Celebrating 25 Years of World Wetlands Day

Celebrating 25 Years of World Wetlands Day

Editor

Richard C. Smardon

MDPI • Basel • Beijing • Wuhan • Barcelona • Belgrade • Manchester • Tokyo • Cluj • Tianjin

Editor
Richard C. Smardon
State University of New York
USA

Editorial Office
MDPI
St. Alban-Anlage 66
4052 Basel, Switzerland

This is a reprint of articles from the Special Issue published online in the open access journal *Land* (ISSN 2073-445X) (available at: https://www.mdpi.com/journal/land/special_issues/world_wetlands_day).

For citation purposes, cite each article independently as indicated on the article page online and as indicated below:

LastName, A.A.; LastName, B.B.; LastName, C.C. Article Title. *Journal Name* **Year**, *Volume Number*, Page Range.

ISBN 978-3-0365-5547-8 (Hbk)
ISBN 978-3-0365-5548-5 (PDF)

Cover image courtesy of Gordon Perkins.

Contents

About the Editor

Richard C. Smardon

Richard C. Smardon has worked in academic, government plus private practice positions before coming to the SUNY College of Environmental Science and Forestry. He is a SUNY Distinguished Service Professor Emeritus at SUNY/ESF. He has a Ph.D. in Environmental Planning from the University of California, Berkeley and a Masters in Landscape Architecture and Bachelors from the University of Massachusetts, Amherst. He has edited/written eight books, the most recent being *The Renewable Energy Landscape* (2017), *Revitalizing Urban Waterway Communities; Streams of Environmental Justice* (2018) *Education for Sustainable Human and Environmental Systems* (2019) and *Selected Papers from the 6th Fabos Conference on Landscape and Greenway Planning*. His new peer-reviewed Special Issue entitled "Celebrating 25 Years of World Wetlands Day" builds on his 2009 book Sustaining the Worlds Wetlands: Setting Policy and Resolving Conflicts. He co-produced the 2019 and 2021 Visual Resource Stewardship conferences and co-edited the 2017 Visual Resource Stewardship conference proceedings https://www.fs.fed.us/nrs/pubs/gtr/gtr_nrs-p-183.pdf.

Preface to "Celebrating 25 Years of World Wetlands Day"

The purpose of this Special Issue is to celebrate 25 years of "World Wetlands Day". There is no other ecosystem that has its very own Ramsar Convention or such a challenge impacting ecosystem sustainability. Papers for this Special Issue provide an overview of wetland status and function within different regions of the world. Of special interest are papers that address wetland ecosystems and human health and wellbeing [1], as well as key international wetland management challenges and actors [2]. A "Universal Declaration of the Rights of Wetlands" has been proposed [3]; therefore, we need innovative solutions for wetland management and maintenance as illustrated in this Special issue.

Tram Chim Nature Reserve Vietnam—A Ramsar wetland Photo by Thang Vo

On February 2nd of each year since 1997, the World Wetlands Day has been celebrated to raise awareness about the vital role that wetlands play for the Earth's ecosystems, as well as the benefits provided to humanity. This day, February 2nd, also marks the date of the adoption of the Convention on Wetlands in the Iranian city of Ramsar on the shores of the Caspian Sea. The theme of World Wetlands Day 2022 focuses on wetlands action for people and nature and so encourages actions to restore and stop their loss. Sub-themes for 2021 included wetlands as a water source and maintaining water quality for usage.
Past World Wetland Days themes are listed below:

2022 Wetlands Action for People and Nature
2021 Intrinsic Link between Wetlands and Freshwater
2020 Wetlands and Biodiversity

2019 Wetlands and Climate Change
2018 Wetlands Making Cities Livable
2017 Wetlands for Disaster Risk Reduction
2016 Wetlands for Sustainable Livelihoods
2015 Wetlands for Our Future Benefits
2014 Wetlands and Agriculture
2013 International Year of Water Cooperation
2012 Wetlands and Tourism
2011 Wetlands and Forests
2010 Wetlands, Biodiversity and Climate Change
2009 River Basins and Their Management
2008 Healthy Wetlands, Healthy People
2007 Fish for Tomorrow
2006 Livelihoods at Risk
2005 There's Wealth in Wetland Diversity—Don't Lose It!
2004 From the Mountains to the Sea – Wetlands at Work for Us
2003 No Wetlands—No Water!
2002 Wetlands: Water, Life and Culture
2001 Wetland World—A World to Discover!
2000 Celebrating our Wetlands of International Importance
1999 People and Wetlands—The Vital Link
1998 Water for Wetlands, Wetlands for Water
1997 First Worlds Wetlands Day

Special Issue Overview

The papers in this Special Issue of LAND consist of three review papers, ten research articles and one perspective paper. Edward Maltby's [4] review paper provides us with an overview of the paradigm shift of how we value and assess wetlands over time. Ballut-Dajud et al. [5] provide us with a worldwide perspective on factors affecting wetland loss. Finally, Jan Vymazal provides us with a historical overview of the development of water quality treatment wetlands in Europe, and North America.

The research papers can be grouped into four groups: 1) use of remote sensing to analyze stability [6] and dynamic factors [7] affecting wetlands; 2) factors affecting the wetlands' ability to store carbon [8,9]; 3) assessment of wetlands effect on water quality [10,11]; and 4) understanding the historical use and value of wetlands [12], farmer's attitudes about wetland management [13], and how we can value wetland ecosystem services [14]. Finally, Bryzek et al. [15] remind us that as wetland researchers and managers, we should minimize damage to wetlands even through field monitoring work.

All these articles illustrate the need for wetland preservation, conservation and in some cases restoration so they can provide valuable ecosystem services to us as well as maintain biodiversity and function. A recent review article by Finlayson et al. [16] has documented that even wetlands designated as Ramsar Convention wetlands continue to suffer from impacts that degrade both function and biodiversity.

Acknowledgements: The guest editor wishes to thank all the contributing authors and the Land journal editorial office for helping to produce this Special Issue. Richard Smardon, PhD, SUNY Distinguished Service, Professor Emeritus, Departments of Environmental Studies and Landscape Architecture, SUNY College of Environmental Science and Forestry, Syracuse, NY, USA.

Richard C. Smardon
Guest Editors

References

1. Millennium Ecosystem Assessment (MEA) Ecosystem and Human Well-Being: Wetlands and Water Synthesis. World Resources Institute, Wash DC 2000.
2. Smardon R.C. Sustaining the Worlds Wetlands: Setting policy and Resolving Conflict. Springer Dordrecht Heidelberg London New York 2009 https://link.springer.com/book/10.1007/978-0-387-49429-6.
3. Davis G.T., Finlayson C. M., Pritchard D.E., Davidson N. C., Gardner R.C. Moomaw W.R., Okuno E., Whitacre J.C. Towards a Universal Declaration of the Rights of Wetlands. Marine and Freshwater Research 2020.https://doi.org/10.1071/MF20219.
4. Maltby E. The Wetlands Paradigm Shift in Response to Changing Societal Priorities: A Reflective Review By Land 2022, 11(9), 1526; https://doi.org/10.3390/land11091526.
5. Ballut-Dajud G.A., Sandoval Herazo L. C., Fernandez-Lambert G., Marin-Muniz J. L., Lopez Mendez M. C., Betanzo-Torres E. A. Factors Affecting Wetland Loss: A Review by Land 2022, 11(3), 434; https://doi.org/10.3390/land11030434.
6. Vymazal J. The Historical Development of Constructed Wetlands for Wastewater Treatment. Land 2022, 11(2), 174; https://doi.org/10.3390/land11020174.
7. Chen Y, Li G, Li L., He L., Ma P. The Effects of Tidal Flat Reclamation on the Stability of the Coastal Area in the Jiangsu Province, China, from the Perspective of Landscape Structure. Land 2022, 11(3), 421; https://doi.org/10.3390/land11030421.
8. Machiwa H., Mango J., Sengupta D. Zhou Y. Using Time-Series Remote Sensing Images in Monitoring the Spatial–Temporal Dynamics of LULC in the Msimbazi Basin, Tanzania Land 2021, 10(11), 1139; https://doi.org/10.3390/land10111139.
9. Aguilos M., Brown C., Minick K., Lie O.J., Hardesty D., Kerrigan M., Noormets A. Millennial-Scale Carbon Storage in Natural Pine Forests of the North Carolina Lower Coastal Plain: Effects of Artificial Drainage in a Time of Rapid Sea Level Rise Land 2021, 10(12), 1294; https://doi.org/10.3390/land10121294.
10. Spangler D. M., Tyler A. C., McCalley C. K. Effects of Grazer Exclusion on Carbon Cycling in Created Freshwater Wetlands Land 2021, 10(8), 805; https://doi.org/10.3390/land10080805.
11. Dattamundi S., Chanda S., Scinto L.J. Microbial Respiration and Enzyme Activity Downstream from a Phosphorus Source in the Everglades, Florida, USA Land 2021, 10(7), 696; https://doi.org/10.3390/land10070696.
12. Langan B., Twiss M.R. Surface Water Quality Differs between Functionally Similar Restored and Natural Wetlands of the Saint Lawrence River Valley in New York Land 2021, 10(7), 676; https://doi.org/10.3390/land10070676.
13. Dobbie M. J. Typing Colonial Perceptions of Carrum Carrum Swamp: The Expected and the Surprising. Land 2022, 11(2), 311; https://doi.org/10.3390/land11020311.
14. Vallizadeh N., Bayat S. E., Hayati D., Viira A-H., Tanaskovik V., Kurban A. Understanding Farmers' Intention towards the Management and Conservation of Wetlands Land 2021, 10(8), 860; https://doi.org/10.3390/land10080860.
15. Gu X., Liu G., Yu J., Wang, H., Yang Y., Zhang P. Changes in Ecosystem Service Value in the 1 km Lakeshore Zone of Poyang Lake from 1980 to 2020 Land 2021, 10(9), 951; https://doi.org/10.3390/land10090951.
16. Bryzek J.A., DeSilva S., Von Haugg C. L., Hartman D., McCall J.E., Veselka W., Anderson J. T. Obligations of Researchers and Managers to Respect Wetlands: Practical Solutions to Minimizing Field Monitoring Impacts. Land 2022, 11(4), 481; https://doi.org/10.3390/land11040481
17. Finlayson C. M., Fennessy S., Grillas P., Kumar R. Commemorating 50th anniversary of the Ramsar Convention on Wetlands. Marine and Freshwater Research 2022, 1-v https://doi.org/10.1071/MFn22161.

land

Review

The Wetlands Paradigm Shift in Response to Changing Societal Priorities: A Reflective Review

Edward Maltby

Department of Wetland Science, Water and Ecosystem Management, School of Environmental Sciences, University of Liverpool, Liverpool L69 3BX, UK; e.maltby@liv.ac.uk

Abstract: This paper reviews some of the key influences that wetlands have had on the development of human society together with the history of wetland use, conservation and management in the context of changing human interactions from prehistoric to modern times. It documents the origins of the Ramsar Convention and the changes in the criteria for defining wetlands of international importance from an emphasis on migratory birds to those of wider functional importance contributing to community well-being. This led to a significant increase in the number of signatories from developing countries The change in scientific emphasis from ecology to ecosystems (and ecosystem services) is identified as a key element of the wetland paradigm shift, which has occurred in the last half century and renewed the recognition of the importance of the natural capital of wetlands. It represents a change in research agenda from what wetlands are to what wetlands do. Modification of the Ramsar wise use concept is documented, and evolution of wetland assessment methods is traced in relation to policy development and the need for a strong science evidence base to improve decision-making connected with wetland conservation and management. The author also addresses the significance of wetland economic valuation and biodiversity issues, transboundary water management with particular reference to the marshlands of Mesopotamia (southern Iraq), conflict, and human livelihood issues. Examples are given of the drive towards wetland restoration in different countries, and at different scales, with awareness of the extraordinarily high costs associated with major schemes such as the Florida Everglades which may prohibit replication in other parts of the world. Adoption of the Ecosystem Approach and the "Wholescapes" concept are seen as important in the future management of wetland ecosystems. The wide-ranging interactions within the structure of a new wetland paradigm are summarized diagrammatically. An examination of current societal priorities and challenges resulting from the nexus of issues arising from food production, energy, water, and environmental change and health suggests both significant threats to wetlands, but also some opportunities for these ecosystems to play a part in sustainable solutions contributing to human well-being. The paper concludes with an endorsement of a new World Charter for wetlands but emphasizes the vital importance of partnership working and the key engagement of local communities to make any new initiative for enhanced protection and management of wetlands to work on the ground. Key challenges facing wetland science are identified, but it is the realization that healthy wetland ecosystems are a significant contributor to human and societal well-being that underpins the paradigm shift in research, management and policy needs.

Keywords: wetland management history; Ramsar convention; wise use; wetland assessment methods; wetland valuation; wetlands paradigm shift; ecosystem approach; wholescapes; sustainable development; climate change; World Charter for wetlands

Citation: Maltby, E. The Wetlands Paradigm Shift in Response to Changing Societal Priorities: A Reflective Review. *Land* **2022**, *11*, 1526. https://doi.org/10.3390/land11091526

Academic Editors: Richard Smardon and Shiliang Liu

Received: 31 May 2022
Accepted: 26 July 2022
Published: 9 September 2022

1. Introduction: Change Is a Normal and Persistent Feature of Earth, Human and Societal History

Earth environmental conditions have evolved and been impacted by natural processes and events throughout geological time. Humans also have altered them, more or less dramatically, by deliberate actions or unintended consequences of activities since prehistory.

What is not necessarily normal or predictable is the scale, location, and speed of change. Yet we now have unprecedented levels of knowledge from ever-increasing bodies of research which can help us to adapt to, reduce, and avoid undesirable consequences of change.

Wetlands have played a seminal role in Earth history and in some of the major environmental changes that have helped shape its present-day geological structures, waters, and atmospheric composition. Human perceptions of and relationships with wetlands have changed significantly over time and from place to place. Altered perceptions resulting from advances in scientific as well as public understanding in the last century have been reflected, at least in part, in significant policy shifts at local, national, and international levels. They have been the focus of sharp conflicts between different sectors of society with contrasting views of how they should be used. Examples of the intensity of feelings are illustrated from the UK and Australia in Figure 1.

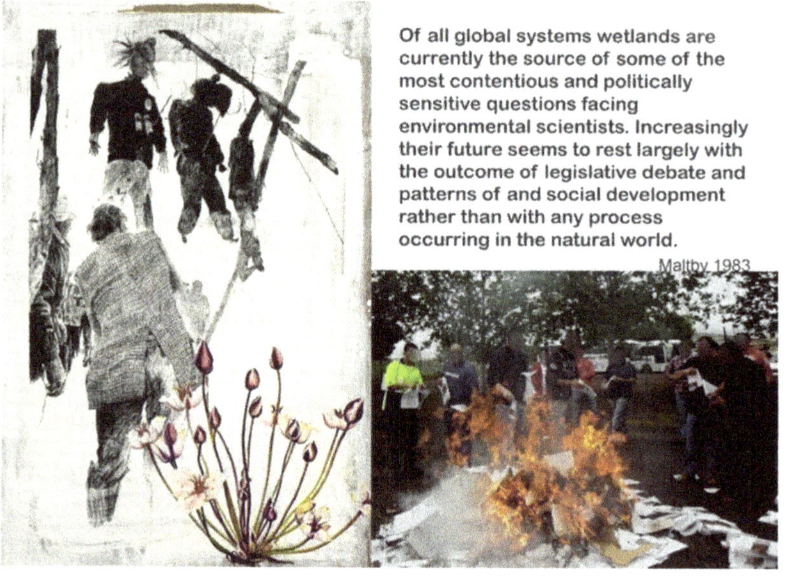

Figure 1. Hanging of the effigies of conservationists on the Somerset Levels, UK by farmers opposed to plans to rewet drained wetlands in 1983; Source: Jeremy Purseglove 2017 [1]. Farmers in the Murray–Darling river basin, Australia, concerned about loss of water for irrigation, burning copies of the 2010 water plan. Source: AAP/Gabrielle Dunlevy.

Over a period of some 50 years of research and advisory work, including to the Ramsar Convention, the present author has witnessed some of the extraordinary changes in our understanding of wetland ecosystems and the views that individuals and society have of them. This article offers a personal digest of some of the notable changes in perspective that have occurred, especially during this snapshot of empirical experience and attempts to interpret their significance for future generations.

Apologies are offered for the self-indulgent emphasis, but it is hoped that the experience might be insightful and in defense there are numerous excellent texts by others that are available for consultation and more comprehensive coverage.

Examples of the ways in which wetlands have been a cultural driving force are given in Maltby [2]. Setting aside their probable intimate role in the origins of life itself, wetlands have stimulated turning points in human culture by virtue of their prominence in key temporal stages or locations of Earth history. Of particular significance, given the current attention on global warming, are the tropical peat-swamp forests (such as those currently occurring in Southeast Asia (Figure 2) which were to become the extensive coal deposits of the Carboniferous period some 250 million years ago [3]. The vulnerability of tropical peat deposits to degradation and loss due to drainage and fire is dramatically illustrated in one of the few remaining peat domes of the Mekong Delta (Figure 3).

Figure 2. The complexity and rich biodiversity of intact peat swamp forest in Indonesia and examples of the simplification resulting from clearance and conversion to agriculture. Source: Edward Maltby.

Figure 3. Vulnerability of tropical peatlands illustrated by the effects of fire on one of the last remaining peat masses in the Mekong Delta at U Minh Tong, Vietnam. Source: Edward Maltby.

Continental drift repositioned these vast carbon stores into higher latitudes where from the mid-eighteenth century they powered the Industrial Revolution, first in Europe and then North America. What followed was arguably one of the most significant changes in societal organization, trade and economics which would have a progressive influence not only on human society but also on our environment. Industrialization and increased national and international trading accelerated urbanization of the human population and increased the development of transport networks and hubs such as ports and centres of manufacturing. Inevitably this led to the conversion of large areas of wetland to alternative uses that has continued to more recent times (Figure 4) In effect, widespread loss and degradation of wetlands throughout the so-called developed world was the price paid (invariably never fully recognized) for rapidly accumulated economic wealth. The fact that the wealth was not shared well with a growing population and its generation also resulted in severe pollution and human health issues is now more seriously viewed in the context of our remaining 'natural capital'.

Figure 4. Clearance and burning of bottomland hardwood forests to convert to agricultural land in Louisiana, United States. Source: Edward Maltby.

Ironically the use of the fossil carbon stores of the wetland ecosystems from earlier geological times became a powerful direct driver of wetland loss. The wider significance of the losses was rarely recognized by the entrepreneurs and participating communities preoccupied with financial wealth creation and direct socio-economic well-being.

There was limited recognition of the adverse consequences of water and atmospheric pollution and no appreciation of the impact that the progressive release of carbon dioxide would have on climate change. It is a further irony that the reconversion to carbon dioxide of carbon previously captured by ancient wetlands should be a major contributor to accelerated global warming. The importance of the biogeochemical coupling between wetlands and larger Earth systems is now increasingly recognized as a key consideration in environmental management [2].

2. From Pre-History to Historical Attitudes

2.1. Intrinsic Dependence to Progressive Detachment

Significant evolution of the human brain is attributed by the "aquatic ape" theory to the fatty acid-rich diet available to hominins living in, and dependent on, the wetland margins of lakes, rivers and the sea [4]. Wetlands continued to play important roles in the evolution of prehistoric communities and were the sites of some of the earliest stages in the development of tool-producing hominins. Excavations of Mesolithic and Neolithic post-glacial lake marginal settlements across Europe have revealed the highly dependent relationships between humans and the wetland ecosystems [5]. Figure 5 illustrates reconstructions of community life and dependence on the wetland resources from archaeological excavations of one such settlement at Starr Carr in North Yorkshire, UK.

Figure 5. Early prehistoric communities were completely reliant on lake and river marginal wetlands throughout Europe for food, shelter, and security. Source: https://www.google.com/search?q=Star+Carr+reconstruction&sa=X&rlz=1C1ONGR_enGB1009GB1009&biw=1280&bih=569&sxsrf=ALiCzsaiGhDJ8Nv2D2DP (accessed on 26 March 2013).

Even whilst this prehistoric dependency persisted in the post-glacial communities of Europe, cradles of civilization were developing based largely on the innovative management of the waters of fertile floodplains such as those of the Nile, Tigris-Euphrates (Mesopotamia), Indo-Gangetic and North China Plains. Utilisation of the natural flood cycle created food security and the creation of agricultural wealth. It demanded community organization and cooperation together with rules and laws [6]. Cities arose from an increasingly prosperous "hydraulic civilization" [7] with food surpluses and trade supporting population growth that made possible the differentiation of roles in society that are familiar today [2]. Archaeological evidence from the city of Ur in Mesopotamia (present day Iraq) has revealed continuity in some of the basic units of floodplain life such as the reed-house (mudhif) or long canoe (mashuf) from Sumerian times to the present day—a cultural connection of more than 5000 years (Figure 6).

Some of the earliest examples of written language come from Sumer, including the origins of the biblical account of the Creation and the Flood of Noah which may be from the epic of Gilgamesh in Sumerian literature. Such a bequest to the culture and philosophy of subsequent generations is hard to over-state.

Figure 6. Present-day Marsh Arabs (Madan) in Southern Iraq (former Mesopotamia) with traditional reed houses (mudhif) often on floating islands and long canoe (mashuf) in a channel through marshland reflooded after 2003. Source: Edward Maltby.

The wider implications of altered hydrology of the floodplain were well understood in ancient Mesopotamia, and recognition of the enrichment of drainage waters with salt from irrigated lands was recorded as early as 4400 BC on clay tablets in the temple of Lagash. These recorded a thirty-fold increase in salinity of one particular agricultural field in just one year [8]. So was heralded the serious water quality issues which may arise from conversion of natural flood-cycle wetlands to agriculture.

Nevertheless, historical attitudes to wetlands progressively detached them from a central role in community life support. The depiction of swamps and bogs in fictional literature as sources of mythical and alien species, as well as numerous hazards to human health and safety, was highlighted by Mitsch and Gosselink [9]. Such portrayal undoubtedly influenced generations of schoolchildren throughout the developed world and became embedded in future attitudes [10]. Their drainage and transformation to other non-wetland uses were seen as laudable "public-spirited" objectives and testimony to the power of technological developments and human ingenuity [11]. We will see this view substantially reversed at the start of the twenty-first century (Figure 7).

Throughout history the benefits that we presume were appreciated by early human cultures and more certainly current traditional users of wetlands were either ignored or dismissed as less significant by more powerful sectoral interest groups [2]. Celebration of major drainage achievements such as those which empoldered the complex Rhine–Meuse–Scheldt delta from medieval to modern times; the drainage of the English Fens [12]; the more recent drainage of the Hula swamps in Israel (Figure 8); and the desiccation of large tracts of the Florida Everglades [13,14] were based on the winning of rich agricultural land, flood protection of encroaching settlement and eradication of disease vectors. The pressures for conversion have been especially strong in developing countries where the attraction of foreign earnings have often overwhelmed conservation interests (Figure 9).

European colonization of the United States resulted in widespread efforts to drain the swamps which were originally owned by the Federal government; petitions to Congress were brought seeking compensation for "improvements" undertaken by States. The resulting legislation—the Federal Swamp Land Acts of 1849—50 and 1860—was intended to reduce flood hazard, improve sanitation and reclaim land for agriculture [15]. For centuries "the drainage of wetlands has been seen as a progressive public-spirited endeavour. the very antithesis of vandalism" [11]. Farmers and other landowners were encouraged by the availability of generous government grants and tax concessions in the UK and elsewhere to convert wetlands to more productive agriculture and other land uses such as forestry.

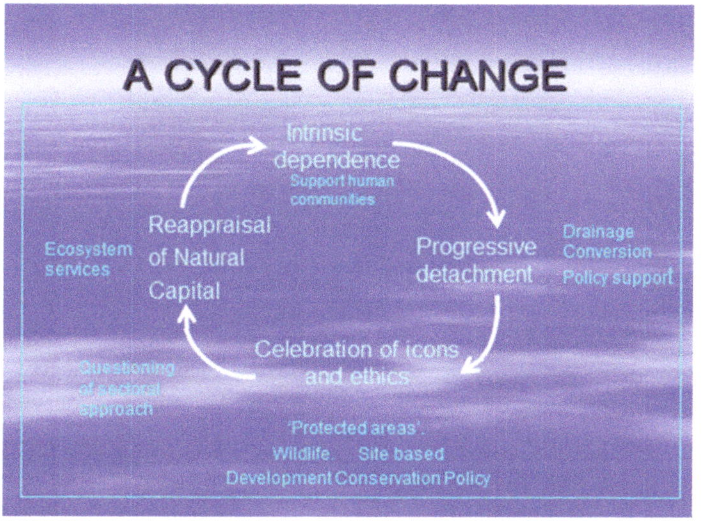

Figure 7. A simplified and generalized representation of the circle of change of the human interaction with wetlands. Source: Edward Maltby.

Figure 8. Peat fires at depth were a common result of the drainage of the Hula valley swamps in Israel. The temperature probe in the residual ash reveals prolonged sub-surface burning. Source: Edward Maltby.

Figure 9. The rich biodiversity of stream, river and lake marginal wetlands throughout Africa, such as this example from Ghana, are under constant threat from unsustainable water extraction and landuse pressures to meet increasing economic needs and poverty alleviation. Source: Edward Maltby.

2.2. Celebration of Ecology and the Rise of Conservation Icons and Policy

Early ecological research focused on habitats and species which only much later became grouped under the umbrella term "wetlands". Studies were of bogs, fens, carr woodland, mires, muskeg swamp, marshes or other habitat descriptors without any reference to "wetland". Figure 10 illustrates just some of the range of such habitats and species.

The term *wetland* was first used in 1953, in a report by the U.S. Fish and Wildlife Service (USFWS) that provided a framework for the first official use of the term in a later publication concerning waterfowl habitat in the United States [16]. It may be that the specific link to waterfowl habitat, as well as the limited interaction between scientists in the United States and those from other countries at this time, explains the reluctance to immediately accept the term beyond North America. Indeed, numerous well-respected European scientists refused for many years to use the term in their empirical research and publications, primarily because they argued that it was too generic and grouped together such highly diverse ecological systems as to become taxonomically confusing. Nevertheless, it became embodied in the only multinational treaty to identify with a single group of ecosystems—the Ramsar Convention.

Early emphasis in research was on fundamental ecological questions. Such can be exemplified by the investigation of the origins, diversity and associated flora and fauna of peat-forming systems, e.g., Moore [17], Moore and Bellamy [18]. They are a pertinent choice given the current switch of focus from basic ecology to their delivery of important ecosystem services and especially their role in greenhouse gas dynamics and climate change. The "formation, development and maintenance of peat depends on an envelope of environmental conditions representing local- and/or regional-scale characteristics in which climate is of critical but not sole importance" (cf. templates of formation) [18]. It has long been stated that peat-forming ecosystems in the UK uplands were originally primarily the result of climate-topographic interactions [18]. A correlation has been made between the 1200 to 1250 mm isohyet (line of equal precipitation) and the extent of blanket peat [19,20], whereas Rodwell [21] reported the threshold for blanket peat formation as at least 160 rain days combined with annual rainfall greater than 1200 mm. Lindsay et al. [22] correlated the change in size, proportion and pattern of open pools and hollows in blanket bog ecosystems with the number of rain days and mean temperature. Much less attention was given originally by ecologists to the potentially overwhelming importance of prehistoric human activities to the onset and subsequent development of upland peat. It is now accepted that anthropogenic activity has played an important role in the development of blanket peat over a substantial area of the UK [17,23,24] through clearance of woodland and hydrological alterations at a time when climate overall was favourable for the growth of *Sphagnum* mosses [25].

(a)

(b)

Figure 10. *Cont.*

(c)

(d)

Figure 10. Cont.

(e)

(f)

Figure 10. Selection of the extreme diversity of wetland types and species: (**a**) muskeg in Alaska showing "drunken" stand of Black spruce (*Picea mariana*) on peat and permafrost; (**b**) the distinctive microrelief patterning of hummock and pools in "hollows" of blanket bog and the extraordinary height of stands of Phragmites in the Perishan wetland of Iran; (**c**) examples of carnivorous plants—(**1**) sundew and (**2**) pitcher plants—adapted to source nutrients from insects to supplement nutrient deficiencies in the peaty substrate, in both temperate and tropical climates (**3**) Human bodies uniquely preserved in the acidic waterlogged peat provide evidence of diet, associated environmental conditions as well as cause of death; (**d**) fringing mangroves in coastal Louisiana; (**e**) cypress swamp (*Taxodium distichum*) in Florida; (**f**) Hummocks of *Spagnum magellanicum* in Tierra del Fuego bog. Source: Edward Maltby.

Somewhat ironically, the actual trigger for peat development might have been human-induced alteration of vegetation and/or soil conditions such as the development of an impermeable iron pan in podzols resulting in hydrological change [26] without necessarily requiring climate change. Dimbleby [27] had drawn attention in the UK to human-induced podzolisation in the Bronze Age stimulating the development of peaty soils, whereas Maltby and Caseldine [28] provided direct pedogenic, pollen and 14C evidence for the possible high speed of change from brown soils to those accumulating a strongly acid peat surface in the Bronze Age (ca. 3500 BP) on Bodmin Moor (Figure 11).

Figure 11. Original brown earth type forest soil preserved beneath a Bronze Age burial structure with a thin iron-pan podzol and peaty surface developed in the surface of the structure and peat accumulated in the surrounding area; Colliford, Bodmin Moor, southeast England. Source: Edward Maltby.

Research highlighted the particular significance of hydrology and water movement in determining the stability, size and slope of the peat mass [29,30]. The natural limits to peat growth, the overall carbon budget and the physical stability of the peat mass was elegantly demonstrated by Clymo [31,32].

Such early ecological research across diverse habitat types played a vital role in our understanding of what wetlands were, how they developed, and why they occurred where they did; this does not mean that we have all the answers to some of the basic ecological questions. The accumulation of deep peat under the remarkably dry climatic conditions of the Falkland Islands, for example, is difficult to explain on the basis of the understanding of peat bog formation in the Northern Hemisphere [33]. Figures 12 and 13 illustrate the remarkable depth and formations of peat on East Falkland, notwithstanding the relatively dry climate.

The sequential accumulation of organic matter and sediment provided palaeoecologists with a time machine from which to document environmental change as well as evolution of the wetland itself.

Figure 12. View of Falkland Island peat bank (bog) between Mt Harriet and Goat Ridge. Source: Edward Maltby.

Figure 13. Exposed eroding face of the peat 'bank' at Goat Ridge, Falkland Islands. Source: Edward Maltby.

Maltby [2] identified characteristics of habitats now described as wetlands which spawned examples of important areas of ecological understanding, notably:

1. Change over time through ecological succession—frequently they are transient features in the landscape and in many cases may be regarded as authors of their own destruction;
2. Zonation both within and at the boundary of the wetland is a common feature of species distribution and substrate characteristics;
3. Temporal variations through seasonal or episodic hydrological or flood "pulsing" cycles—see especially Junk et al. [34] or individual events.

Initially research was essentially site-based and has contributed to a wealth of taxonomic data and understanding of species-habitat relationships.

Inevitably the focus on wetlands as parts of the landscape worthy and in need of protection also became site based as well as their importance for particular species and especially birds. Indeed, this was also reflected in the central commitment to the Ramsar Convention which required signatories to list at least one wetland site of International Importance.

The growing concern for wildlife and the vulnerability of their habitats had already begun to emerge at the end of the nineteenth century. In the UK the Royal Society for the Protection of Birds (RSPB) was formed in 1889 and was highly influential in the passing of the Bird Protection Act in 1954 (much earlier a Wild Birds Protection Act had been passed in 1872).

Particularly significant in the United States was the founding in 1886 of the National Audubon Society in response to the unprecedented levels of hunting, particularly of wetland birds for their plumage that was much sought after to satisfy the demands of the millinery trade responding to the fashion requirements of an increasingly wealthy part of Society. For some 50 years from about 1870, plume hunters invaded the Everglades and in particular decimated colonies of the Snowy Egret that was much favoured by the fashion houses of the day. In the mid-1880s up to five million birds a year were being harvested just for their plumage [35]. The Florida Audubon Society requested the American Ornithologists Union to hire a game warden to prevent the illegal killing. Guy Bradley was the first such warden in South Florida who was tragically killed in 1905 trying to arrest a notorious plume hunter. His death further inspired the conservation movement and was an additional stimulus for future legislation. Although birds were a key driver of the conservation movement, there are many plants and other animals that are unique and charismatic features of wetlands. The American alligator is used as an example of the specialized fauna, not least because of its ancient origins in the Oligocene period some 30 million years ago (Figure 14).

The year 1905 was also significant in the Netherlands where the Society for Preservation of Nature Monuments was established in response to plans by Amsterdam City Council to use a nearby wetland as a rubbish dump for the rapidly expanding city; the Society bought the wetland and saved it from loss. Today Natuur Monumenten is one of the largest landowners of nature reserves and parks in the Netherlands (where of course much of the country historically was wetland).

Whilst wetland species and specific iconic sites were often the *cause celebre* for the early conservationists, a more general environmental movement achieved prominence only from the 1960s with many crediting Rachel Carson's book *Silent Spring*, published in 1962 [36], as an influential catalyst. Her message was important in demonstrating the devastating impacts of pesticides on wild non-target species often via the links provided by water. Recognition of the coupling among environmental contaminants, food chains and the water cycle was a precursor of the subsequent paradigm shift that would take us to the present day.

In the middle of the twentieth century the conservation focus was not only site and species based, but primarily concerned with the preservation of rarity, uniqueness or particularly good examples of nature. Following its much earlier connection with the inspiration for conservation, Everglades National Park was officially opened in 1947 when land and funding was eventually secured and became the largest designated subtropical wilderness reserve on the North American continent. It became an International Biosphere Reserve in 1976, a World Heritage Site in 1979 and a Ramsar site in 1987, to confirm its position as one of the world's iconic wetlands.

Figure 14. Example of American alligators in their natural wetland habitat of Louisiana. Source: Edward Maltby.

Mid twentieth-century threats to wetlands resulting from sectoral policy conflicts also managed to galvanize communities around the conservation banner. In the UK the Forestry Commission was established by government to enhance the much-depleted post-First World War forest stock. It later proposed (in the 1950s) to plant coniferous trees on The Chains, the most extensive tract of blanket bog on Exmoor. At the same time there was substantial government grant aid to support the conversion of peat and organic soils to more productive farmland. In the case of Exmoor this led to the formation (in 1958) of the Exmoor Society, an effective non-governmental organisation (NGO) and lobby group dedicated to maintaining the "traditional" moorland landscape (Figure 15) and its associated rural livelihoods and public access. Afforestation on Exmoor was prevented, the blanket bogs became a celebrated conservation icon, and the Exmoor Society a beacon for policy change.

Policy innovations included management agreements with landowners who were compensated to prevent alteration of the moorland landscape. It is ironical that much of the moorland so cherished at the present time was covered in mixed deciduous woodland in prehistoric times; clearance, especially during the Bronze Age by human communities, served as a catalyst for ecological and pedogenic change resulting in peat development. For immediately succeeding generations this impact might have been considered as a first environmental planning disaster of that period because of the decline in sources of food, shelter and heating [25].

The blanket peats of Caithness and Sutherland in Scotland were not as fortunate as The Chains, and 67,000 ha (17%) of the peatland were either planted or approved for planting [37] as a result of tax incentives or Forest Grant Scheme approval [22]. The tax incentives have since disappeared and the Forestry Commission now has guidelines to protect blanket bog from further afforestation [38].

Figure 15. The Chains blanket bog, Exmoor, UK; dominated over large tracts by the deciduous grass *Molinia caerulea* which has contributed to active peat accumulation and was threatened by afforestation. Source: Edward Maltby.

Individual site-based conservation is enshrined in the various scales of protected area policies worldwide. Within the UK the Wildlife and Countryside Act of 1981 empowered the government conservation body, Natural England, to identify and protect Sites of Special Scientific Interest (SSSIs) on the basis of wildlife, geology or landform. These include numerous wetlands, but protection by law of the site does not necessarily ensure protection from external impacts.

The Ramsar Convention in 1971 provided a seminal change in the conservation strategy for wetlands by recognizing the importance for migratory birds of not just single sites, but of the network connections of wintering, breeding and feeding (especially re-fueling and resting sites) often over hundreds if not thousands of kilometres and across multiple sovereign territories. In recognition of the rapid and accelerating losses of wetland habitats, especially throughout Europe and North America, the movement to secure conservation and where possible restoration of the remaining resource at the global scale gained traction. The Ramsar Convention became an iconic standard bearer for the conservation and sound management of wetlands. It emerged from the efforts of passionate ornithologists, notably Luc Hoffman from Tour du Valat research station in the Camargue, France, Geoffrey Matthews from The Wildfowl and Wetlands Trust in Slimbridge, UK, and Escandar Firouz, former Minister of Environment in Iran. The key motivation was recognition that it was necessary to safeguard the essential international connectivity of habitat requirements of migratory waterfowl. Single site conservation and management alone was not sufficient to maintain the populations of migratory species which weremuch revered by bird-watchers throughout the developed world. The Ramsar Convention on Wetlands of International Importance Especially as Waterfowl Habitat was established in 1971 at a meeting in Ramsar on the shores of the Caspian Sea (Iran) and became unique as being the only international agreement to cover a specific single group of ecosystem type.

3. From Ecology to Ecosystems and Ecosystem Services

The last 50 years has witnessed a major shift in scientific research from the description of wetland habitats and their relationships with plants and animals to attempting to understand how wetland ecosystems actually work, as part of larger environmental systems and in support of human well-being and exemplified in Figures 16 and 17.

Figure 16. Examples of the services realized by communities in the natural Melaleuca wetlands of the Mekong Delta, Vietnam; (**a**) fish trap; (**b**) cut Melaleuca stems; (**c**) sorting wood for different uses; (**d**) distillation of essential oils from leaves. Source: Edward Maltby.

Figure 17. Catching even small fish in the wetlands of the Mekong Delta (Vietnam) can be an important addition to family nutrition. Source: Edward Maltby.

The emphasis on migratory waterfowl, which was the original driver of the Ramsar Convention, was seen by many observers as an indulgence on the part of richer nations that poorer countries could ill-afford [2]. Confronted with the immediate and pressing challenges of poverty, food and clean water shortages, lack of economic development and the burden of foreign debt, many developing nations were reluctant to become signatories to the Convention. They could see little advantage in undertaking management obligations perceived as primarily benefitting the conservation ethos of bird watchers in already wealthy countries. This position was to change dramatically after the 1987 Conference of Parties held in Regina, Canada [39]. Two developments were largely responsible for the change in perception and an encouragement of greater interest and commitment from the developing world.

1. Changes to the criteria for the listing of wetlands of international significance to increase the recognition of their wider functional importance beyond bird habitat;
2. Elaboration of the "wise use" commitment to emphasise the contribution of wetlands to human welfare and sustainable development.

A summary table of the changes in criteria from the Conference of Parties in Cagliari in 1980 and Montreux in 1989, either side of the Regina Conference, is given in [2]. Not all delegates at Regina initially supported the changes, with some concern that they might change responsibility for the Convention from government departments of nature conservation to those dealing with natural resources (E. Maltby personal observation).

Revision of the criteria was influenced by a rapidly growing body of scientific evidence revealing the significance of the hydrological and biogeochemical as well as ecological functioning of wetlands. These translated to benefits such as flood control, water quality and fisheries support to which people could more easily relate. Modifications to the criteria focused on how wetland ecosystems functioned, rather than their habitat description and ecological characterization orientated towards traditional nature conservation values such as rarity, uniqueness, and particularly good examples of an ecosystem type. The switch in emphasis from habitat and ecological research to ecosystem processes and functioning was led from the United States where the significance of "wetland functions and values" was linked to state wetland regulatory laws starting in 1963 [40] and from 1972 at the Federal scale to the requirements of section 404 of the Clean Water Act [41].

Internationally, the World Wide Fund for Nature (WWF) and World Conservation Union (IUCN) initiated the Wetland Conservation Programme 1985–1987. The programme set up a scientific advisory group, which *inter alia* provided technical advice and guidance to the Ramsar Convention. This included developing the first definition of "wise use" [42] and active participation in influential discussions at the Regina COP in which the present author served as chair of the session of contracting parties reviewing criteria for designating sites of international importance. The Programme also commissioned the preparation of a new text to draw international attention to the real importance of wetlands to Society. *Waterlogged Wealth, Why waste the world's wet places?* [15] attempted to capture the essence of the changing wetland paradigm, which had moved to a new emphasis beyond fundamental ecology to functioning which underpinned a wide range of benefits enjoyed by human communities.

One indicator of the influence of this change in emphasis is found in the number of developing countries signing up to the Ramsar Convention. Pre-Regina there were only 13 developing countries out of 37 signatories. Post-Regina the numbers had changed to 123 out of 167 in 2018 and are currently 124 out of 170. The proportional change from 35% to 73% is a clear reflection of the fundamental change in understanding and perception by governments worldwide that wetland conservation and sound management was not just for the benefit of the populations of rich nations, but was an important tool to underpin the improved well-being of developing nation communities.

3.1. Elaboration of the Wise Use Concept

Signatories to the Ramsar Convention are required to nominate at least one wetland of international importance but also undertake to promote the "wise use" of all wetlands within their territories. The original 1987 definition was formulated by the IUCN Wetlands Programme Advisory Committee: "The wise use of wetlands is their sustainable utilization for the benefit of mankind in a way compatible with the maintenance of the natural properties of the ecosystem". Sustainable utilization was defined as "human use of a wetland so that it may yield the greatest continuous benefit to present generations while maintaining its potential to meet the needs and aspirations of future generations" [42]. Clarification and fuller interpretation of the wise use obligation coincided with the revision of the criteria for recognition of wetlands of international importance. The Regina COP can be regarded as a key milestone in the global wetland research and policy shifts in emphasis including from birds to people, habitat to ecosystems as well as nature conservation to environmental quality and human well-being.

A wide gap has often existed between accepting the definition and obligations of wise use and applying its principles and objectives to specific wetlands, or in particular countries. Maltby [43] indicated that in general the application of wise use would require:

1. Identification of wetland functions and values;
2. Integration of compatible uses where possible;
3. Separation of incompatible uses;
4. Zoning and environmental planning;
5. Catchment management;
6. Appropriate employment, social and economic strategies to relieve the ecosystem of damaging pressures.

The continued loss and degradation of wetlands across all signatory countries to Ramsar brings into question the effectiveness, not only of the wise use obligation but of the Convention itself, in securing the world's wetland resources. Whilst regrettably ineffective in preventing wetland loss globally, the Convention has been a key factor in raising the importance and political awareness of iconic wetlands, which are symbolic of the wider wetland resource. Such has been the case of the Florida Everglades threatened by water quality issues and hydrological changes resulting largely from agricultural development and urbanization in its wider catchment. The legal basis for intervention by the Federal Government (advised by the present author) to try and halt their progressive degradation was supported in no small measure by the triple international designation of Everglades National Park which, in addition to being a Ramsar site of international importance, is a World Heritage Site and an International Biosphere Reserve (S. Ponzoli, pers. comm.). The attorney and expert scientist fees associated with the Everglades lawsuits alone ran to many millions of dollars and the costs of remedial works orders of magnitude more. Kadlec [13] and Richardson [14] give authoritative summaries of the main issues and subsequent actions. It is unlikely that any other country in the world could afford to tackle a prosecution and mount a restoration plan on the scale seen in the Everglades. The effectiveness of such a message remains to be seen, but the over-riding conclusion is that it makes economic as well as ecological sense to avoid wetland degradation in the first place and such a strategy can be aided by following wise use guidance.

The wise use definition under the Convention was revised at the 2005 Conference of Parties (COP 9) to "Wise use of wetlands is the maintenance of their ecological character achieved through the implementation of ecosystem approaches, within the context of sustainable development". This modification reflected the influence of not only the Convention's mission statement, but also Millenium Assessment terminology [44], elaboration of the "ecosystem approach" under the Convention on Biological Diversity [45,46], and the Brundtland Commission's definition of sustainable development [47].

3.2. The Rise of a Functional Approach and Assessment Procedures

Concomitant with the change in emphasis from what wetlands *a* to what wetlands *do* has been the requirement to better understand and assess wetland functioning.

"A functional approach to wetland assessment is one that acknowledges that wetlands can perform work at a variety of scales in the landscape, which result in significant direct and/or indirect benefits to people, wildlife and the environment" [48]. It effectively broadens consideration of wetlands from a view as conservation icons to recognizing their wider utilitarian importance resulting from multiple ecological, biogeochemical, and physical processes and their natural dynamics (Figure 18).

The need for a different perspective on wetlands which recognized the functions they performed (and the values resulting) was recognized initially in the United States where state wetland regulatory laws, starting in 1963, stimulated the first assessment methodologies [40]. Functional assessment subsequently became central to the US Federal permitting process, which regulates wetlands under Section 404 of the Clean Water Act (33 US Code 1344) and coupled to a "no net loss" policy. Conceptual development of assessment in the US is described by Brinson [41], and Smith [42] provides examples in practice. In developing countries, the underpinning science-based evidence is much more limited (see examples in Roggeri) [49]. Hawson et al. [50] examine approaches in Canada.

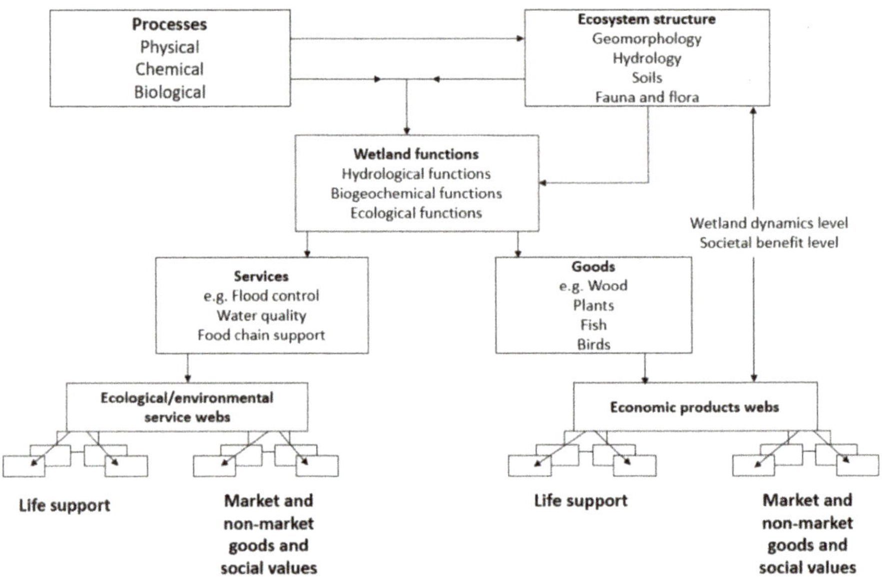

Figure 18. Physical, chemical, and biological processes lie behind the provision of ecosystem services [51].

The early literature from the United States was highly influential in stimulating further research, e.g., Sather and Smith [52], but whilst pioneering, was also simplistic by grouping together functions and values (without clear separation between the two) and not making sufficient distinction between the ability of different wetlands or parts of the same wetland to perform particular functions. It became increasingly clear that not all wetlands perform the same functions or perform the same functions to the same extent.

Physical, chemical and biological processes, individually or in combination, within the diverse structures of different wetland ecosystems, control different patterns and quantities of functioning such as hydrological (e.g., flood control), biogeochemical (e.g., nutrient retention) and ecological (e.g., habitat provision) functions. Functions result in the provision of different goods and services valuable to people such as flood risk reduction, pollution reduction, and food chain support (Figure 18). These and other outcomes are now commonly referred to as "ecosystem services", recognising where and how they impart human benefits.

"Assessment of functioning is a key pre-requisite to making management decisions that affect delivery, by wetlands, of specific or particular combinations of ecosystem services.

This knowledge is critical to provision of the evidence necessary to underpin strategic and policy decisions. Effective functional assessment provides an essential tool for those individuals, regulatory bodies, and other government or non-government organisations that make informed decisions for appropriate wetland management" [53].

Bartoldus [54] identified 40 different wetland assessment procedures for the United States alone, reflecting inter alia the diversity of ecosystems, species targeted, time/effort required, costs, outputs, expertise and user needs. The rationale has been driven by policy and the need to better inform decision-makers of the public values provided by wetland functioning which may be lost or impaired by development. The inevitable legal disputes involving the rights of individuals and property ownership have led to the concept in the US of "jurisdictional wetlands" with a need to delineate wetland areas (and their functions) on the basis of both legally as well as scientifically verifiable criteria [54].

Tools in the United States for "rapid" assessment have tended to treat a wetland as a single functional unit. The most widely used method initially was the Wetland Evaluation Technique (WET) developed from the work of Adamus [55] and Adamus et al. [56]. It has been the basis for training and use by the US Army Corps of Engineers (who together with the US Environmental Protection Agency are responsible for wetland permitting under the Clean Water Act legislation) and other regulatory programmes [41].

Subsequently the HGM (Hydrogeomorphic) Approach was developed to overcome criticisms of WET [57]. It was initially designed to estimate change in wetland condition by means of a quantitative comparison of altered/impacted wetlands with those that had not been altered and considered as "reference" wetlands using ecosystem functions as the basis for evaluation. Brinson [41] was intimately involved in the development of the HGM concept and provides a critical analysis of the HGM approach. He cautions on two major limitations in practice. First, it "does not provide decision-makers with complete information to determine the full consequences of degrading a wetland" and second, "only a rapid level of assessment is provided and the building of a reference system is expensive". For an up-to-date summary of recommended approaches in the United States see [39].

With financial support from the European Commission, an empirically based methodology of wetland functional assessment was developed based on trans-European multidisciplinary research [58]. The approach in Europe differed from that adopted in North America, partly because there was no regulatory framework for wetlands and partly because of the generally smaller scale of wetlands and the often intimate association with agricultural land use systems. The basic unit of assessment developed by the Maltby-coordinated international team was the "hydrogeomorphic unit" (HGMU), defined as "areas of homogeneous geomorphology, hydrology and/or hydrogeology, and under normal conditions, homogeneous soil/sediment" [58]. The mapping of HGMUs allows prediction of the variation in wetland process across often complex European wetland ecosystems and associated landscapes, based on the recognition of easily identifiable or measured controlling variables [58].

Its application has never been mandated by regulatory agencies in Europe because of the absence of specific wetland policy requirements. It remains as a tool available to predict the likelihood of a wetland, or part of a wetland, performing a particular function or combination of functions and resulting ecosystem services, together with assessment of the effects of wetland alteration without the need for expensive and time-consuming empirical research [58].

One important further application has been to link with Earth observational data to facilitate the mapping of wetland functions using remote sensing [59]. This is just one example of the very recent expansion of geospatial technologies such as GIS, remote sensing and spatial modelling that are increasingly used for wetland applications including mapping, assessment and simulations.

3.3. Re-Discovery of the Natural Capital Value of Wetlands through Economic Determinations

Society today ultimately is no less dependent on the globe's wetland resources than were the prehistoric communities whose livelihoods were inextricably linked with these ecosystems. Such realization has emerged at least in part from the idea of "natural capital", which recognizes the importance of the natural assets of the planet and their underpinning of economic development as well as wider human well-being. Costanza et al. [60] emphasized the disproportionate contribution of wetlands to the natural capital provided by the world's ecosystems. One estimate is that only 1% of global terrestrial rainfall flows through wetlands but the yields of ecosystem goods and services are disproportionately higher that this [60–63]. Hence, wetlands multiply the value of rainfall compared with other natural and man-made ecosystems [2]. Balmford et al. [61] concluded that conversion of wild habitats to other uses, such as mangroves to aquaculture, was always harmful in overall economic terms. They cite as one example the analysis by van Vuuren and Roy [64], who reported that for freshwater marshes in Canada, the total economic value was more than twice as high when the wetlands remained intact rather than converting them to agriculture.

The Millennium Ecosystem Assessment engaged between 2001 and 2005 over 1360 experts worldwide to assess the consequences of ecosystem change on human well-being. Their findings provide a state-of-the-art scientific appraisal of the condition and trends in the world's ecosystems and the services they provide, as well as the scientific basis for action to conserve and use them sustainably [44]. The key messages from the wetlands and water synthesis report provide insight into the condition, rate of change in the global wetland resources, and captured the new emphasis on the services provided, the value of which often far exceeded alternative land uses. (Box 1).

Box 1. Key messages from the MEA wetlands and water synthesis report [44].

- Wetland ecosystems (including lakes, rivers, marshes, and coastal regions to a depth of 6 m at low tide) are estimated to cover more than 1280 million hectares, an area 33% larger than the United States and 50% larger than Brazil. However, this estimate is known to under-represent many wetland types, and further data are required for some geographic regions. More than 50% of specific types of wetlands in parts of North America, Europe, Australia, and New Zealand were destroyed during the twentieth century, and many others in many parts of the world degraded.
- Wetlands deliver a wide range of ecosystem services that contribute to human well-being, such as fish and fiber, water supply, water purification, climate regulation, flood regulation, coastal protection, recreational opportunities, and, increasingly, tourism.
- When both the marketed and nonmarketed economic benefits of wetlands are included, the total economic value of unconverted wetlands is often greater than that of converted wetlands.
- A priority when making decisions that directly or indirectly influence wetlands is to ensure that information about the full range of benefits and values provided by different wetland ecosystem services is considered.
- The degradation and loss of wetlands is more rapid than that of other ecosystems. Similarly, the status of both freshwater and coastal wetland species is deteriorating faster than those of other ecosystems.
- The primary indirect drivers of degradation and loss of inland and coastal wetlands have been population growth and increasing economic development. The primary direct drivers of degradation and loss include infrastructure development, land conversion, water withdrawal, eutrophication and pollution, overharvesting and overexploitation, and the introduction of invasive alien species.
- Global climate change is expected to exacerbate the loss and degradation of many wetlands and the loss or decline of their species and to increase the incidence of vector-borne and waterborne diseases in many regions. Excessive nutrient loading is expected to become a growing threat to rivers, lakes, marshes, coastal zones, and coral reefs. Growing pressures from multiple direct drivers increase the likelihood of potentially abrupt changes in wetland ecosystems, which can be large in magnitude and difficult, expensive, or impossible to reverse.

Box 1. *Cont.*

- The projected continued loss and degradation of wetlands will reduce the capacity of wetlands to mitigate impacts and result in further reduction in human well-being (including an increase in the prevalence of disease), especially for poorer people in lower-income countries, where technological solutions are not as readily available. At the same time, demand for many of these services (such as denitrification and flood and storm protection) will increase.
- Physical and economic water scarcity and limited or reduced access to water are major challenges facing society and are key factors limiting economic development in many countries. However, many water resource developments undertaken to increase access to water have not given adequate consideration to harmful trade-offs with other services provided by wetlands.
- Cross-sectoral and ecosystem-based approaches to wetland management—such as river (or lake or aquifer) basin-scale management, and integrated coastal zone management—that consider the trade-offs between different wetland ecosystem services are more likely to ensure sustainable development than many existing sectoral approaches and are critical in designing actions in support of the Millennium Development Goals.
- Many of the responses designed with a primary focus on wetlands and water resources will not be sustainable or sufficient unless other indirect and direct drivers of change are addressed. These include actions to eliminate production subsidies, sustainably intensify agriculture, slow climate change, slow nutrient loading, correct market failures, encourage stakeholder participation, and increase transparency and accountability of government and private-sector decision-making.
- Major policy decisions in the next decades will have to address trade-offs among current uses of wetland resources and between current and future uses. Particularly important trade-offs involve those between agricultural production and water quality, land use and biodiversity, water use and aquatic biodiversity, and current water use for irrigation and future agricultural production.
- The adverse effects of climate change, such as sea level rise, coral bleaching, and changes in hydrology and in the temperature of water bodies, will lead to a reduction in the services provided by wetlands. Removing the existing pressures on wetlands and improving their resiliency is the most effective method of coping with the adverse effects of climate change. Conserving, maintaining, or rehabilitating wetland ecosystems can be a viable element to an overall climate change mitigation strategy.
- The MA conceptual framework for ecosystems and human well-being provides a framework that supports the promotion and delivery of the Ramsar Convention's "wise use" concept. This enables the existing guidance provided by the Convention for the wise use of all wetlands to be expressed within the context of human wellbeing and poverty alleviation.

The cross-fertilisation between the expert guidance to the Ramsar Convention and the scientific expertise involved in the MEA was significant in ensuring that the latest thinking could be engaged within the working practices and implementation strategy of the Convention.

Hard on the heels of the MEA was the UK National Ecosystem Assessment (UK NEA)—the first ecosystem assessment to be carried out at the national scale [65]. Figure 19 depicts the Conceptual Framework of the UK NEA, and Figure 20 shows how ecosystem services are incorporated.

The UK NEA originated partly as a need to provide an evidence base on which to adopt an ecosystem approach in government policy as required by commitments to international agreements—notably the Convention on Biological Diversity as well as the Ramsar Convention. It built on the conceptual framework established by the global MEA; most importantly, it provided a direct feed into new UK government policies relating to environment and including wetlands. It aimed to provide a comprehensive overview of the state of the natural environment and a new way of expressing its wealth. The assessment was carried out under eight Broad Habitats in the UK, which meant that wetlands were considered in three of these: freshwaters including open waters, wetlands and floodplains; mountains, moorlands and heaths; and coastal margins.

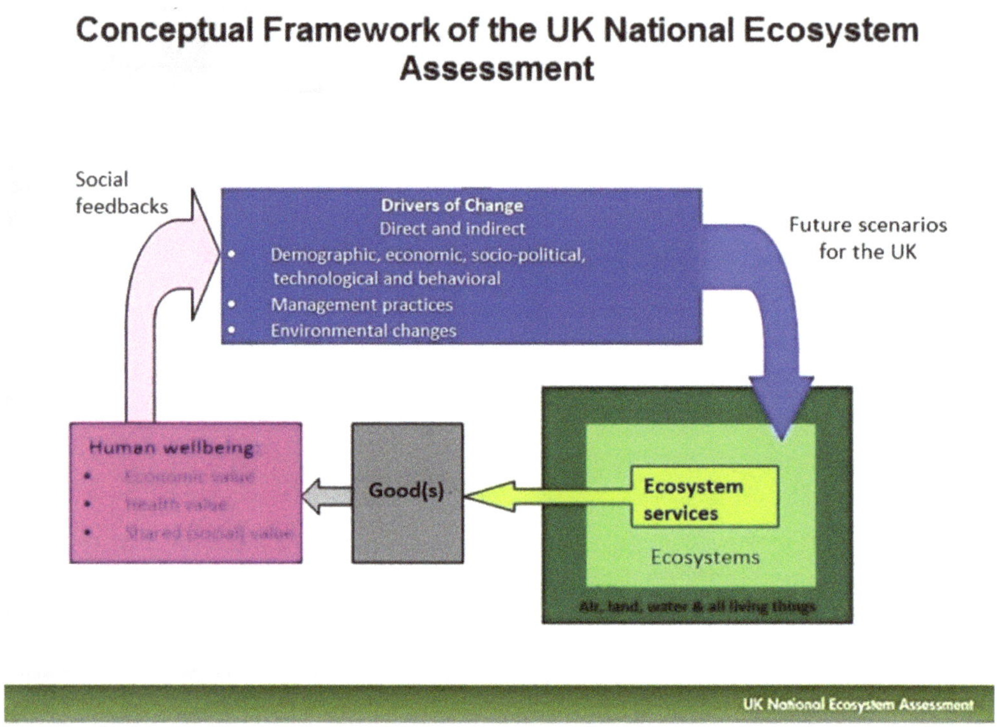

Figure 19. Conceptual framework of the UK National Ecosystem Assessment. Source: UK National Ecosystem Assessment 2011 UNEP-WCMC Cambridge.

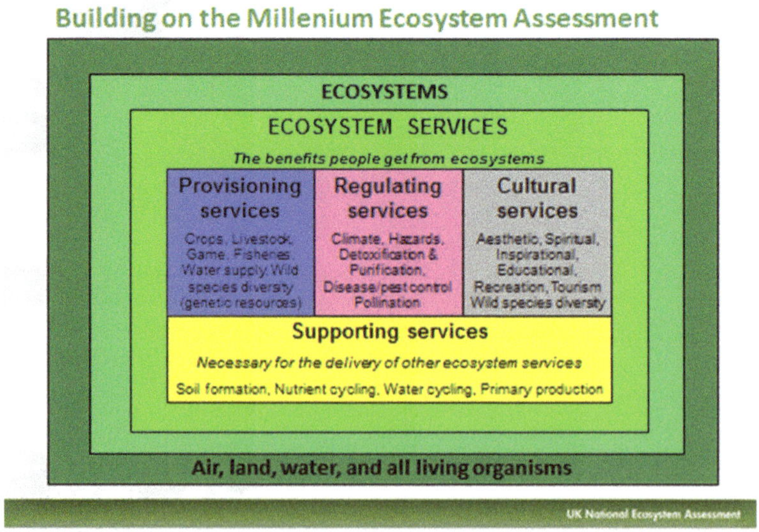

Figure 20. Incorporation of Ecosystem services terminology within the UK NEA. Source: UK National Ecosystem Assessment 2011 UNEP-WCMC.

Of particular significance in the assessment is that each key finding was assigned a level of scientific certainty which provides decision-makers with a level of confidence often previously lacking in the use of research findings for policy innovations. The over-arching conclusion from the freshwaters (including open waters, wetlands) and floodplains assessment was "a need to reappraise our view of the importance of Freshwater ecosystems and their critical position in policy, management and a sustainable economy. This involves recognizing the multiple benefits, potential cost-benefits and wide range of public and private interests which can be supported simultaneously ... through a more holistic approach linked to their pivotal role in delivering ecosystem services. In turn, this recognition will arise from a more practical implementation of the ecosystem approach to integrate the sustainable management of land, water and living resources" [66], and so congruent with both the Ramsar Convention and the Convention on Biological Diversity.

The need for more holistic and integrated thinking has strongly influenced policy, at least in the UK.

The Government's 25-year Plan for the Environment sets out its comprehensive and long-term approach to protect and enhance natural landscapes and habitats. This includes working with nature and using natural capital to benefit communities.

A parallel initiative under the auspices of the Natural Capital Initiative has developed the concept of "Wholescapes", which emphasized the importance of partnership working as essential to integrate sound management of the land, its freshwater, the coast and open seas [67]. The UK National Ecosystem Assessment and follow-on Natural Capital Committee reports have catalogued the severe loss of natural capital in the UK. This is due partly to failure of land and water management practices, because individual sectors are often too narrowly focused and miss the benefits that can be achieved from working across traditional institutional and geographical boundaries. There is increasing awareness that better partnership working can support local economies, improve livelihoods, and enhance quality of life that are all consistent with meeting the objectives within the Sustainable Development Goals (particularly Goal 17 but also 11, 14 and 15). Figure 21 illustrates in a simplistic way how the "wholescapes" concept is inclusive and connective from upland freshwater catchments to the open sea and superimposes the delivery of ecosystem services on the wide range of habitats more traditionally defined.

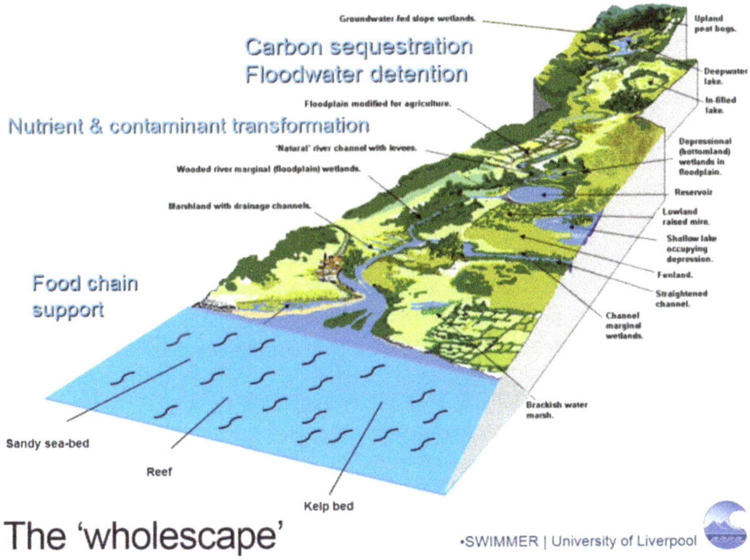

Figure 21. Schematic representation of the Wholescape concept [53].

The most recent changes in the focus on wetlands have concentrated particularly on their roles at scales well beyond individual sites such as within whole freshwater catchments, extensive coastal zones, major biomes and the balance of radiatively active gases in the atmosphere with a particular emphasis on their contribution to the challenges of climate change. The need to uncouple the converted uses of freshwater catchments, such as for food production or urbanization from negative impacts downstream, has brought the wide range of ecosystem services provided by wetlands into strong consideration. A new case for the conservation of peatlands based on their functioning and provision of services contributing to human well-being was outlined by Maltby [25]. Subsequently, Maltby and Acreman [67] have explained the transformation in how modern society values the benefits of natural ecosystems and highlighted the pathfinder role that wetland research has played in this paradigm shift.

Currently wetlands are at the heart of Nature-Based Solutions (NBS) to tackle issues ranging from flooding hazard and poor water quality to the many environmental challenges arising from global climate change.

There is now an unprecedented level of societal awareness of and concern over the possible far-reaching consequences of climate change. Potential impacts include increased extreme events of storms, flooding, drought and wildfires, as well as progressive sea-level rise and biodiversity loss. Wetlands can provide mitigation in respect of all these impacts as well as playing a vital role in the dynamics of greenhouse gases. An assessment of Nature-Based Solutions for Climate Change in the UK coordinated by the British Ecological Society sets out measures that can be used in "Freshwaters, Peatlands and Coastal and Marine systems" as well as other natural systems [67].

A key conclusion from the Freshwaters assessment is that NBS are best and most effectively delivered by local-based partnerships. The role of "Trusted Intermediaries" is vital to facilitate local support, attract resources, and foster engagement of local communities and facilitate Citizen Science. This finding is consistent with the success of the Rivers Trust movement throughout the UK in implementing a wide range of wetland restoration actions with the support of volunteers from the local communities.

3.4. *Upstream and Downstream Thinking*

Changing farming and land management practices aims to reduce the movement of sediments, pesticides and animal waste into rivers. In southwest England, moorland restoration is helping to achieve these aims and improve the condition of rivers, such as the Dart and Exe. This reduces downstream water treatment costs, defers large capital investments, and lowers household water bills. This "Upstream Thinking" approach has been facilitated through a partnership of South West Water, Exmoor National Park, Devon's and Cornwall's Wildlife Trusts, and the Westcountry Rivers Trust (https://wrt.org.uk/project/upstream-thinking/ (accessed on 26 March 2022)).

The long-term goal is to see partnerships amongst and between government, civil society and businesses that operate at the whole scale—linking, where appropriate, land, the coast and sea. Although wholescape is based on geography (the bio-physical scale), its application needs a transformation in human behaviour to affect a cultural change also at this scale. Successful implementation will also require the parallel application of the ecosystem approach and the new thinking that this necessitates (Figure 22).

3.5. *The Economics of Ecosystems and Biodiversity (TEEB)*

Paralleling the new emphasis on natural capital and the ecosystem services which result from wetland functioning is the recognition of the real economic values of wetland ecosystems and water. The Economics of Ecosystems and Biodiversity (TEEB) is a global initiative focused on "making nature's values visible"; its principal objective is to mainstream the values of biodiversity and ecosystem services into decision-making at all levels. It aims to achieve this goal by following a structured approach to valuation that helps decision-makers recognize the wide range of benefits provided by ecosystems and biodiversity, demonstrate their values in economic terms and, where appropriate, capture those values in decision-making. The findings from the part of the programme of work dealing with wetlands capture the essence of the challenges facing policymakers worldwide (see Box 2).

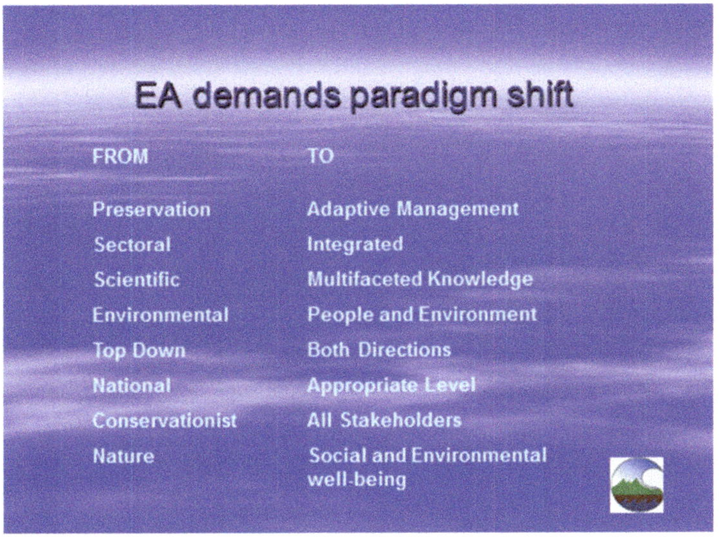

Figure 22. Summary of the key changes in thinking required to effectively apply the ecosystem approach [46].

Box 2. Key Messages from TEEB Wetlands and Water 2013.

1. The "nexus" between water, food and energy is one of the most fundamental relationships—and increasing challenges—for society.
2. Water security is a major and increasing concern in many parts of the world, including both the availability (including extreme events) and quality of water.
3. Global and local water cycle are strongly dependent on wetlands.
4. Without wetlands, the water cycle, carbon cycle and nutrient cycle would be significantly altered, mostly detrimentally. Yet policies and decisions do not sufficiently consider these interconnections and interdependencies.
5. Wetlands are solutions to water security—they provide multiple ecosystem services supporting water security as well as offering many other benefits and values to society and the economy.
6. Values of both coastal and inland wetland ecosystem services are typically higher than for other ecosystem types.
7. Wetlands provide natural infrastructure that can help meet a range of policy objectives. Beyond water availability and quality, they are invaluable in supporting climate change mitigation and adaption, support health as well as livelihoods, local development and poverty eradication.
8. Maintaining and restoring wetlands in many cases also lead to cost savings when compared to manmade infrastructure solutions.
9. Despite their values and despite the potential policy synergies, wetlands have been, and continue to be, lost, or degraded. This leads to biodiversity loss—as wetlands are some of the most biodiverse areas in the world, providing essential habitats for many species—and a loss of ecosystem services.
10. Wetland loss can lead to significant losses of human wellbeing, and have negative economic impacts on communities, countries and business, for example through exacerbating water security problems.
11. Wetlands and water-related ecosystem services need to become an integral part of water management in order to make the transition to a resource efficient, sustainable economy.
12. Action at all levels and by all stakeholders is needed if the opportunities and benefits of working with water and wetlands are to be fully realised and the consequences of continuing wetland loss appreciated and acted upon.

3.6. International Recognition of the Significance of Trans-Boundary Water Control on Wetlands, Political Conflict and the Effects on Human Culture and Livelihoods

One of the outcomes of the war in Iraq was media attention and increased public awareness of the vulnerability of one of the world's most ancient wetlands, and the unique Madan culture associated with the floodplain marshes and shallow water bodies of the former Mesopotamia.

The Mesopotamian marshlands are fed by the Tigris and Euphrates rivers mainly through the flood pulse generated by the annual snow melt from the mountains of Turkey, and to some extent Iran. Numerous tribes of marsh dwellers, Marsh Arabs or "Madan" provide the cultural link with Sumer, one of the world's earliest civilisations. In a pioneering interdisciplinary international study, funded by the NGO Amar Appeal, attention was drawn to the importance of the wetland ecosystems in supporting the livelihoods as well as cultural traditions of the Madan [68,69]. Their significance also at the regional and wider international scale was illustrated in terms of the connectivity to important bird flyways, linkages of aquatic species between the marshes and the Gulf, and as home for rare and endemic species [68,70].

Until recently the marshes occupied an area of some 25,000 km^2 but about 90% of the wetlands existing in the 1970s had disappeared by 2000 [69]. Further assessment in 2003 indicated that just 7% of the original area remained (Figure 23). It was estimated that the entire wetland would disappear within 10–20 years [68], and in 2003 by UNEP within 5 years.

Figure 23. Landsat imagery of the Mesopotamian marshes showing the progressive desiccation from the 1970s to 2000 [68,69].

Dam construction and other engineering works upstream in Turkey and Iran, as well as Iraq itself, provided the Saddam regime with the opportunity to accelerate the drainage of the marshlands through deliberate diversions within the wetlands themselves. Driven by the desire to suppress the traditional opposition from Madan tribes and the provision of refugia for political opponents, the Saddam administration may have displaced up to half a million people, with many killed and the remainder becoming "environmental refugees" [69]. The impact, both on human misery as well as wildlife and cultural heritage, was seized upon by the global media and politicians creating strong public awareness and engagement of national and international NGOs as well as governments.

The uncoordinated restoration of the marshes started in 2003, very soon after the fall of the Saddam regime. Individual communities took the initiative to breach local structures that were preventing normal flooding and wetland regeneration. More organized restoration of hydrological flows followed, and in Spring 2005 UNEP reported recovery of more than 50% of the marshland present in the 1970s (see https://postconflict.unep.ch/publications/UNEP_IMOS.pdf (accessed on 30 March 2022)). There was a remarkably swift re-establishment of many species [71,72], reinforcing the adage "just add water and nature will take care of the rest" (Figures 24 and 25).

The post-war Iraq government, new NGOs such as Nature Iraq and the international aid community (especially US AID) invested heavily in planning and assessment initiatives. From personal observations in the field and more recent reports of others, the restoration of what has been regarded as a "human and environmental catastrophe" is technically feasible [2,73]. Most recently Crisp [73] describes occurrence of the deepest water conditions in the marshes for some 20 years due in part to the increased release of Tigris and Euphrates flows from Turkey together with higher rainfall. In response, there has been significant recovery of important bird and fish species. More than 250,000 people have returned since 2003 but most have occupied surrounding towns and villages rather than the reed houses and floating islands associated with the traditional Madan culture.

Figure 24. Mesapotamian wetlands. Desiccated landscape. Reflooded impoundment with limited biodiversity. Restored wetland within the natural floodplain landscape with traditional canoe, reed housing and animal husbandry. Source: Edward Maltby.

Figure 25. Electrofishing in restored Mesopotamian marshland often replaces more traditional methods, and without sound management and community cooperation may threaten the sustainability of stocks. Source: Edward Maltby.

This likelihood was forecast in interviews carried out by this author during the US AID funded international mission in 2004. There is no doubting the importance of wetland restoration in the opinion of many of the local inhabitants to recover the historic land and waterscape as well as support some livelihoods, environmental quality, and biodiversity. However, there is wide consensus that this should not deny people access to improved health care, more formal education, better transport facilities and energy resources.

The challenge is to achieve the most appropriate balance of restored "natural" ecosystems and sustainable development with the support of the local population, together with awareness of its significance to the wider regional and interests. This may require innovative financial mechanisms to compensate or incentivise communities, in a way like the shift in policy in the UK and elsewhere towards payment for ecosystem services, recognizing that their provision to the benefit of others generally comes at a cost to the local custodians or owners of the resource.

Sustainability of the right balance will inevitably require progressive river basin dialogue among Turkey, Iran and Iraq. It will require comprehensive engagement of diverse stakeholders with often conflicting demands for water and other resources in the region (Figure 26).

The principles of the Ecosystem Approach offer guidance on how to achieve an integrated solution to maintaining sustainable land, water and living resources [47]. There is a potentially important role for the Ramsar Convention to catalyse the ongoing process.

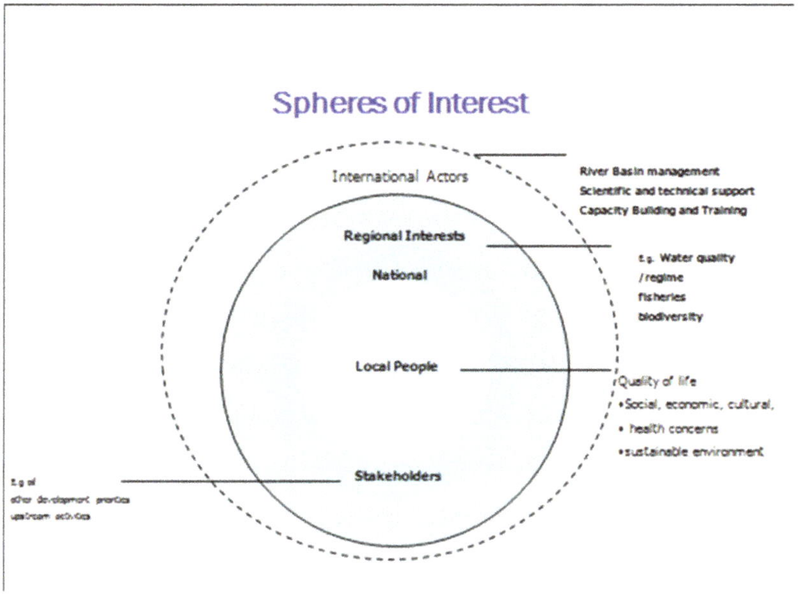

Figure 26. Spheres of influence of stakeholders as presented by the author to the round table discussions among Iraq, Iran and international delegates at the UN Geneva 2003 [74].

3.7. Increase in the Political Priority of Wetland Restoration

A significant element of the paradigm shift in wetland management in recent years has been an increase in momentum of wetland restoration at a range of scales and for a variety of purposes. The overriding driver for the restoration effort has been recognition that such restoration is required to underpin economic and social well-being.

Notably in the United States, The Comprehensive Everglades Restoration Plan (CERP) was authorized by Congress in 2000 as a plan to "restore, preserve, and protect the south Florida ecosystem while providing for other water-related needs of the region, including water supply and flood protection". At a cost of more than USD 10.5 billion and with a 35+ year timeline, this is the largest hydrologic restoration project ever undertaken in the United States, if not the world [75]. It is a further irony that the need for such restoration is to support the population growth, urban expansion, and economic development that have actually been responsible for reduction of the historic extent of Everglades. The Water Resources Development Act (2007) further authorized funds towards restoration of the Florida Everglades. It also recommended the establishment of a Coastal Louisiana Ecosystem Protection and Restoration Task Force that would make recommendations and propose strategies to protect, repair, restore and maintain the ecosystems of the Louisiana coastal zone. This built on The Coastal Wetlands Planning, Protection and Restoration Act, (CWPPRA), the federal legislation enacted in 1990 that is designed to identify, prepare, and fund the construction of coastal wetlands restoration projects. Since its inception, 210 coastal restoration or protection projects have been authorized, benefiting approximately 100,000 acres in Louisiana [76]. Federal, State and university partners have dedicated considerable research effort into the understanding of the underlying causes of coastal wetland loss in Louisiana, currently estimated at about 75 square kilometers per year, and determining the optimum solutions to mitigate and/or reverse such dramatic losses and the resulting economic and human misery [77]. There has been considerable scientific debate over the most sustainable approach to restoration and management. This has focussed in large measure on the restoration of sediment nourishment to the Mississippi Delta. The review by Allison and Meselhe provides some insight into the range of scientific opinion [78]. Current approaches at individual project locations include one or more of the following actions: marsh creation and restoration, shoreline protection, hydrologic restoration, beneficial use of dredged material, terracing, sediment trapping, vegetative planting, barrier island restoration, and bank stabilization techniques.

It is particularly noteworthy that in recent years China has made considerable advances in wetland conservation and protection which now includes over half of the nation's extensive resource. The *People's Daily* newspaper reports wetland expansion of more than 200,000 hectares from 2016–2020, and a new wetland protection system comprising mainly national and more local wetland parks and wetland nature reserves [79]. The enhanced status of wetlands is enshrined in the Wetland Protection Law, passed by the national People's Congress in December 2021 and which entered into force in June 2022 [80]. The new law includes a specific section on wetland restoration in which the principle "natural restoration first" is emphasised partly in recognition that some restoration projects have actually caused more ecological harm. In addressing how to make the law work in practice, Wang Xinyi and Sheng Xiaoying from Friends of Nature, have stressed the importance of public participation in making the law work, especially because of the crucial knowledge of local communities. The potential benefits in creating jobs and prosperity to local residents is a powerful argument for sound management, and where necessary wetland restoration. Almost 10,000 villagers close to the Longji terraced fields national wetland park in South China's Guangxi Zhuang autonomous region have benefitted from Park revenue and local government subsidies. According to the National Forestry and Grassland Administration (NFGA) China's wetland parks contributed CNY 53.6 billion (USD 8.21 billion) to regional economic growth and created 47,000 new jobs. In 2019 China's national wetland parks received 385 million visitors [79].

In Europe, the most recent five year environmental report [81] portrays a somewhat dismal picture of the state of the continent's waters, with only 40% of surface water bodies achieving good ecological status and wetlands widely degraded; this is especially the case for floodplains, resulting in a "critical impact on the conservation status of wetland habitats and the species that depend on them". The report strongly advocates a river basin scale and ecosystem-based approach to future management, with restoration including natural water retention measures and buffer strips to parallel economic measures such as smart water pricing, more efficient irrigation, and so-called precision agriculture. Restoration of wetlands is a common theme supported by policy at the European and individual country levels with the objective of reconnecting rivers to their floodplains a high priority [82,83].

4. Structure of a New Paradigm

An attempt is made in Figure 27 to capture the complexity of the new perspectives of wetlands amidst the present challenges facing Society. The paradigm shift, which is now merging nature, economic growth and prosperity within the same conceptual framework, is being embraced increasingly in policy initiatives worldwide.

4.1. Current Priorities and Future Challenges

Civil society faces arguably unparalleled new threats that may lead to severe and accelerated degradation of our remaining wetland resources but could also offer great opportunities for wetlands to play a renewed role in sound planetary management. The nature of these threats can be grouped under the headings of **F**ood, **E**nergy, **W**ater, **E**nvironment (FEWE). Associated with all of these is the global biodiversity crisis which is a fundamental constraint to actually achieving sustainable economies, livelihoods and human well-being. To what extent can wetlands be seen as the nexus of the threats and opportunities associated with **FEWE**?

Human conflicts, plus forced and unforced migration, are creating refugee crises at various scales and have raised public as well as political awareness of the stark reality and far-reaching implications of climate change; these are just some expressions of exacerbated concerns for sustained human well-being.

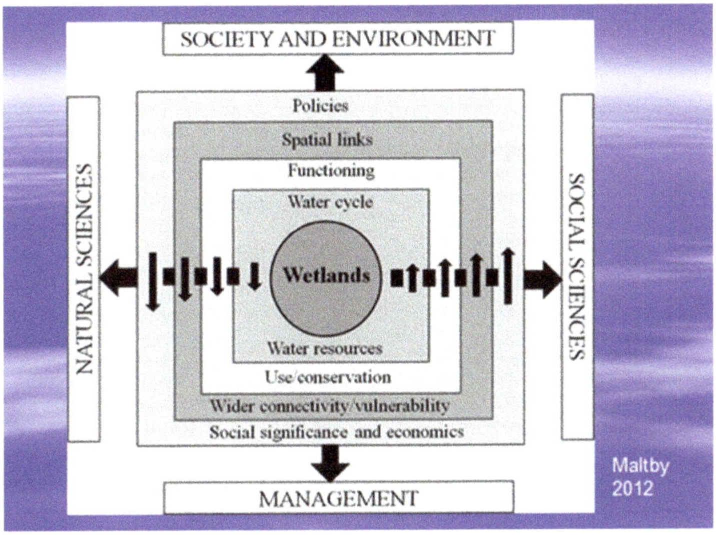

Figure 27. Outline of the new wetland paradigm with multiple linkages among the natural world and human society [2].

Current food shortages and dramatic price inflation arising from the war in Ukraine are likely to renew pressures to convert wetlands and other natural ecosystems to agriculture. A notable historical precedent and example of such a threat comes from the expansion of soybean cultivation in the southern United States from the late 1950s. Between 1959 and 1964, 400,000 hectares in the Mississippi Delta region were drained and cleared almost exclusively for soybean, and in north Louisiana alone forested wetlands disappeared at 45,000 hectares a year [15]. The economics of wetland forest clearance and drainage were given an unanticipated twist in the early 1970s by events off the Pacific coast of South America. Years of over-fishing were compounded in 1972 by a change in the upwelling pattern of the cold, nutrient-laden Peru current resulting in collapse of the anchovy industry, one of the most important sources of protein meal. The demand for soybean increased as a replacement and prices soared. Despite the enormous costs of clearance and drainage, the anchovy crash made cultivation of even difficult wetland terrain financially attractive. Between 1973 and 1975 over 200,000 hectares of wetlands on the coastal plain of North Carolina were cleared. Large agricultural corporations with substantial outside investment converted enormous tracts of intact pocosin(literally 'swamp on a hill') wetland into agriculture. The pocosins once covered almost a million hectares of North Carolina, but by 1980 only 21,800 hectares remained with considerable, but then largely ignored, loss of wetland ecosystem services [84]. Maltby [15] records numerous other examples where changing economics and policy decisions such as relating to taxation (as was the case with commercial afforestation of the peat bogs of the Scottish Flow country) [37], have led to dramatic wetland losses.

At the present time there are major concerns for the global supply of sunflower oil because of the conflict in Ukraine. It remains to be seen if this results in renewed pressures for the conversion of more peat swamp forests in Southeast Asia to palm oil plantations, where already deep concerns surround existing rates of expansion and the loss of habitat for iconic species such as the Orang Utan together with wider ecosystem services [85] (Figure 28).

There is little doubt that the world food and other manufacturers will be closely monitoring the need for alternative ingredients or new country sources, should supplies from Ukraine and elsewhere continue to dwindle and prices rise further. Developing countries, in particular, facing mounting international debt and possibly famine, may find it difficult to resist the pressures which may lead to further wetland degradation and loss in the bid to realize short-term economic returns and meet the problem of food commodity shortages.

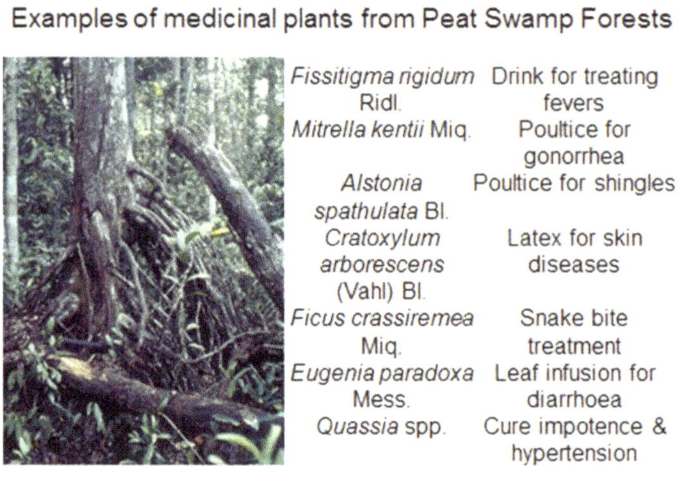

Figure 28. An illustration of the frequently ignored services provided by Peat Swamp Forests in Southeast Asia. After Chai et al. [86]. Photograph Edward Maltby.

Even if wetlands are not converted directly to support additional food production, intensification of farming will inevitably result in greater pressures on catchment hydrology, especially downstream water quality and biodiversity resulting in adverse changes in ecological character and wider ecosystem functioning. The dramatic extent of potential off-site impacts has been well demonstrated in the case of the Everglades and has been a major concern of the Ramsar Convention (Figures 29 and 30).

Figure 29. Wetland stresses and impacts on ecosystem structure and functioning in Everglades. Source: Edward Maltby.

Figure 30. Wetland conversion to agriculture outside Everglades National Park—the Everglades Agricultural Area. Source: Edward Maltby.

It is possible that the use of modified wetlands, such as Storm Treatment Areas (STAs), aimed to reduce the nutrient load from the Everglades Agricultural Area (Figure 30) reaching and altering the wetland ecosystems of the Everglades National Park, can be a potential solution but at costs that may prove prohibitive for application elsewhere. At smaller scales, wetlands may be used effectively as buffer zones to protect adjacent aquatic ecosystems from pollution and ecological damage [87]. Whilst policy has often encouraged the use of buffer zones immediately adjacent the water bodies to be protected, such locations may not be the most effective because of the pattern of runoff and nutrient pollution pathways (Figure 31).

Whilst wetland buffer zones may be highly effective in reducing nutrient pollution and protecting water quality, there may be a downside contribution to global warming resulting from the incomplete process of denitrification releasing nitrous oxide rather than nitrogen gas (Figure 32). Decisions to use wetlands control of nutrient pollution will involve an assessment of "trade-offs" and cost-benefit analysis that goes well beyond short-term economic expedients. This will always depend on the availability of an empirically sound and verifiable evidence base to which the scientific community has contributed already a great deal, but will benefit from increasingly targeted policy-relevant research.

In the face of such major threats to wetlands it is even more important to emphasise the many ways in which these ecosystems actually contribute to the wider sustainable supply of food—a key part of the provisioning services they provide.

It may be necessary now to follow the example set by the IPCC reports, and assemble the evidence regarding the benefits from the healthy functioning of the wetland resource and the consequences of their loss and/or degradation, in such a way that policy makers within governments can take a much more informed, urgent and serious view of what is at stake. The Ramsar Convention could reasonably take a strong lead in such an initiative and subsequent "Wetland Days" may assist in generating supportive public awareness, building on already highly relevant themes of previous years.

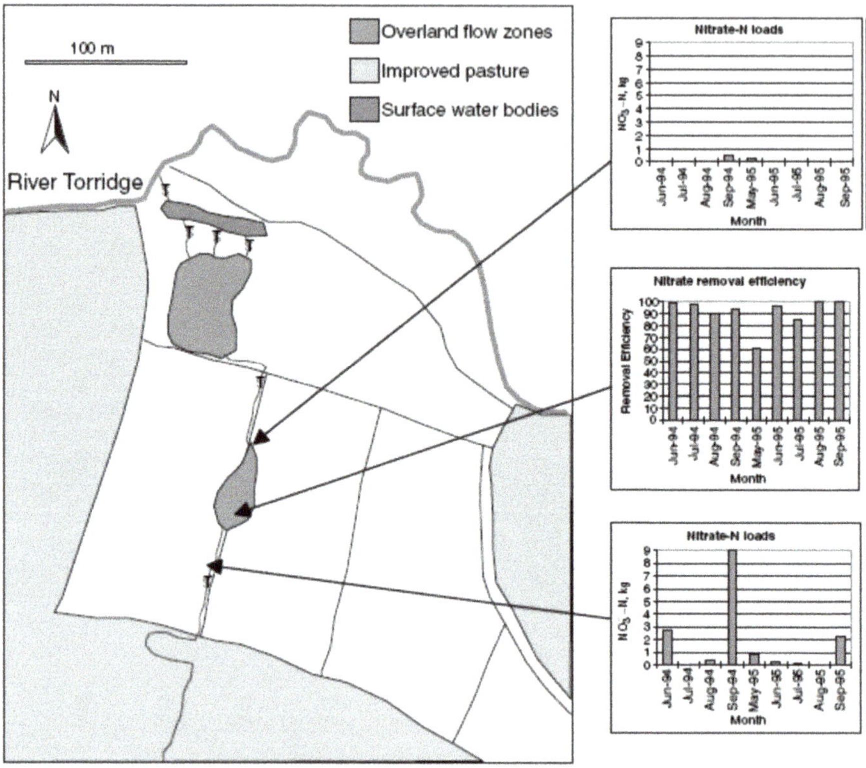

Figure 31. Incongruity between wetland science and policy in which river marginal wetlands running parallel to the channel may not necessarily offer the most effective buffers against nitrate pollution of freshwaters. The graphs show reduced levels of nitrate at locations along the runoff flow line from the source in agricultural fields and at right angles to the alignment of the channel. The grey shaded areas are particularly effective wetland hotspots for nitrate reduction especially by denitrification [88].

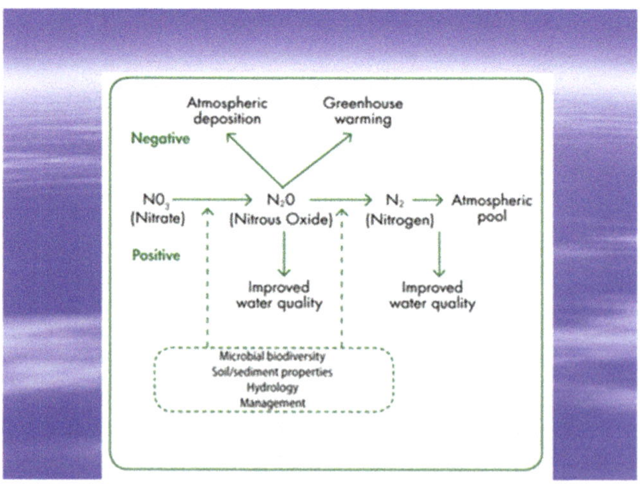

Figure 32. Example of trade-offs between water quality maintenance and global warming [89].

4.2. Energy

The rationale for a drive away from hydrocarbons and the development of alternative renewable energy sources has become generally well-accepted in the face of increasingly incontrovertible evidence for their role in global warming. The argument has gained additional force because of the 2022 spike in world energy prices and an increasing imperative for nations to become less dependent on potentially less-reliable supplies from other countries.

The switch to "green" energy sources is of paramount importance, but care must be exercised where there is the possibility of wetland loss and/or degradation that could outweigh any gains. This might be the case in inappropriate tidal barrage schemes, hydropower dams, or selecting wetland areas for solar or wind power if potentially adverse impacts on ecosystem functioning are either ignored or subjugated to a lesser importance.

4.3. Water

The world has a finite freshwater resource, and the lack of adequate, safe and reliable supply is a major contributor to poverty and deprivation for a significant proportion of the global population. Wetlands are a vital link in the hydrological cycle, controlling flows as well as water quality and mediating the essential connectivity from atmosphere through the land to the sea. Modification of hydrology has been an ever-increasing feature of human history. There are increasing examples of the restoration of pre-existing hydrological conditions including reconnecting rivers to their floodplains [82], managed realignment to accommodate rising sea-levels, and re-wetting peatlands. It is unlikely, however, that future pressures which result in over-exploitation of water resources will cease. In addition to increasing populations, urbanization, and industrial and agricultural pressures, the premium in real estate values attracted by water-side locations will continue to threaten river, lake and coastal wetlands even though in many countries such new development is now severely restricted [90].

Over-exploitation of groundwaters has been a long-standing threat to iconic wetlands dependent on those aquifers, and the Tablas de Damiel in Spain offers a salutary example. Despite its designated and "protected" status, the wetland was severely degraded by groundwater abstraction from the surrounding cultivated area via the La Mancha aquifer in the 1970s. Desiccation combined with high temperatures led to combustion of the peat substrate and extensive fire damage to the ecosystem. The lesson is that even wetland "jewels" included within a nation's conservation network are not necessarily safeguarded from external damaging demands for water. Llamas 1988 [91] describes the conflicts over water use and still the emergency remedial measures to save the wetland are on-going in the face of limited financial resources. Additionally, over-use of coastal aquifers by increasing urban populations, as well as sea-level rise, are causes of salt-water intrusion which can extend far inland resulting in ecological as well as economic damage [92,93].

Desiccation of major wetlands such as Lake Chad in Africa [94] and the Aral Sea in Central Asia [49], in addition to the case of Mesopotamia, are clear signposts to the reality of potentially more extensive losses due to poor management of water resources.

Regional rainfall and hydrology will be affected by climate change. Increases in extremes and unpredictability of droughts and floods are likely consequences, and in this case, wetlands can be an invaluable resource adding a high level of natural resilience to freshwater catchments against the adverse consequences of climate change. Particular benefits will include cost effective management of flood peaks and maintenance of base flows.

4.4. Environmental Change and Health

Concerns surrounding climate change have brought into sharp focus the importance of the natural environment and its "natural capital" in mitigating, arresting or even reversing the effects and/or rate of global warming. Wetlands offer some of the most potent tools to meet the challenges posed by the current climate crisis. Of particular importance is their role as sinks or stores of carbon, which otherwise would add to the atmospheric load; these are well-documented.

Somewhat less explored are the physical and mental health benefits to be gained from engagement with the wetland ecosystem. There is at least some evidence that investment in improving access to wetlands may be cost effective in relation to the returns in human health benefits and reduced costs to traditional health care providers [95].

Responding to the so-called climate crisis in terms of the wetland resource is not invariably straightforward, and an area of concern must be the potential for an increased threat from disease vectors either in type or geographical extent. The complexity and difficulties in predicting the effects of climate change on important disease vectors have been examined by Rocklov and Dubrow [96]. Analysis is urgently required of the potential role of wetlands in facilitating the increased risk of water-borne or other disease vectors under various climate change scenarios and strategies developed for prevention or amelioration.

4.5. World Wetlands Day and the Future

The themes of a quarter of a century of World Wetlands Day, celebrated in this special series, well reflect the new wetland paradigm in which human well-being is a central focus. This in no way reduces the importance of biodiversity and the migratory birds which were the original primary focus of the Ramsar Convention. Biodiversity is a good indicator of ecosystem health, which in turn reflects the ability to function in ways that can deliver the services essential for sustainable livelihoods. Healthy wetland ecosystems are an important part of Earth's natural capital.

5. A New World Charter—Conclusions

Despite the increasing evidence base for the vital natural and economically important roles played by wetlands, the resource continues to decline and degrade. The pressures on wetlands imposed by Society's responses to current and possible future social, economic, and environmental challenges are unlikely to diminish. Frustrated by the lack of effective policies to protect wetlands, a transdisciplinary team have proposed a Universal Declaration of the rights of Wetlands [97]. (Box 3).

Box 3. The proposed Universal Declaration of the rights of Wetlands.

1. The right to exist
2. The right to their ecologically determined location in the landscape
3. The right to natural, connected, and sustainable hydrological regimes
4. The right to ecologically sustainable climatic conditions
5. The right to have naturally occurring biodiversity, free of introduced or invasive species that disrupt their ecological integrity
6. The right to integrity of structure, function, evolutionary processes and the ability to fulfil natural ecological roles in the Earth's processes
7. The right to be free from pollution and degradation
8. The right to regeneration and restoration.

They conclude by urging "all governments, from local to national, as well as international organisations to support this Declaration and provide mechanisms and funding for implementation and enactment". They "specifically encourage the Contracting Parties (countries) to the Ramsar Convention on Wetlands to seek ways to embrace the Declaration and to incorporate the rights of wetlands into their national procedures and operational processes".

Such an initiative is to be applauded, but we must remember that it is people who use and abuse wetlands and it will be at the individual or community level that their values or "rights" will be safeguarded for future generations and the welfare of the planet. Wetland scientists can contribute significantly to this effort, and the ways in which this might be achieved are summarized in Figure 33.

Figure 33. Some Key Challenges for wetland science. Source: Edward Maltby.

Human evolution, prehistoric community survival and the development of civilisations owe much to the natural capital provided by wetlands. Failure to maintain recognition of such significance led to the progressive demise of these ecosystems worldwide, and throughout history. Whilst the rise of conservation ethics and the emergence of ecology as a respected scientific discipline did much to underpin their importance, it was the Ramsar Convention that was of key significance in raising awareness internationally. Notwithstanding the role of Ramsar as a standard bearer of the wetland conservation movement, it has been a renewed recognition of the importance of their natural capital, manifest as wide-ranging ecosystem services, that has reshaped the perspective of Society. Realisation of the roles of wetlands in underpinning human well-being has come full-circle and is the essence of the paradigm shift for wetland scientists, managers and policy-makers alike.

Funding: This research received no external funding.

Informed Consent Statement: Informed consent was obtained from all subjects whose identifiable image appears in the figures.

Data Availability Statement: There are no data sets to support this review.

Acknowledgments: The author would like to thank the many wetland science colleagues worldwide who have generously shared knowledge, insight and friendship that has made his own scientific journey so rewarding. The research referred to would have been impossible without the support of my family and successive teams of research assistants and students together with a plethora of funding bodies who showed confidence in their investment. The present contribution has benefitted from helpful recommendations from anonymous reviewers and the editorial staff of the Journal have provided exceptional and much appreciated support in finalizing the manuscript.

Conflicts of Interest: The author declares no conflict of interest.

References

1. Purseglove, J. *Taming the Flood: Rivers, Wetlands and the Centuries-Old Battle Against Flooding*; William Collins: London, UK, 2017.
2. Maltby, E. The Changing Wetland Paradigm. In *The Wetlands Handbook*; Maltby, E., Barker, T., Eds.; Wiley-Blackwell: Oxford, UK, 2009.
3. Siefferman, R.G. La systeme des grandes tourbieres equatoriales. *Ann. Geogr.* **1988**, *544*, 642–666.
4. Morgan, E. *The Aquatic Apes*; Souvenier Press: London, UK, 1982.

5. Coles, B.; Coles, J. *People of the Wetlands, Bogs, Bodies and Lake-Dwellers*; Thames and Hudson: London, UK, 1989.
6. Steward, J.H. Cultural causality and law: A trial formulation of the development of early civilization. *Am. Anthropol.* **1949**, *51*, 1–27. [CrossRef]
7. Mitchell, W.P. The hydraulic hypothesis: A reappraisal. *Curr. Anthropol.* **1973**, *14*, 532–534. [CrossRef]
8. Wooley, C.L. *The Sumerians*; Norton: New York, NY, USA; London, UK, 1965.
9. Mitsch, W.J.; Gosselink, J.G. *Wetlands*, 3rd ed.; Wiley: New York, NY, USA, 2000.
10. Anderson, S.; Moss, B. How wetland habitats are perceived by children: Consequences for children's education and wetland conservation. *Int. J. Sci. Educ.* **1993**, *15*, 473–485. [CrossRef]
11. Baldock, D. *Wetland Drainage in Europe*; IIED/IIEP: London, UK, 1984.
12. Darby, H.C. *The Changing Fenland*; Cambridge University Press: Cambridge, UK, 1983.
13. Kadlec, R.H. Wetlands for Contaminant and Wastewater Treatment Chapter 20. In *The Wetlands Handbook*; Maltby, E., Barker, T., Eds.; Wiley: New York, NY, USA; London, UK, 2009.
14. Richardson, C.J. The Everglades: North America's subtropical wetland. *Wetl. Ecol. Manage* **2010**, *18*, 517–542. [CrossRef]
15. Maltby, E. *Waterlogged Wealth, Why Waste The Worlds Wet Places*; Earthscan: London, UK, 1986.
16. Shaw, S.P.; Fredine, C.G. *Wetlands of the United States, Their Extent, and Their Value for Waterfowl and Other Wildlife*; Circular 39; U.S Fish and Wildlife Service, U.S Department of Interior: Washington, DC, USA, 1956; 67p.
17. Moore, P.D. Origin of blanket mires. *Nature* **1975**, *256*, 267–269. [CrossRef]
18. Moore, P.D.; Bellamy, D.J. *Peatlands*; Elek Science: London, UK, 1974.
19. Godwin, H. *The Archives of the Peat Bogs*; Cambridge University Press: Cambridge, UK, 1981.
20. O'Connell, M. Origin of Irish lowland blanket bog. In *Ecology and Conservation of Irish Peatlands*; Doyle, E.J., Ed.; Royal Irish Academy: Dublin, Ireland, 1990; pp. 49–71.
21. Rodwell, J.S. (Ed.) *British Plant Communities, Volume 2. Mires and Heath*; Cambridge University Press: Cambridge, UK, 1991.
22. Lindsay, R.; Charman, D.J.; Everingham, F.; O'Reilly, R.M.; Palmer, M.A.; Rowell, T.A.; Stroud, D.A. *The Flow Country: The peatlands of Caithness and Sutherland*; Joint Nature Conservation Committee: Peterborough, UK, 1988.
23. Merryfield, D.I.; Moore, P.D. Prehistoric human activity and blanket mire initiation on Exmoor. *Nature* **1974**, *250*, 439–441. [CrossRef]
24. Moore, P.D. The influence of prehistoric cultures on the initiation and spread of blanket bog in upland Wales. *Nature* **1973**, *241*, 350–353. [CrossRef]
25. Maltby, E. Effects of climate change on the societal benefits of UK upland peat ecosystems: Applying the ecosystem approach. *Clim. Res.* **2010**, *45*, 249–259. [CrossRef]
26. Taylor, J.A.; Smith, R.T. The role of pedogenic factors in the initiation of peat formation and the classification of mires. In Proceedings of the 6th International Peat Congress, Duluth, MN, USA, 17–23 August 1980.
27. Dimbleby, G.W. *The Development of British Heathlands and Their Soils*; Oxford Forestry Memoirs 23; Clarendon Press: Oxford, UK, 1962.
28. Maltby, E.; Caseldine, C.J. Prehistoric soil and vegetation development on Bodmin Moor, southwestern England. *Nature* **1982**, *297*, 397–400. [CrossRef]
29. Ivanov, K.E. *Water Movement in Mirelands*; Academic Press: London, UK, 1981.
30. Ingram, H.A.P. Size and shape in raised mire ecosystems: A geophysical model. *Nature* **1982**, *297*, 300–303. [CrossRef]
31. Clymo, R.S. The limits to peat bog growth. *Philos. Trans. R. Soc. Lond. B* **1984**, *303*, 605–654.
32. Clymo, R.S. Models of peat growth. *Suo* **1992**, *43*, 127–136.
33. Smith, R.; Clymo, R. An extraordinary peat-forming community on the Falkland Islands. *Nature* **1984**, *309*, 617–620. [CrossRef]
34. Junk, W.J.; Bayley, P.B.; Sparks, R.E. The flood pulse concept in river-floodplain systems. *Can. Spec. Publ. Fish. Aquat. Sci.* **1989**, *106*, 110–127.
35. McIver, S.B. *Death in the Everglades: The Murder of Guy Bradley, America's First Martyr to Environmentalism*; University of Florida Press: Gainesville, FL, USA, 2003.
36. Carson, R. *Silent Spring*; Houghton Miffin Company: New York, NY, USA, 1962.
37. Stroud, D.A.; Teed, T.M.; Pienkowski, M.W.; Lindsay, R.A. *Birds, Bogs and Forestry—The Peatlands of Caithness and Sutherland*; Nature Conservancy Council: Peterborough, UK, 1987.
38. Patterson, G.; Anderson, R. *Forests and Peatland Habitats: Guideline Note*; Forestry Commission: Edinburgh, UK, 2000.
39. Ramsar. Regina COP. 1987. Available online: https://www.ramsar.org/search?f[]=field_tag_body_event%3A366&f[]=field_tag_body_event%3A561 (accessed on 26 June 2009).
40. Larson, J.S. Methodologies for Wetland Assessment Chapter 21. In *The Wetlands Handbook*; Maltby, E., Barker, T., Eds.; Wiley: New York, NY, USA; London, UK, 2009.
41. Brinson, M.M. The United States HGM (Hydrogeomorphic) Approach Chapter 22. In *The Wetlands Handbook*; Maltby, E., Barker, T., Eds.; Wiley: New York, NY, USA; London, UK, 2009.
42. Smith, R.D. Wetlands Assessment in Practice: Development and Application in the United States Regulatory Context, Chapter 24. In *The Wetlands Handbook*; Maltby, E., Barker, T., Eds.; Wiley: New York, NY, USA; London, UK, 2009.
43. Hollis, G.E.; Holland, M.M.; Maltby, E.; Larson, J.S. Wise Use of Wetlands. *Nat. Resour.* **1988**, *24*, 2–13.

44. Maltby, E. *Wise Use and Conservation of Wetlands. Evidence to the House of Commons Select Committee on European Legislation Document: Commission Communication (16343) 8564/95 COM (95)189*; House of Commons Library: London, UK, 1996.
45. Millennium Ecosystem Assessment (MEA). *Ecosystem and Human Well-Being: Wetlands and Water Synthesis*; World Resources Institute: Washington, DC, USA, 2000.
46. Maltby, E. Ecosystem Approach: From Principles to Practice. In *The Norway/UN Conference on the Ecosystem Approach for Sustainable Use of Biological Diversity. Proceedings of the Trondheim Conference, Norway September 1999*; Schei, P.J., Sandlund, O.T., Strand, R., Eds.; Norwegian Institute for Nature Research and Norwegian Directorate for Nature Management: Trondheim, Norway, 1999; pp. 30–40.
47. Maltby, E. Using the Ecosystem Approach to implement the CBD. In *Report of the International Workshop on 'Further Development of the Ecosystem Approach'*; Korn, H., Schliep, R., Stadler, J., Eds.; German Federal Agency for Nature Conservation: Bonn, Germany, 2002; pp. 104–109.
48. Brundtland Commission. *Our Common Future*; Oxford University Press: Oxford, UK, 1987.
49. Roggeri, H. Wetland Evaluation in Developing Countries, Chapter 25. In *The Wetlands Handbook*; Maltby, E., Barker, T., Eds.; Wiley: New York, NY, USA; London, UK, 2009.
50. Hanson, A.; Swansin, L.; Ewig, D.; Grabas, G.; Meyer, S.; Ross, L.; Watmough, M.; Kibey, J. Wetland Ecological Functions Assessment: *An Overview of Approaches*; Canadian Wildlife Service Technical Report Series No. 497; Atlantic Region. 2008. Available online: https://publications.gc.ca/collections/collection_2010/ec/CW69-5-497-eng.pdf (accessed on 26 June 2012).
51. Maltby, E.; Hogan, D.V.; Immirzi, C.P.; Tellam, J.H.; van der Peijl, M.J. Building a new approach to the investigation and assessment of wetland ecosystem functioning. In *Global Wetlands: Old World and New*; Mitsch, W.J., Ed.; Elsevier: Amsterdam, The Netherlands, 1994; pp. 637–658.
52. Sather, J.M.; Smith, R.D. *An Overview of Major Wetland Functions and Values Report for US Fish and Wildlife Service FWS/OBS/-84 (citing US Corps of Engineers 1972)*; US Fish and Wildlife Service: Washington, DC, USA, 1984.
53. Maltby, E.; Barker, T.; Linstead, C. Development of a European Methodology for the Functional Assessment of Wetlands. In *The Wetlands Handbook*; Maltby, E., Barker, T., Eds.; Wiley-Blackwell: Oxford, UK, 2009.
54. Bartoldus, C.C. *A Comprehensive Review of Wetland Assessment Procedures: A Guide for Wetland Practitioners*; Environmental Concern Inc.: St. Michaels, MD, USA, 1999.
55. Adamus, P.R. *A Method for Wetland Functional Assessment. Volume II FHWA Assessment Method Rep No FHWA-IP-82-24*; Federal Highway Administration US Department of Transportation: Washington, DC, USA, 1983.
56. Adamus, P.R.; Stockwell, L.T.; Clarain, E.J.; Smith, R.D.; Young, R.E. *Wetland Evaluation Technique (WET) Volume II Operational Draft TRY-87*; US Corps of Engineers Waterways Experiment Station: Vicksburg, MS, USA, 1987.
57. Smith, R.D.; Ammann, A.; Bartoldus, C.; Brinson, M.M. *An Approach for Assessing Wetland Functions Using Hydromorphic Classification, Reference Wetlands and Functional Indices*; Technical Report TR WRP-DE-10; US Army Corps of Engineers, Waterways Experiment Station: Vicksburg, MS, USA, 1995.
58. Maltby, E. (Ed.) *Functional Assessment of Wetlands: Towards Evaluation of Ecosystem Services*; Woodhead Publishing Ltd.: Sawston, UK, 2009.
59. Rapinel, S.; Hubert-Moy, L.; Clement, B.; Maltby, E. Mapping wetland functions using Earth observation data and multicriteria analysis. *Environ. Monit. Assess.* **2016**, *188*, 641. [CrossRef] [PubMed]
60. Costanza, R.; d'Arge, R.; de Groot, R.; Farber, S.; Grasso, M.; Hannon, B.; Limburg, K.; Naeem, S.; O'Neill, R.V.; Paruelo, J.; et al. The value of the world's ecosystem services and natural capital. *Nature* **1997**, *387*, 253–260. [CrossRef]
61. Balmford, A.; Brouer, A.; Cooper, P.; Costanza, R.; Farber, S.; Green, R.; Jenkins, M.; Jefferiss, P.; Jessamy, V.; Madden, J.R.; et al. Economic reasons for conserving wild nature. *Science* **2002**, *297*, 950–953. [CrossRef]
62. Falkenmark, M. Meeting water requirements of an expanding world population. *Philos. Trans. R. Soc. Lond. B* **1997**, *352*, 929–936. [CrossRef]
63. Falkenmark, M.; Rockstrom, J. *Balancing Water for Humans and Nature. The New Approach in Ecohydrology*; Earthscan: London, UK, 2004.
64. Van Vuuren, W.; Roy, P. Private and social returns from wetland preservation versus those from conversion for agriculture. *Ecol. Econ.* **1993**, *8*, 289–305.
65. UKNEA (UK National Ecosystem Assessment). *Preliminary Synthesis and Progress Report on Status and Trends*; UK National Ecosystem Assessment, UNEP World Conservation Monitoring Centre: Cambridge, UK, 2010.
66. Maltby, E.; Ormerod, S.; Acreman, M.; Blackwell, M.; Durance, I.; Everard, M.; Morris, J.; Spray, C. *Freshwaters: Openwaters, Wetlands and Floodplains. The UK National Ecosystem Assessment: Technical Report*; UNEP-WCMC: Cambridge, UK, 2011; pp. 295–360.
67. Maltby, E.; Acreman, M.; Maltby, A.; Bryson, P.; Bradshaw, J. Wholescape Thinking: Towards Integrating the Management of Catchments, Coast and the Sea through Partnerships—A Guidance Note. [Online] Natural Capital Initiative. 2019. Available online: http://goto.rsb.org.uk/rsbocly7 (accessed on 10 January 2021).
68. Maltby, E. (Ed.) *An Environmental and Ecological Study of the Marshlands of Mesopotamia*; Amar Appeal Trust: London, UK, 1994.
69. Maltby, E. *Approaches to the Re-Establishment of the Freshwater-Marine Continuum through the Mesopotamian-Marshes: Regime, sectoral and Transboundary Challenges*; ROPME and UNEP: Geneva, Switzerland, 2005; pp. 7–9.
70. Partow, H. *The Mesopotamian Marshlands: Demise of An Ecosystem Report United Nations Environment Program*; UNEP: Geneva, Switzerland, 2001.

71. Richardson, C.J.; Hussain, N.A. Restoring the Garden of Eden: An ecological assessment of the marshes of Iraq. *Bioscience* **2006**, *56*, 477–489. [CrossRef]
72. Alwash, S. *Eden Again: Hope in the Marshes of Iraq*; Talbet House Publishing: Fullerton, CA, USA, 2013.
73. Crisp. Available online: https://www.positive.news/environment/how-a-disaster-zone-in-iraq-became-a-conservation-success-story/ (accessed on 20 March 2022).
74. Maltby, E. Re-establishment of Wetland Functioning. In *The Wetlands Handbook*; Maltby, E., Barker, T., Eds.; Blackwells: Oxford, UK, 2009; pp. 721–728.
75. Comprehensive Everglades Restoration Plan (CERP). Available online: https://www.nps.gov/ever/learn/nature/cerp.htm#:~{}:text=The%20CERP%20was%20authorized%20by,line%2C%20this%20is%20the%20larges (accessed on 31 May 2022).
76. The CWPPRA Legislation. Available online: https://www.lacoast.gov/new/About/Default.aspx (accessed on 31 May 2022).
77. Louisiana Coastal Wetlands: A Resource At Risk. Available online: https://pubs.usgs.gov/fs/la-wetlands/#:~{}:text=%22The%20swamps%20and%20marshes%20of,billion%20per%20year%20seafood%20industry (accessed on 31 May 2022).
78. Allison, M.A.; Meselhe, E.A. The use of large water and sediment diversions in the lower Mississippi River (Louisiana) for coastal restoration. *J. Hydrol.* **2010**, *387*, 346–360.
79. China Makes Major Progress in Protecting Its Wetlands. Available online: http://english.www.gov.cn/news/topnews/202103/26/content_WS605d77e6c6d0719374afb882.html#:~{}:text=China%20has%20made%20g (accessed on 24 May 2022).
80. How to Make China's Long-Awaited Wetlands Protection Law Work. Available online: https://chinadialogue.net/en/nature/how-to-make-chinas-long-awaited-wetlands-protection-law-work/ (accessed on 31 May 2022).
81. The European Environment—State and Outlook 2020: Knowledge for Transition to a Sustainable Europe. Available online: https://www.eea.europa.eu/soer/2020 (accessed on 24 May 2022).
82. Blackwell, M.S.A.; Maltby, E. *Ecoflood Guidelines: How to Use Floodplains for Flood Risk Reduction*; European Commission D.G. Research: Brussels, Belgium, 2006; p. 144.
83. Natural Water Retention Measures. Available online: https://ec.europa.eu/environment/water/adaptation/ecosystemstorage.htm (accessed on 22 May 2022).
84. Richardson, C.J. Pocosins: An Ecological Perspective. *Wetlands* **1991**, *11*, 335–354.
85. Meijaard, E. The Importance of Swamp Forest for the Conservation of the Orang Utan (*Pongo pygmaeus*) in Kalimantan, Indonesia. In *Tropical Peatlands*; Riley, J.O., Page, S.E., Eds.; Samara Publishing Ltd.: Cardigan, UK, 1997; pp. 243–254.
86. Chai, P.P.K.; Lee, B.M.H.; Ismawi, H.O. *Native Medicinal Plants of Sarawak Report FB1*; Forestry Department: Sarawak, Malaysia, 1989.
87. Blackwell, M.S.A.; Hogan, D.V.; Pinay, G.; Maltby, E. The Role of Buffer Zones for Agricultural Run-Off. In *The Wetlands Handbook*; Maltby, E., Barker, T., Eds.; Wiley-Blackwell: Oxford, UK, 2009.
88. Blackwell, M.S.A.; Hogan, D.V.; Maltby, E. The use of conventionally and alternatively located buffer zones for the removal of nitrate from diffuse agricultural run-off. *Water Sci. Technol.* **1999**, *39*, 157–164. [CrossRef]
89. Elmquist, T.; Maltby, E. *Biodiversity, Ecosystems and Ecosystem Services. Ecological and Economic Foundations. Chapter 2*; Earthscan: London, UK, 2010.
90. Waterfront Property in Demand as Buyers Weigh Quality of Life during Covid Pandemic. Available online: https://www.knightfrank.com/research/article/2020-09-23-waterfront-view-2020 (accessed on 23 May 2022).
91. Llamas, M.R. Conflicts between wetland conservation and groundwater exploitation: Two case histories in Spain. *Environ. Geol. Water Sci.* **1988**, *11*, 241–251.
92. Werner, A.D.; Simmons, C.T. Impact of sea-level rise on saltwater intrusion in coastal aquifers. *Groundwater* **2009**, *47*, 197–204. [CrossRef]
93. Alfarrah, N.; Walraevens, K. Groundwater over-exploitation and seawater intrusion in coastal areas of arid and semi-arid regions. *Water* **2018**, *10*, 143. [CrossRef]
94. Coe, M.T.; Foley, J.A. Human and natural impacts on the water resources of the Lake Chad Basin. *J. Geophys. Res.* **2001**, *106*, 3349–3356.
95. Maund, P.R.; Irvine, K.N.; Reeves, J.; Strong, E.; Cromie, R.; Dalliner, M.; Davies, Z.G. Wetlands for Wellbeing: Piloting a Nature-Based Health Intervention for the Management of Anxiety and Depression. *Int. J. Environ. Res. Public Health* **2019**, *16*, 4413. [CrossRef]
96. Rocklöv, J.; Dubrow, R. Climate change: An enduring challenge for vector-borne disease prevention and control. *Nat. Immunol.* **2020**, *21*, 479–483. [CrossRef] [PubMed]
97. Davis, G.T.; Finlayson, C.M.; Pritchard, D.E.; Davidson, N.C.; Gardner, R.C.; Moomaw, W.R.; Okuno, E.; Whitacre, J.C. Towards a Universal Declaration of the Rights of Wetlands. *Mar. Freshw. Res.* **2020**, *72*, 593–600. [CrossRef]

land

MDPI

Review

Factors Affecting Wetland Loss: A Review

Gastón Antonio Ballut-Dajud [1,†], Luis Carlos Sandoval Herazo [1,†], Gregorio Fernández-Lambert [1], José Luis Marín-Muñiz [2], María Cristina López Méndez [1] and Erick Arturo Betanzo-Torres [1,*,†]

1 Wetlands and Environmental Sustainability Laboratory, Division of Graduate Studies and Research, Tecnológico Nacional de México/Instituto Tecnológico de Misantla, Km 1.8 Carretera a Loma del Cojolite, Misantla 93821, Veracruz, Mexico; gaston.ballut@unisucre.edu.co (G.A.B.-D.); lcsandovalh@itsm.edu.mx (L.C.S.H.); gfernandezl@itsm.edu.mx (G.F.-L.); mclopezm@itsm.edu.mx (M.C.L.M.)

2 Academy of Sustainable Regional Development, El Colegio de Veracruz, Xalapa 91000, Veracruz, Mexico; soydrew@hotmail.com

* Correspondence: eabetanzot@itsm.edu.mx

† These authors contributed equally to this work.

Abstract: Despite occupying an area no greater than 8% of the earth's surface, natural wetland ecosystems fulfill multiple ecological functions: 1. Soil formation and stabilization support, 2. Food, water, and plant biomass supply, 3. Cultural/recreational services, landscape, and ecological tourism, 4. Climate regulation, and 5. Carbon sequestration; with the last one being its most important function. They are subject to direct and indirect incident factors that affect plant productivity and the sequestration of carbon from the soil. Thus, the objective of this review was to identify the incident factors in the loss of area and carbon sequestration in marine, coastal, and continental wetlands that have had an impact on climate change in the last 14 years, globally. The methodology consisted of conducting a literature review in international databases, analyzing a sample of 134 research studies from 37 countries, organized in tables and figures supported by descriptive statistics and content analysis. Global results indicate that agriculture (25%), urbanization (16.8%), aquaculture (10.7%), and industry (7.6%) are incident factors that promote wetlands effective loss affecting continental wetlands more than coastal and marine ones. Regarding carbon sequestration, this is reduced by vegetation loss since GHG emissions raise because the soil is exposed to sun rays, increasing surface temperature and oxidation, and raising organic matter decomposition and the eutrophication phenomenon caused by the previous incident factors that generate wastewater rich in nutrients in their different activities, thus creating biomass and plant growth imbalances, either at the foliage or root levels and altering the accumulation of organic matter and carbon. It is possible to affirm in conclusion that the most affected types of wetlands are: mangroves (25.7%), lagoons (19.11%), and marine waters (11.7%). Furthermore, it was identified that agriculture has a greater incidence in the loss of wetlands, followed by urbanization and industry in a lower percentage.

Keywords: anthropogenic activities; climate change; terrestrial ecosystems; environmental impacts; greenhouse gases

Citation: Ballut-Dajud, G.A.; Sandoval Herazo, L.C.; Fernández-Lambert, G.; Marín-Muñiz, J.L.; López Méndez, M.C.; Betanzo-Torres, E.A. Factors Affecting Wetland Loss: A Review. *Land* **2022**, *11*, 434. https://doi.org/10.3390/land11030434

Academic Editor: Richard C. Smardon

Received: 12 February 2022
Accepted: 14 March 2022
Published: 17 March 2022

1. Introduction

The ecosystemic value of wetlands is the set of functions, characteristics, or processes that indirectly or directly contribute to human well-being [1]. The most stand out functions are: wildlife habitat, water supply, and carbon sequestration [2,3]. They additionally produce food, medicine, and recreational uses [4], along with water purification and filtration of agricultural pollutants [5]. In addition, they counteract the effects of climate change through atmospheric CO_2 sequestration that they capture and store in the long term either naturally or through man-made sinks [6].

It is important to consider that wetlands represent between 5% and 8% of the earth's surface [7,8], constituting 29.83 million km^2 distributed in Asia (9.2 million), South America

(7.95 million) and North America (5.65 million), and representing 78% of the global percentage [9]. Africa has 19% (5.6 million) [10]; especially vegetated coastal wetlands, that store organic and inorganic carbon in greater quantities [11].

Despite the benefits provided by wetlands, the dense populations that develop in coastal areas along with their added poverty, are both responsible for 1% to 3% of annual deforestation [12]. It is estimated that this population around wetlands, favored with its benefits, would reach half the world population located within a radius of 100 km from the coast [13] as a result of the damages caused to their vegetation. Davidson [14] made a balance of area losses by continents since 1900 and adjusted the deforestation rate from 64% to 71%, being higher in coastal wetlands.

One of the main environmental services provided by natural wetlands is carbon sequestration that takes place both in the soil and biomass. In sediments, it is in a 2 to 3 ratio compared to biomass storage [15]. In addition, Marin-Muñiz et al. [16], from their research on several marshes and swamps from the Gulf of Veracruz in Mexico, determined that accumulated carbon was higher in swampy soils than marshes, (0.92 ± 0.12 KgC/m^{2}Yr and 0.31 ± 0.08 KgC/m^{2}Yr, respectively) in a ratio of 3 to 1.

Riparian and coastal wetlands can sequester 50 times more carbon in the soil than other land forest systems [17]. Other more conservative authors such as Donato et al. [18] manifest that in the first 30 cm of soil in wetlands such as mangroves, twice more carbon can be stored in one hectare than boreal forests. This thickness or layer is the most susceptible to changes in land use. On the other hand, Adame et al. [19], citing the Intergovernmental Panel on Climate Change (IPCC), state that mangroves can store two or three times more carbon than tropical and temperate forests.

The largest soil organic carbon reserves are found in tropical wetlands. Köchy et al. [20] and Villa and Bernal [21] confirm that they can store up to one-third of global carbon. Mangroves store 218-ton C/year from the atmosphere, a key component of the so-called 'blue carbon' (Cui et al.) [22] referring to the carbon exchanged by habitats near the coast with vegetation [23].

Therefore, determining natural wetland conditions and their evolution over time is important for the global context since the value of ecosystem services per unit area would be known, including areas where vegetation losses develop. They are desirable to restore and protect, as well as to enhance through priority policies [24] due to their quality of being natural carbon sinks.

Vegetation loss implies loss of its functions not only in the ecosystem services for man, but also in the breakdown of the world carbon cycle, and in the water and nutrient cycles [25]. Furthermore, global warming as a result of climate change born from the industrial revolution puts future human survival at risk and has become a challenge to mitigate gas emissions and conserve ecosystems such as wetlands that absorb CO_2 [26].

In this context, the objective of this review was to identify the incident factors of loss of area and carbon sequestration in marine, coastal, and continental wetlands that have had an impact on climate change in the last 14 years at a global level.

2. Materials and Methods

2.1. Information Sources

This research is of a qualitative type, made up of articles integrated with a database of 134 publications from the last decade, adding some research from the second last one to further strengthen the search. This temporality was determined following what was suggested by von Uexkull and Buhaug [27] indicating that in the last decade, scientific research has abundantly been carried out in relation to conflicts associated with climate, considering variable incidents in losses. Such is the case of natural events such as floods, cyclones, pests, or diseases because mangroves and wetlands show resistance to this kind of disturbance [12].

The review was structured by articles, book chapters, books, and theses, and was published in both English (93%) and Spanish (7%). For the search, Google Scholar (26%),

Sciencedirect (10%), 1findr (19%), Springer (11%), Proquest (19%), and EBSCO (15%) were used. The selected articles were reviewed by academic peers, who guaranteed the quality of the collected data. When starting the search, keywords were used at phase one such as: carbon fixation in wetlands, carbon in tropical wetlands, carbon in tropical coastal wetlands, carbon sequestration in tropical coastal wetlands, wetland boundaries, and remote sensors in coastal wetland vegetation. At phase two, more specific words such as: coastal wetlands, carbon sequestration, remote sensing, and others. Finally, in Spanish, pérdida de humedales [loss of wetlands] and actividades antropogénicas y naturales [anthropogenic and natural activities], four inclusion and exclusion criteria were used to review the literature and obtain the final database, as shown in Figure 1.

2.2. Information Analysis

This document was organized with all the extracted data highlighting activities that affect vegetation loss and decrease carbon sequestration capacity in marine, coastal, and continental wetlands, considering its geographical position, and organizing it in tables by continent, country, wetland location, incident factor, indicator, and percentage of affected area.

Descriptive statistics were applied, constructing frequency histograms by incident factor by country, continent, and type of affected wetland.

According to the Ramsar International Convention, a classification for wetlands was followed in five levels: system, subsystem, class, subclass, and wetland type. This review was focused on two levels: the system including the marine, coastal, and continental ones; and the wetland type, here, the mangroves, lagoons, marshes, rivers, marine waters, peatlands, swamps, and river deltas. Some wetlands considered natural reserves and those within national parks were also included.

2.3. Statistical Analysis

For mapping locations of sites studied in the investigations, the software ArcGIS, Version 10.0 was used. In addition, content analysis was the technique used, and an analysis guide was used as an instrument for the information by continents regarding the incident factors, countries, and types of wetlands. The Origin Pro Software, version 2021 (OriginLab Corporation, Northampton, MA, USA) was used to elaborate the frequency histograms. Figure 1 shows the methodology in a flowchart.

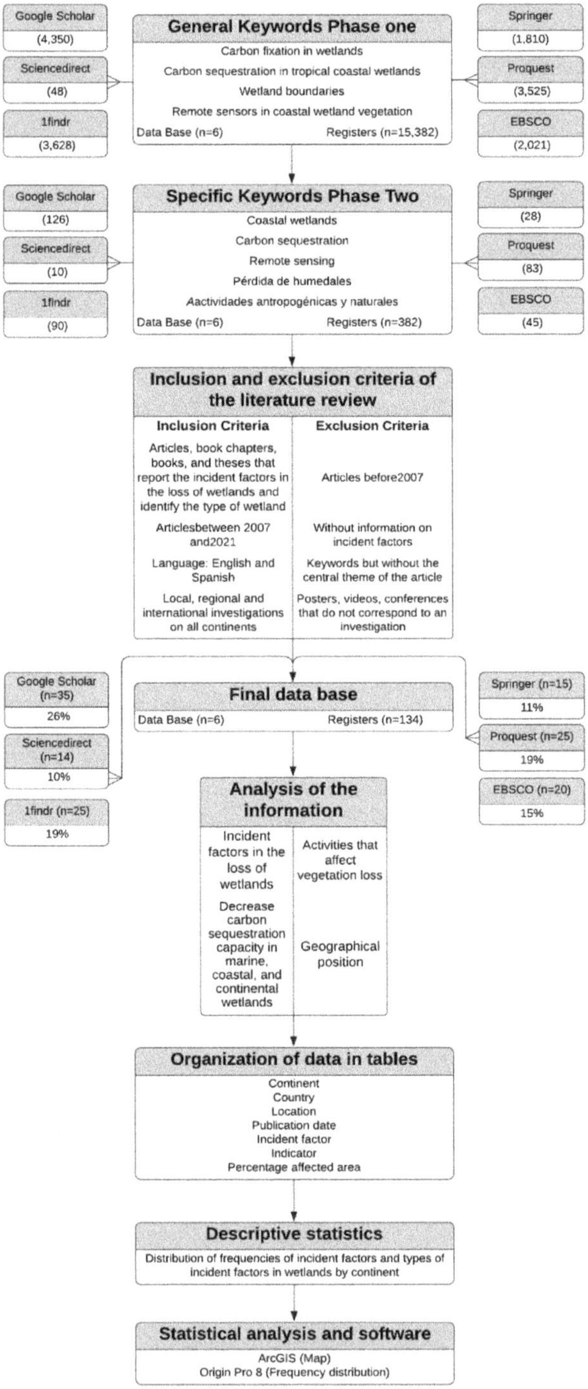

Figure 1. Methodology for the literature review.

3. Results and Discussion

Figure 2 shows the spatial location of the number of investigations reviewed that highlighted the incident activities in the loss of coverage and reduction of carbon sequestration capacity in marine, coastal, and continental wetlands. These include 134 documents structured by countries and continents, where the United States of America and China hold first place with the most publications on the subject, followed by Australia, Mexico, India, and Brazil.

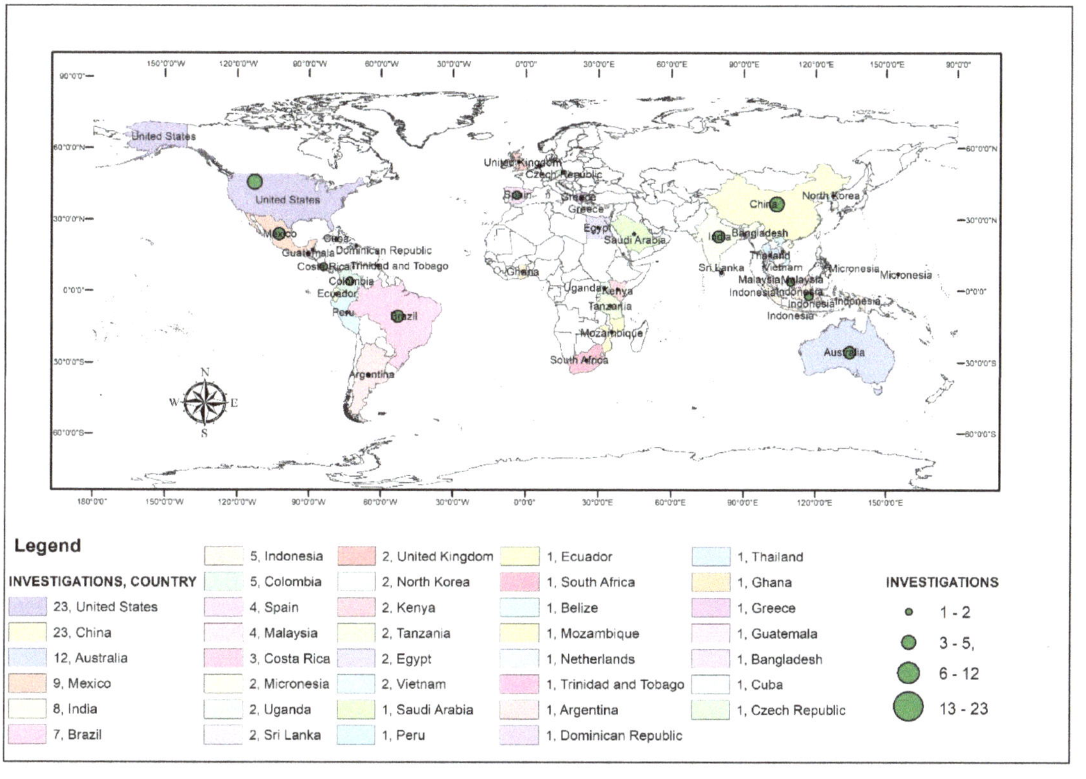

Figure 2. Number of investigations reviewed globally.

This information makes it possible to visualize the distribution and importance required for research on natural wetlands, to gain a better understanding of the current state of the effects on wetlands in the world. Although it is possible that there are other unreported or unavailable research studies, it is very likely that the incident factors reported are similar. The collected information became the sample that was classified by continent for analysis, which allowed us to find the following results.

3.1. Research on the African Continent

Table 1 presents 10 studies found in Africa where location and wetland type, impact period, incident factor in vegetation loss and sequestration, loss indicator, sequestration and impact percentage are analyzed, showing a pattern of continental behavior and helping to build an analysis. It is evident that although few investigations were recorded, the eastern part of the continent allows us to analyze the following:

Table 1. Data from the African continent on wetlands in coastal, marine, and continental zones (n = 10).

Country	Site and Type of Wetland or Flooded Area	Affectation Time (Years)	Incident Factor in Loss of Vegetation and Sequestration	Loss and Sequestration Indicator	Affectation Percentage	Author
Kenia	Gazi Bay Mangroves	Yearly	Logging	Vegetation Loss	0.7	[28]
Kenia	Gazi Bay Seagrasses	Last 140	Fishing with nets	Degradation	N.A.	[29,30]
Egypt	Coastal Area Governorate of Kafr Elsheikh-Nile River Delta,	N.A.	Salinity	Productivity	N.A.	[31]
Egypt	Burullus Lake	Last two centuries	Soil erosion, agricultural soil drainage.	High Sedimentation and Eutrophication	62.5	[32]
Ghana	Densu Delta, Sakumo II and Muni-Pomadze	1985, 2002, 2017	Flood, marin erosion	LDD, NDVI	N.A.	[33]
Tanzania	Tanzania, Kenia and Mozambique Coastal Zone	2000–2016	Urbanization, agriculture, livestock, lumber industry	Vegetation Loss	N.A.	[34]
Tanzania and Mozambique	Rufiji and Zambeze	N.A.	Illegal Logging and Coastal Erosion	Vegetation Loss	N.A.	[35]
Uganda	Kirinya Wetland and Nakivo	1950–	Agriculture	Vegetation Loss	N.A.	[36]
Uganda	Wetland Naigombwa	N.A.	Agriculture	Vegetation Loss	N.A.	[37]
South Africa	Mkuse Floodable Plain	N.A.	Agriculture, Dams	Vegetation Loss, Flood and Sediment Control	N.A.	[38]

N.A.: Not Available. LDD: Landscape Deviation Degree. NDVI: Normalized Difference Vegetation Index.

First, it was found that in the Gazi-Kenya Bay, mangrove forests are used by inhabitants for construction and firewood, authorized by the government for being the only forests. This has brought erosion consequences that finally lower sequestration capacity, experimentally observed in a small-scale plot [28]. On the other hand, Githaiga et al. [29] and Juma et al. [30] indicated that Gazi-Kenya Bay contains some seagrasses that have been degraded by daily fishing with trawls and purse seines by artisanal fishermen. These pastures have not been studied on their vulnerability and carbon sequestration capacity. However, it was determined that pastures with sediments and vegetation sequestered more carbon than others with a scarcity of these.

Second, in another latitude, the coastal area of the Nile River is also affected by salinity when the tide rises, implying a low nitrogen and carbon content since vegetation is scarce. Restoration and ecosystem management with crops was recommended to improve sequestration and mitigate climate change [31]. Authors such as Eid and Shaltout [32], emphasized that Lake Burullus is a Ramsar site and before being declared as such it had lost almost 62.5% of its area due to erosion and sediment deposit loaded with allochthonous carbon and pollutants resulting from agricultural land drainage in the Nile River Delta.

Furthermore, in Ghana, an ecosystem health study was conducted in three coastal Ramsar wetlands, Sakumo II, Densu Delta, and Muni-Pomadze; using structure, function, and resilience indicators such as the Landscape Deviation Degree (LDD), Normalized Difference Vegetation Index (NDVI), and Normalized Difference Water Index (NDWI). The result showed that in 1985, 2002, and 2017, the LDD indicator increased due to urban development activity as the main fragmentation cause in the three wetlands, and the NDVI behaved inversely due to the reduction of areas covered by vegetation. The NDWI was variable due to flooding processes and erosion caused by the sea [33].

In Tanzania, coastal ecosystems not only protect but also serve as habitats for population and fauna. This last function has been threatening animals, transforming their vegetation, especially by man. Thus, these changes along with their dynamics, and socioeconomic drivers were studied between 2000 and 2016. It was evidenced that urban development, agriculture, livestock, and the logging industry are the main activities affecting ecosystems [34].

Comparatively, Lagomasino et al. [35] conducted research in four deltas of two continents on carbon in mangroves, its losses and gains. In the African Deltas, Rufiji (Tanzania), and Zambeze (Mozambique), they identified that incident factors in vegetation loss were illegal logging of mangrove forests and coastal erosion, respectively. Remote sensing techniques were used in the study to determine the changes. The other two deltas will be observed in the Asian continent sections.

On the other hand, agricultural intrusion into Ugandan wetlands, especially Kirinya and Nakivudo, very close to Lake Victoria, has replaced native endemic papyrus vegetation with Cocoyam cultivation. The impact suffered by these incident changes in carbon sequestration and CO_2 emission was determined using the Eddy Covariance Technique [36]. Therefore, it is very important to increase statistics of carbon sequestration worldwide. Were et al. [37] studied the Naigombwa wetland located in Iganga, Uganda, where the estimate was determined since a large part of freshwater wetland is being transformed into rice fields. It was found that in the natural wetland carbon sequestration was higher than in the transformed area and it was recommended to use other options of sites for rice cultivation and not in the wetlands.

Finally, the Mkuze floodplain is the largest wetland area in Kwazulu-Natal, South Africa which has been affected by agriculture and construction of the Pongola dam to regulate hydrology, consequently varying the thickness and width of peat deposits as well as sediments along the tributary valley towards Lake Mpanza [38].

Figures 3 and 4, shows eleven (11) incident factors that affect wetlands and the types of wetlands affected respectively in Africa. The descriptive analysis is premature to infer what happens continentally, but it is possible to affirm that there is a trend of incident factors of agricultural activities in the first place, such as farming, fishing, livestock (40%), urbanization, and industry. The latter being directly correlated with each other. Nevertheless, they represent 20% of the analyzed sample.

On the other hand, direct anthropogenic factors alternate; understood as those where pressure is exerted by man on ecosystems first-hand, with indirect factors, negative externalities, consequences, or collateral damage that were classified as coastal erosion, salinity, and flood.

The effects in Africa are mainly due to agricultural activities since this continent is the second most populated but at the same time the poorest on the planet, being rich in natural minerals such as platinum, diamonds, chromium, and so on. It also has a high number of countries without access to the sea and cities with high population density. Its development has been slowed down in terms of continental trade, due to its large desert and jungle area that prevent people from transit [39].

The interpretation given to the previous results can be focused on two aspects: In the first instance, the two figures show a trend in data behavior in Africa, because, with only ten investigations, they represent 7.46% of the total research. However, in the second one, this information indicates a consistent trend of what is happening with wetlands due to the lack of an appropriate conservation policy and extreme poverty. Therefore, farmers do not consider the effects on wetlands from their productive activities.

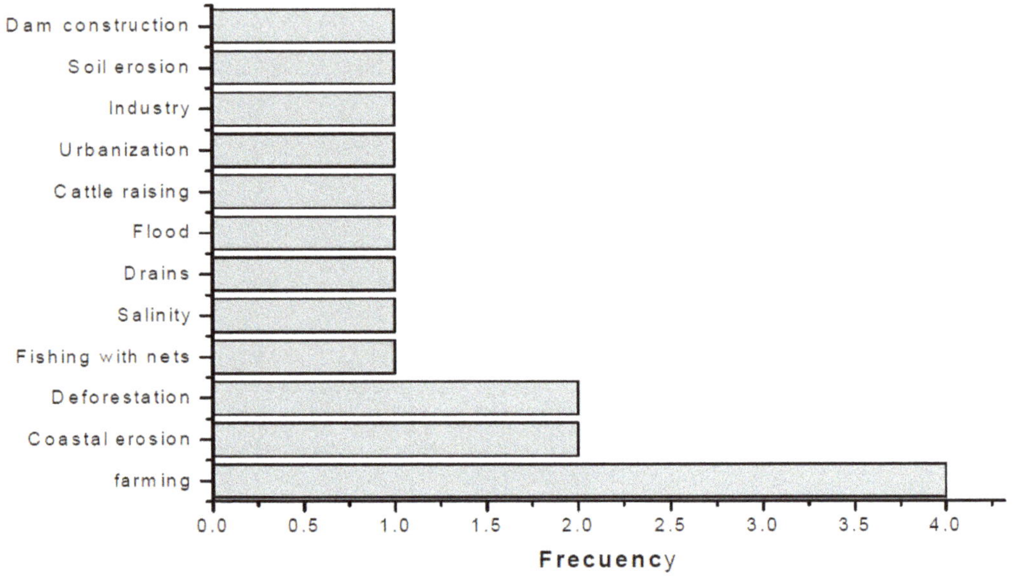

Figure 3. Frequency distribution of incident factors of wetland loss in Africa.

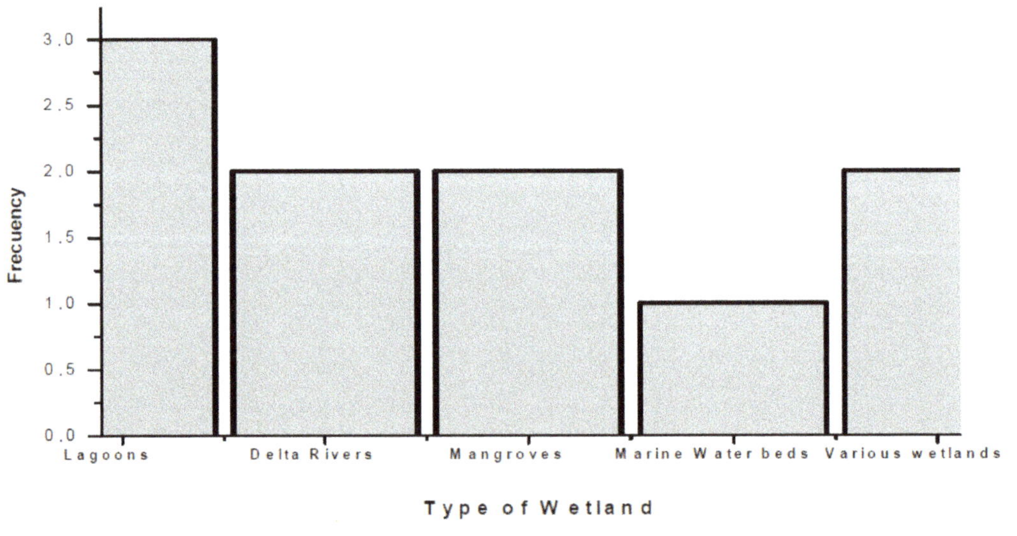

Figure 4. Frequency distribution by type of affected wetland in Africa.

3.2. Research in the American Continent

In Table 2, fifty-five (55) studies found in the American continent are described, occupying the first place among all the continents with 41.04% of the total sample, making it an important number, covering research from the United States to Argentina.

Table 2. Data from the American continent on wetlands in coastal, marine, and continental zones (n = 55).

Country	Site and Type of Wetland or Flooded Area	Affectation Time (Years)	Incident Factor in Loss of Vegetation and Sequestration	Loss and Sequestration Indicator	Affectation Percentage	Author
USA	Wetlands of Albermarle Strait, North Carolina	N.A.	Agriculture	Drainage	N.A.	[40]
USA	Estuaries from Delaware Bay and the Indian River Florida	1970–	Rising Sea Levels	Vegetation Loss	N.A.	[41]
USA	Tidal Marshes Estuaries-National Research Reserve, Delaware	1780–	Agriculture and Urbanization	Vegetation Loss	54	[42]
USA	Seagrasses from the Virginia Coastal Reserve	N.A.	Mud Mold and a Hurricane.	Vegetation Loss	N.A.	[43]
USA	Timberlake, Albemarle Peninsula, North Carolina	1900–1980	Deforestation Agriculture	Vegetation Loss	N.A.	[44]
USA	Salt Marsh in Rowley Massachusetts	N.A.	Waste Water	Eutrophication	N.A.	[45]
USA	San Francisco Bay	20th century	Agriculture and Urbanization	Vegetation Loss	90	[46]
USA	New England Coast	Last 30 years	Overgrazing	Vegetation Loss	N.A.	[47]
USA	Georgia Coast	–2100	Rising Sea Levels	Vegetation Loss	20	[48]
USA	Elkhorn Slough Wetland California	Last century	Agriculture	Eutrophication	N.A.	[49]
USA	Coasts	1960–1970	Oil Spill	Pollution	N.A.	[50]
USA	New York	N.A.	Agriculture	Respiration and Nitrogen Mineralization	N.A.	[51]
USA	Indiana and Illinois	150–200	Agriculture	Drainage	90	[52]
USA	Coastal Wetlands of Louisiana	Last two centuries	Flood Control Levees	Degradation	N.A.	[53]
USA	Continental Wetlands Louisiana	200	Agriculture	Drainage	80	[54]
USA	Barataria Bay, Lousiana		Coastal Sinking	Rising Sea Levels, Erosion		[55]
USA	Ohio Freshwater Wetlands		Agriculture	Erosion	N.A.	[56]
USA	Marshes, California	100	Agriculture, Livestock	Drainage	N.A.	[57]
USA	Everglades Boglands	19th Century	Waste water	Eutrophication	N.A.	[58]
USA	Everglades Wetlands, Great Dismal Swamp	200	Agriculture, Urbanization and Fire	Vegetation Loss	N.A.	[59]
USA	Indian River Lagoon Vero Beach and Fort Pierce Florida	At the end of the 20th Century	Reservoir Construction	Drainage	variables	[60]

Table 2. *Cont.*

Country	Site and Type of Wetland or Flooded Area	Affectation Time (Years)	Incident Factor in Loss of Vegetation and Sequestration	Loss and Sequestration Indicator	Affectation Percentage	Author
USA	Mangroves in Tampa Bay	1950–1990	Urbanization	Vegetation Loss	21	[61]
USA	Naples Bay Mangroves, Florida	2005	Urbanization	Vegetation Loss	70	[62]
Mexico	Nuxco Sub-basin Mangroves, Municipality of Tecpan de Galeana	1981–2015	Agriculture and Urbanization	Vegetation Loss	50	[63]
Mexico	Terminos Lagoon	Last Years	Waste water and Garbage	Disturbance	N.A.	[64]
Mexico	Alvarado Lagoon System	Colonial Period	Agriculture and Livestock	Change in vegetation	variable	[65]
Mexico	Sinaloa Marshes	Last Three (3) Decades	Aquaculture	Vegetation Loss	variable	[66]
Mexico	Marshes and Swamps State of Veracruz	N.A.	Livestock, Petrochemistry and Urbanization.	Vegetation Loss	N.A.	[67]
Mexico	Gulf of México Mangroves	100	Sub-freezing and Increased level of the substrate due to sediments	Vegetation Loss vegetal	variable	[68]
Mexico	La Encrucijada Biosphere Reserve Chiapas	N.A.	River dredging and Fires	Degradation	N.A.	[69]
Mexico	California Gulf Mangroves	N.A.	Urbanization	Population Growth	50	[70]
Mexico	Biosphere Reserve of Sian Ka'an Yucatán Peninsula	N.A.	Climate Change	Sea Level, Roads, tourism	N.A.	[71]
Belize and Guatemala	Milpa, blue creek and Zotz. Wetlands.	Recent Years	Agriculture	Carbon Isotope Variation	N.A.	[72]
Honduras	Fonseca Gulf Mangroves	1985–2013	Aquaculture	Vegetation loss	5800 ha	[73]
Costa Rica	Estereo Brook Basin	Last ten years	Urbanization, Waste water	Vegetation Loss and Eutrophication	N.A.	[74]
Costa Rica	Palo Verde Freshwater W	30	Livestock	Grazing	N.A.	[24]
Costa Rica	Earth University, Parismina River Basin	N.A.	Agriculture	Vegetation Loss	N.A.	[75]
Cuba	Caguanes National Park	19th Century	Agriculture and livestock	Deforestation	N.A.	[76]
Dominican Republic	Providence of Montecristi Mangroves	1983–1993	Aquaculture	Vegetation loss	N.A.	[77]
Colombia	Magdalena River	2007–2012	Livestock, Agroindustry, Mining, Energy and Urban Expansion	Population Growth	24	[78]

Table 2. *Cont.*

Country	Site and Type of Wetland or Flooded Area	Affectation Time (Years)	Incident Factor in Loss of Vegetation and Sequestration	Loss and Sequestration Indicator	Affectation Percentage	Author
Colombia	Uraba Gulf Mangroves	Two (2) decades	Deforestation, Agrícultural soils and Urban areas.	Population Growth	29.8	[79]
Colombia	Orinoco River Basin and the Caribbean	Last four (4) years	Agriculture	Change in Land Use	variable	[80]
Colombia	Malaga Bay Mangroves and Buenaventura Bay	N.A.	Deforestation, Urbanization, Expansion of ports and docks.	Vegetation Loss	N.A.	[81]
Colombia	El Tunjo Freshwater Wetland	1940–2016	Urbanization	Fragmentation	90	[82]
Ecuador	El Pantanal Wetland, Technical University of Machala	2007–2016	Solid waste and Enclosure	Vegetation Loss	40	[83]
Peru	Santa Rosa Wetland, Lima	N.A.	Agriculture, Porciculture, Livestock and Waste water	Invasive Plants, Eutrophication	N.A.	[84]
Trinidad and Tobago	Coasts of the Islands	N.A.	Port Industry, Agriculture and Urbanization	Ecological Stress and Loss of Cover	N.A.	[85]
Brazil	O Pantanal	N.A.	Natural and Arson Fires	Biomass Burning	N.A.	[86]
Brazil	Cananeia-Iguape Lagoon	165	River Diversion	Intrusion of Macrophyte Species	N.A.	[87]
Brazil	Whale Coast in Bahía	Last 34 years	Eucalyptus Forestry, Agriculture, Urbanization	Vegetation Loss	N.A.	[88]
Brazil	Varzea Clear Water Alluvial Plain	Currently	Livestock	Soil Compaction	N.A.	[89]
Brazil	Atibaia River Basin and Jaguari River	N.A.	Agriculture and Urbanization	Eutrophication and Vegetation Loss	72.4	[90]
Brazil	Jaguaribe River	N.A.	Wastewater and Aquaculture	Eutrophication Vegetation Loss	N.A.	[91]
Brazil	Sepetiba Bay / Rio de Janeiro	N.A.	Metallurgical industry and Urbanization	Heavy Metals	N.A.	[92]
Argentina	Parana River Delta	19	Livestock	Vegetation Loss	58.3	[93]

N.A.: Not Available.

It is important to highlight that from the total investigations in this continent (n = 55), 41.82% were developed in the United States and the rest in Latin America (Mexico, Brazil, and Colombia). Results show that agriculture, urbanization, and ranching occupy the first places with a frequency of double digits in the entire continent. Wetlands in the United States have had more impact.

In contrast, regarding agriculture and urbanization in Latin America: Although Quimbayo Ruiz [40] stated that Latin America and the Caribbean are the most urbanized regions in the world, in this region, the territory has been developed by socio-political appropriations of space degrading ecological issues, so it could be assumed that it is not due to socio-economic activities.

From the ecosystem point of view, it is evident that coastal wetlands (mangroves, lagoons, and marshes) are more affected than continental ones. It is important to highlight that they correspond to the spatial location of the most important and populated cities of the continent. On another level, global efforts have been made to tackle climate change by trying to reduce net carbon emissions, but they have not been enough. Only two countries in Latin America and the Caribbean have mitigation plans [41]. Consequently, conservation and preservation of these ecosystems should be on the political agenda of current leaders, for the planet's sustainability due to its great capacity to sequester carbon and counteract the effects of global warming.

A first aspect to analyze is the repeated entry of saltwater during drought times that abruptly decreases concentrations of dissolved organic carbon in the freshwater coastal wetlands of the Albermarle Strait, reaching a salinity of 12 ppt in the dry season. These conditions are given by agricultural development that has connected the area with drainage channels [42].

It was also found that a SLAMM/HEA model or approach was developed to predict effects and economic costs caused by sea-level rise in the Delaware Bay and Indian River Florida Estuaries, USA since 1970. Finally, it was shown that it can be achieved with a resolution of 10 MT, a large-scale analysis, useful for ecosystem managers [43]. According to St. Laurent et al. [44], organic matter and carbon variability in sediments from two marshes in Delaware were studied. Results showed significant differences in sequestration and even in vegetation. Stream basins that supply wetlands are affected by agriculture and urbanization.

On the other hand, a disease such as mud mold and a hurricane in 1933 extinguished seagrasses in the Virginia-USA Coastal Reserve, causing the closure of the fishing industry. Starting in 2001, new grasslands were planted and were subject to study to determine carbon sequestration at different ages of pastures, resulting in higher values in pastures of 10 years [45]. The restoration of wetlands has been carried out since 2004 subject to deforestation and agriculture in the 20th century. Then, through monitoring with Lidar sensors, biomass in this young vegetation was estimated with limited results, and the use of optical sensors and high spatial resolution was recommended [46].

Furthermore, Moseman-Valtierra et al. [47], consider that some anthropogenic activities deposit low nitrogen concentrations in the saltwater marshes in Rowley, Massachusetts. For this reason, the effects of nitrate in gas production were studied during a time through parcels. It was concluded that anthropogenic additions alter emissions substantially.

According to the point of view of Stralberg et al. [48], the San Francisco Bay in the USA contains marshes that will be threatened by the rise in sea level and supply of sediments limited by works upstream with a possible vegetation loss. However, it was evident that since European colonization, about 90% of marshes were lost to agriculture and urbanization.

Regarding the wetland degradation process, it was determined that on the coasts of New England, not only has overgrazing determined vegetation loss due to population growth, but also burrows of crabs because they weaken the peat and cause erosion with tides [49].

On the Georgia coasts, properties of soil, carbon sequestration, and accretion in freshwater forests from three rivers affected by increases in sea level were studied. It was concluded that the accelerated level will decrease the forests and expand saline marshes, and the conversion into marshes will improve carbon sequestration [50].

Now, Siciliano et al. [51], used three approaches to determine changes in nutrient enrichment in the Elkhorn Slough wetland, California. This served to measure how an estuary is affected by agricultural activities where amounts of nutrients are discharged to its waters causing eutrophication. He also used hyperspectral images to detect the spatial changes that vegetation underwent when using two spectral indices. They found a relationship between reflectance, chlorophyll, and nutrients.

Similarly, researchers analyzed the impacts that coastal ecosystems can suffer in their flora and fauna due to the spills of different types of oil, either due to transport or production in the high seas, especially the one caused in 2010 off the coast of the Gulf of Mexico. The spilled oil reached the coasts by high tides or winds, stagnated in the vegetation and soil, and caused damage to fish and wildlife due to its chemical toxicity [52].

In contrast, the agricultural activity in the Marlens Tract wetland, located in the Montezuma Wild Management Area in New York State, has partly affected its 99 ha area, reducing its spatial heterogeneity in soil and microbial properties [53].

In the Corn Belt located in the USA Midwest (Indiana, Illinois), Craft et al. [54] argued that almost 90% of wetlands have been drained in 150 to 200 years, for agricultural activities that require large amounts of soil nutrients.

On the other hand, Lane et al. [55] reported that the construction of dams to control floods has been affecting the forested coastal wetlands of Louisiana, USA, largely due to scarcity of sediments and the contribution of fresh water. Failing that, these receive effluents have been treated by wastewater treatment plants. The potential of these wetlands as carbon pools was researched. Louisiana has 40% of the wetlands in the USA and they accumulate 42% of the world's carbon reserve. Furthermore, 80% of losses from the 19th to the 20th century in Louisiana were due to drainage to convert wetlands into agricultural land, which accelerated organic matter loss, carbon oxidation, and its release into the atmosphere as CO_2 [56]. In the same state in Barataria Bay, the coastal subsidence of 5 to 16 mm/year and the increase in sea level of 3.4 mm/year has given rise to the loss of 25.9 km^2 per year of wetlands due to erosion, causing soil carbon to be dissolved and degraded until it is released into the atmosphere [57].

Fennessy et al. [58], studied the variation of carbon sequestration depending on the ecological condition and by eco-region in the USA. Nine (9) freshwater wetlands were taken in the Erie Drift Plain region (Ohio) and 10 in Ridge and Valley (Pennsylvania). Results showed that in the Ohio region soil accretion rates were higher because agriculture dominates there and there was greater sediment carry-over, while in Pennsylvania the region is made up of mountains and wooded areas that limit erosion and sediment transport.

In the Sacramento-San Joaquin Delta located in the central valley of northern California, Hemes et al. [59] studied the carbon that accumulated organic matter in freshwater marshes for more than 7000 years, generating a peat layer with more than 15 m deep rich in carbon, but a large part of it was eliminated in just 100 years due to dams' construction, drains for agriculture and livestock, generating a high emission of greenhouse gases. Therefore, their restoration was proposed.

The nutrient load that the peatlands from the Everglades in the USA receive has allowed phosphorus to be the main cause of changes in the ecosystem, especially in vegetation structure. This element has been deposited with the construction of drainage channels [60]. The Everglades in Florida have historically been subject to agricultural exploitation on the north and urbanization on the east, thus becoming the subject of restoration. In addition, the Great Dismal Swamp is a swamp affected by fire, drainage, and deforestation in the last 200 years [61].

Verhoeven et al. [62] stated that insects are being controlled globally through the Rotational Impoundment management (RIM) approach, which consists of creating reservoirs and pumping water from wetlands, then returning it through sewers, which has caused changes in vegetation due to the nitrogen cycle. It was implemented in Florida until 1985 causing the vegetation cover to decrease from 75% to 30%.

In another research, Dontis et al. [63] deduced that in Tampa Bay, Florida, mangroves have been displacing marshes to a great extent due to climate change. However, wetlands lost an average of 2000 ha between 1950 and 1990 to urban development. In Naples Bay, approximately 70% of mangroves have been lost due to urbanization. Where soil samples were taken, carbon was estimated, and sequestration data were lower than world averages [64].

In Mexico, more than 50% of mangroves in the coastal area of the Nuxco sub-basin from Guerrero State have been lost in 34 years. Geographic information systems were used to prepare thematic maps that identified agriculture, livestock activities, and developers as the main causes of vegetation losses [65].

Furthermore, as highlighted by Cerón et al. [66], the Laguna de Terminos in the Yucatan Peninsula, has been considered a Ramsar site since 2004. Two research sites *Esterero-Pargo* and *Bahamitas* were chosen, both sites affected by human pollution, especially wastewater. Carbon content was determined at different seasonal times and in relation to depth.

In addition, Vázquez-González et al. [67] in relation to current vulnerability and, current and strategic trend, using a scale index for coastal wetlands, found in the lagoon system of Alvarado, Mexico, that the main anthropogenic factors that have affected them have been livestock since the arrival of conquerors and sugar cane agriculture. As a result, they found that vulnerability in the current scenario and current trend increases in all municipalities of the State of Veracruz.

In Mexico, the area with the highest aquaculture shrimp production is the Sinaloa state. Berlanga-Robles et al. [68] found with satellite images that 75% of shrimp farming occurred in marshes and 1% on mangroves, modifying the spatial vegetation pattern in coastal wetlands.

Now, due to livestock, petrochemical, and urbanization activities, coastal wetlands in the State of Veracruz have been transformed, and their capacity to retain water and sequester carbon has decreased. It was determined that marshes and swamps are important for these two functions [69].

The Gulf of Mexico was subject to a historical reconstruction of mangroves in its area, given its significance on carbon sequestration and the effects of climate change it has suffered. It was concluded that in 1983 the sub-freezing in Texas affected approximately 80% of mangroves. On the other hand, those without vegetation (open water), decreased from 1951 to 1967 because the substrate increased due to sediments driven by Hurricane Carla in 1961 [70].

Additionally, Adame et al. [71], in research developed on the Pacific coast of southern Mexico, in Chiapas to be exact, located La Encrucijada Biosphere Reserve (LEBR) with an area of 144,868 ha and different types of wetlands. There, the carbon pool, as well as sequestration rates, were determined by comparing stocks in trees, fallen wood, and soil. The degradation of wetlands was evidenced thanks to a load of sediments resulting from the dredging of the river upstream and fires threatening the potential sequestration estimated at 38 tons of carbon.

Furthermore, in Mexico, Ochoa-Gómez et al. [72] studied in the region of the Gulf of California, the coastal zone including *Bahía de la Paz* that covers an extensive area of 14 mangrove patches with a total area of 270 ha, affected by the expansion of the urban center of La Paz adjacent to the mangroves. This region has the second-highest rate in Mexico in terms of population growth and urban development, reducing the mangrove area in some cases up to 50%.

In the Ramsar site of Sian Ka'an Biosphere Reserve (SKBR), in the Yucatan peninsula, carbon reserve in its wetlands was quantified with three vegetation types, finding that tall mangroves had the largest reserves. It was also concluded that climate change has affected mangroves with the rise of sea levels, road construction, and water pollution due to tourism [19].

On the other hand, in Belize and Guatemala, carbon isotope relationships were analyzed over time since the lowland wetlands have been cultivated by Mayan indigenous people, finding variations in the different soil layers in the Milpa wetlands [73].

Interestingly, in order to increase carbon data in wetlands, a study was carried out in three areas of Honduras: *Laguna de los Micos*, Roatan islands (Guanaja, Utila), and Gulf of Fonseca. The latter affected by aquaculture since 1985, aimed to transform 5800 ha of mangroves. The results did not show significant differences among the three sites [74].

Now, according to Rodríguez-Arias and Silva Benavides [75], the wetlands of the Estereo brook, located in San Ramon, Costa Rica, are being prioritized due to degradation suffered from urban expansion and wastewater generated by the population, largely losing the functions of its ecosystem services in the last 10 years.

Ref. [25] studied 12 wetland communities on two continents with different characteristics to compare carbon sequestration: two freshwater wetlands in the humid tropics in Costa Rica and two wetlands in the dry tropics, one in the Okavango north of Botswana and the other one in Costa Rica. Authors stated for the latter, that in the previous three decades it has been the object of cattle grazing. It was found that humid tropical wetlands had higher carbon content than dry ones and also, the dry ones had similar data, but the humid ones differed significantly. This research was developed at the Earth University, it has an approximate area of 3300 ha and is located in the Parismina river basin. The study described how wetlands are being used internally within the campus, finding that farming activities with plantations of bananas and pineapples have been affecting the basin [76].

In Cuba, since the middle of the 19th century, the cultivation of sugar cane increased, deforesting and draining hectares of forests and swamps. Then, when the Soviet Union ended, these areas were switched to large-scale livestock activity. Through an agreement between American universities, the environmental threats that affect these ecosystems in the Caguanes National Park were determined. The complexity of the research relationships was evident due to political causes, and it was considered that the community should be involved in the wetland's conservation [77].

Kauffman et al. [78], quantified carbon reserves in wetlands in the province of Montecristi, Dominican Republic, including abandoned shrimp ponds since 1993. Results showed that they represented only 11% of the mangroves in their reserve and that they also emitted more CO_2 gases. The construction of dikes for aquaculture blocked the flow of fresh water and tides.

Now, in Colombia, in the Magdalena River basin, two types of wetlands, one fluvial and the other one isolated, were compared in relation to carbon sequestration. The result was higher in the isolated one. On the other hand, in the same research, Pérez-Rojas et al. [79], stated in the discussion, that this basin has been too deteriorated due to having the main river artery of the country and a high settlement of the population in its surroundings. In general, from 2007 to 2012, the area of wetlands decreased by 24% in Colombia, due to livestock, agroindustry, mining, energy activities, and urban expansion.

The Gulf of Uraba region, also in Colombia, is suffering from the excessive deforestation of mangroves attributed to agricultural and urban activities among others, greatly affecting mangrove structure. In terms of biomass, diameter varied; in terms of species, there was a reduction of some and proliferation of others [80].

Furthermore, Ricaurte et al. [81], analyzed the loss of wetlands with remote sensor images. Results highlight that, in the Orinoco and Caribbean regions, the most incidents of anthropogenic activities are oil palm cultivation.

According to Palacios Peñaranda et al. [82], mangroves of the Colombian Pacific, and especially the bay of Malaga and Buenaventura, served as scenarios to quantify carbon reserves, finding that they were similar to other tropical mangroves. They determined that these mangroves are being degraded by deforestation, urbanization, expansion of docks, and ports.

In addition, in El Tunjo wetland located in the Colombian capital of Bogota, Mateus and Caicedo [83] confirmed that it underwent a transformation process from 1940 to 2014, determined through satellite images, where it was evidenced that urban processes affected up to 90% of the area, causing fragmentation.

The study area located at the Technical University of Machala, Ecuador, is a wetland with approximately 12,500 m^2. This has been affected not only by the enclosure of the university campus, but also by the deposit of solid waste, especially construction debris [84].

The Santa Rosa wetland, located on the north coast of Lima, is an environmental lung filtering air and water due to its great biodiversity. It has been affected by activities such as agriculture, pig farming, livestock, and wastewater, developing in it, a large number of invasive species such as *Pistia stratiotes*, which is a floating plant that causes eutrophication [85].

Regarding the Caribbean islands, Trinidad and Tobago has approximately 362 km of coasts. It is considered as the country in the Caribbean with the industry most heavily dependent on oil and natural gas. On its coasts, port industries, agriculture, and urbanization have been developed affecting the ecosystem value of its wetlands, especially the economic valuation of carbon sequestration [86].

In 2014 and over 2 years, Ding et al. [87] sampled three wetlands in different continents, Everglades, Pantanal, and Okavango, in order to determine the effects of natural and provoked fires that burn aerial biomass and its final deposit as black carbon dissolved not only in the soil but also in the atmosphere, water, and sediments; representing large significant amounts.

Similarly, in Brazil, Rovai et al. [88] determined that the hydrological alterations made by man 165 years ago in the Valo Grande canal river modified the deposit patterns and biogeochemistry of the water by diverting the river. Similarly, the mangrove area was reduced due to the intrusion of freshwater macrophyte species in the estuarine system of the Cananeia-Iguape lagoon. With this county being the second largest mangrove area in the world.

Furthermore, the disorderly growth of eucalyptus forestry, agriculture, and urbanization have been occupying the coastal wetlands of *Costa das Baleias*, Brazil for the last 34 years. Therefore, it is urgent to research this area to value and preserve this natural environment to counteract human actions [89]. In the Varzea alluvial plain in the Amazon River, cultivation of cocoa, jute, and rubber occurred between 1940 and 1990. Currently, they are secondary forests that keep livestock among their trees causing soil compaction. Remote sensors were used to understand how the stored carbon suffered variability over time as a consequence of livestock, floods, and forest age [90].

The basin areas of the Atibaia and Jaguari rivers in Brazil have been affected by the eutrophication of agricultural lands, mainly by the cultivation of sugar cane and urbanization, respectively, affecting the physical characteristics of the streams and fish community [91]. Now, in the Jaguaribe River, there is a mangrove area where Ferreira et al. [92], studied the effects on vegetation and soil by herbivorous crabs and also, biomass restoration as a result of deforestation. The main factors that affected the area are wastewater and aquaculture. After 10 years it was found that the restored area sequestered more carbon than the self-recovered one.

Regarding pollutants, there are metallurgical companies that produce Zn that contaminate the Sepetiba Bay with toxic metals generated when processing minerals such as bauxite. Concentrations of these metals affect the population when consuming diets based on shellfish from the bay [93].

Finally, in Argentina in the Paraná River delta, studies were conducted using remote sensors to identify anthropogenic activities that cause area losses. It was concluded that from 1994 to 2013, the main cause was the change of use of the soil for cattle production [5].

Figure 5 shows the 14 incident factors that affect wetlands in the American continent, highlighting agriculture, urbanization, livestock, and wastewater, among others.

On the other hand, and according to the bibliography consulted, it is observed in Figure 6 that the most affected types of wetlands are mangroves, lagoons, and swamps, followed by various types of wetlands and rivers. In America, it is where the greatest evidence of work related to the effects was found, which may indicate the interest in their study, restoration, and of course the economic capacity to finance this type of project, especially in the United States of America, which represented more than 50% of the studies in that continent.

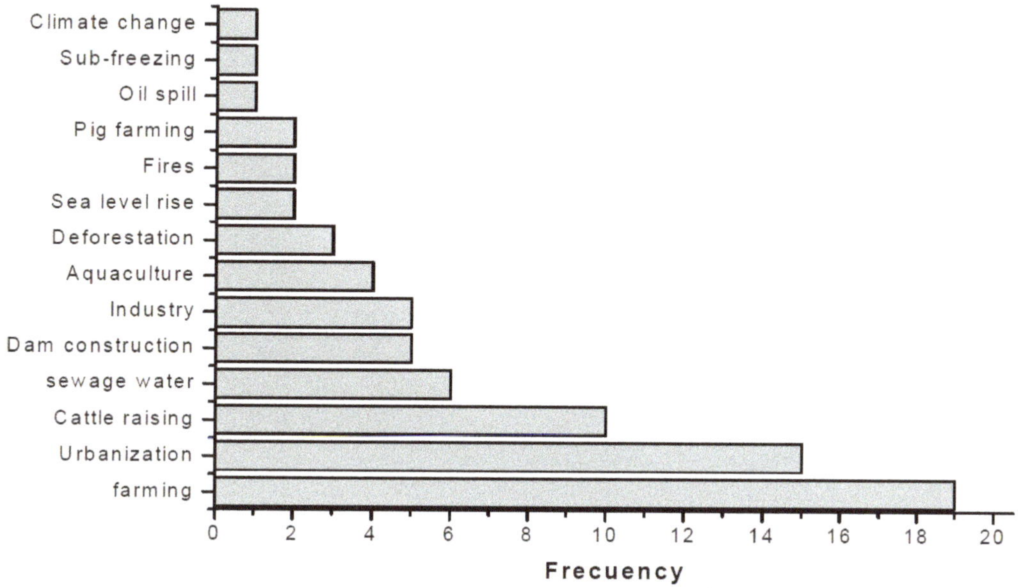

Figure 5. Frequency Distribution of Incident Factors affecting wetlands in America.

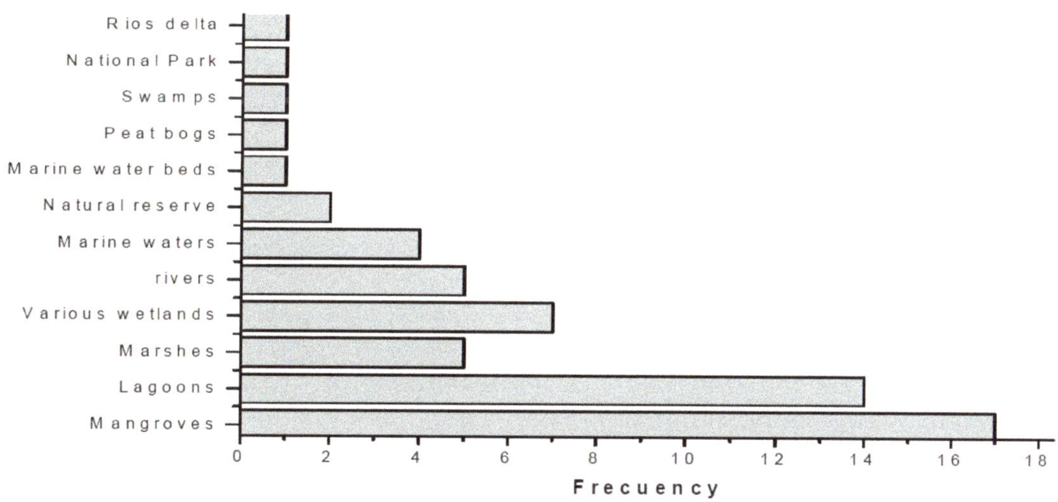

Figure 6. Frequency distribution of the type of wetland affected in America.

3.3. Research in the Asian Continent

Table 3 shows studies found in Asia, with a total of 46 investigations, highlighting China, India and Malaysia, with a highly significant presence of wetlands, where affectations were also reported.

China is the country with the largest area in the Asian continent and the longest length of its coasts, hence the greatest number of research studies (n = 23). It is considered the second most important economy in the world and results from Table 3, and Figures 6 and 7 indicate this. Factors such as aquaculture (the largest producer in the world), agriculture, urbanization, industry, coastal erosion, wastewater, and population growth, and the affected wetlands such as: bays, mangroves, lagoons, estuaries, and river deltas are related to each other in a cause–consequence relationship [94], where population and urbanization have affected its coastal wetlands such as bays and estuaries because three large groups of industrial cities settled in the bays of the most important rivers in the country.

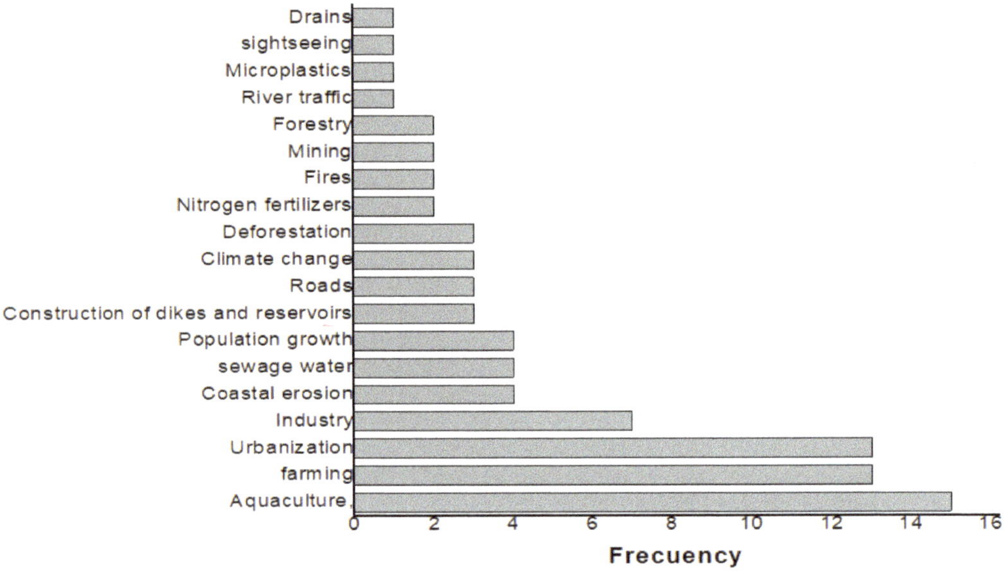

Figure 7. Frequency distribution of incident factors that affect wetlands in Asia.

Exports from Japan, Taiwan, South Korea, and Hong Kong together represent much more than twice those of the third world. Similarly, their growth rate is higher than other countries. Their Gross National Product is comparable to that of the United States of America [95].

Table 3. Data from the Asian continent on wetlands in coastal, marine, and continental zones (n = 46).

Country	Site and Type of Wetland or Flooded Area	Affectation Time (Years)	Incident Factor in Vegetation Loss and Sequestration	Loss and Sequestration Indicator	Affectation Percentage	Author
China	Yellow River	1990–2015	Urbanization	Land Use Cover Change LUCC	15	[17]
China	Xiamen Coast Wetlands	N.A.	Urbanization, Population Growth	Vegetation Loss	N.A.	[96]
China	Mangroves Northeast Coast Hainan Island	1960–	Aquaculture	Vegetation Loss	73	[97]
China	Xincun Bay, Hainan Island	N.A.	Aquaculture	Eutrophication	N.A.	[98]
China	Shanyutan Swamp, Minjiang River Estuary	19th Century	Aquaculture, Agriculture and Wastewater	Carbon Mineralization	N.A.	[99]
China	Sanjiang Plain	Last 50 years	Agriculture	Vegetation Loss	N.A.	[100]
China	Liaoche Rivel Delta	N.A.	Agriculture, Aquaculture, Oil Exploitation, Forestry, Industrial Construction	Vegetation Loss	N.A.	[101]

Table 3. *Cont.*

Country	Site and Type of Wetland or Flooded Area	Affectation Time (Years)	Incident Factor in Vegetation Loss and Sequestration	Loss and Sequestration Indicator	Affectation Percentage	Author
China	Daya Bay	Last Decades.	Domestic, Industrial, Agricultural and Aquacultural Waste Water.	Eutrophication	N.A.	[102]
China	North Hangzhou Bay	Last Years	Urbanization, Agriculture and Roads	Vegetation Loss.	N.A.	[103]
China	Shenzhen Bay	N.A.	Microplastics	Heavy and Organic Metal Pollution	N.A.	[104]
China	Yellow River Delta, Shandong Province	Last Century	Agricultural Fertilization and Fossil Fuel Combustion	Plant Growing.		[105]
China	Chongming Island, Yangtze Estuary	Two Decades	Breakwater Construction Agricultural Use	Puddling below Ground Level (30 cm)	N.A.	[106]
China	Zhangjiang Estuary	N.A.	Shrimp Aquaculture	Variation of Mangrove Height by Nutrients	N.A.	[107]
China	Poyang Lake, Yangtze River	Last Decades	Climate change and intensive human activities	Loss of Water Level, Change in Vegetation	N.A.	[108]
China	Northeast	Last Centuries	Agriculture	Vegetation Loss	N.A.	[109]
China	Hangzhou Bay	2000, 2010	Industry and urbanization	Economic Development	35.81, 15.19	[110]
China	Jiaozhou Bay	N.A.	Aquaculture	High Alkalinity and Salinity	N.A.	[111]
China	Bohai Bay	1979–2014	Urbanization and agriculture	Economic Development	N.A.	[112]
China	Heilongjiang, Jilin and Liaoning Provinces	50	Climate Change	Desiccation	N.A.	[113]
China	Yangtze River	N.A.	Waste water	Decreasing Biomass	N.A.	[114]
China	Sanjiang Plain	50	Agriculture	Drainage	N.A.	[115]
China	Yancheng Natural National Reserve	1988–2006	Agriculture, aquaculture urbanization	Landscape Fragmentation	N.A.	[116]
China	Baiyangdian Freshwater Wetland	1970–	Reservoir Construction	Wetland Drying	N.A.	[117]
Thailand	Phang-nga Bay Phuket Province	Variable	Mining, Agriculture, Aquaculture and Urbanization	Change in Use	N.A.	[118]
Saudi Arabia	Pérsian Gulf	Last Century	Oil Industry, Urbanization, Population Growth	Vegetation Loss	90	[119]
Korea	Marshes Mud Flat	1987–	Industry	Pressure due to Development	22	[120]

Table 3. *Cont.*

Country	Site and Type of Wetland or Flooded Area	Affectation Time (Years)	Incident Factor in Vegetation Loss and Sequestration	Loss and Sequestration Indicator	Affectation Percentage	Author
Korea	Seagrasses Korea Peninsula	Last two or three decades	Construction of Levees, Urbanization and Industrialization	Vegetation Loss	N.A.	[121]
India	Coastal zone Cambay Gulf, Gujarat	N.A.	Industry, Aquaculture, Urbanization and Coastal Erosion	Vegetation Loss	N.A.	[122]
India	Lake wetlands University of Kalyani West Bengal.	N.A.	Industrial Waste Water	Eutrophication	N.A.	[123]
India	Kerala State Coastal Mangroves	Last 50 years	Aquaculture and Coastal Erosion	Vegetation Loss	N.A.	[124]
India	Chilika brackish water lagoon	N.A.	Nitrogen Fertilizers	Eutrophication	N.A.	[125]
India	Kannur, Kerala, and kunhimangalam	Currently	Agriculture, Aquaculture, Urbanization, Roads	Vegetation Loss.	N.A.	[126]
India	Kodungallur-Azhikode Estuary	N.A.	Aquaculture y Agriculture	Organic and Inorganic Waste.	N.A.	[127]
Malaysia	Kalimantan indoor bog, Borneo	Last Decades	Fires and Drainage	Vegetation Loss	N.A.	[128]

N.A.: Not Available. LUCC: Land Use Cover Change.

The northeast coast of Hainan Island, in southern China, has lost around 73% of its mangrove area due to aquaculture activities since 1960, generating high rates of suspended matter and nutrients such as nitrogen, causing eutrophication of waters, and possibly causing damage to reefs, seagrasses, and corals [96]. In another area of China Fan et al. [97] found that the loss of coastal wetlands in Xiamen is due to high urbanization and high population from recent years, which has caused an expansion of reclamation lands and a decrease in land for cultivation.

On another level, Jiang et al. [98] stated that contributions of nutrients generated by shrimp aquaculture in Xincun Bay, Hainan Island, South China Sea, have allowed eutrophication to indirectly reduce the seagrass, sequestering organic carbon, and conversely enhancing labile organic carbon. Carbon mineralization in the Mianjiang River, China has been affected by anthropogenic activities, mainly aquaculture, agriculture, and pollutant discharge, inhibiting the sequestration process [99].

Now, the Sanjiang plain is a region with different types of wetlands, bathed by the Heilong, Wusuli, and Songjhua rivers. Marshes have been converted to farmland for 50 years before this publication, becoming one of the most productive regions in China. They were converted into Ramsar sites and the rate of carbon accumulation in the short and long term was estimated [100].

One of the largest coastal wetlands in Asia is the Liaoche River delta, which has suffered the impact of changes in land use due to different anthropogenic activities, which led to research on how land reclamation affects the storage of organic carbon and total nitrogen. It was identified that: oil pollution had a higher concentration of carbon and nitrogen but does not serve as a fertility test. Sugar cane released both elements quite a lot. In terms of uses, there were also differences in storage [101].

In another place of high economic importance such as Daya Bay, an industrialized and highly-populated area, Zhao et al. [102] identified a discharge process of wastewater product of different anthropogenic activities, causing the eutrophication phenomenon in its waters due to nitrogen. That is why the incidence of some environmental attributes within the denitrification process was researched, studying key enzymes that favorably act to achieve it. It was concluded that temperature and ammonia are key factors in the removal. Furthermore, the mangrove has a 48% efficiency in relation to other types of wetlands.

In the Yangtze Delta, North Hangzhou Bay is the largest economic center in China with a dense population. Demand for land has been such that it has been used for urbanization, agriculture, and road construction. Huang et al. [103], conducted their analysis, thanks to the use of remote sensing and geographic information systems, as well as a weighted linear model used to evaluate the risk of loss and degradation of wetlands.

Moreover, a very serious environmental problem generating pollution is micro-plastics, not only in the fauna transferred through the food chain causing energy depletion, slowing growth, and even causing death, but also in the flora because they function as adsorption vectors and carry heavy metals such as (lead, cadmium, and chromium) and organic pollutants (polychlorinated, pesticides, biphenyls, and so on) affecting the growth of microalgae and macrophytes. Depending on the wetland characteristics, whether they were deposited in it, a study was carried out in Shenzhen Bay, and it was determined that severe pollution in the mangrove accumulates more in the edge strip than in the inside and on the floor [104].

The deposition of nitrogen in the soil of wetlands, a product of anthropogenic activities of fertilization or fossil combustion, affects carbon dynamics and the soil physicochemical properties, being reflected in the growth of plants and soil microorganisms. The delta of the yellow river in the Shandong province does not escape this reality, that is why an experiment was carried out at the mesocosmic level with different amounts of nitrogen. The effect on plant growth and carbon decomposition was evidenced [105].

Additionally, the effects of climate change at the global level and the constant rise in temperature, brings with it damaging and catastrophic effects on land ecosystems, especially in wetlands and in their soil, affecting the microbial activity and the same soil chemical properties that directly affect the wetland vegetation, for which the possible changes in the reserves of Carbon, Nitrogen, Phosphorus of the soil and the related microbial activities were examined after 7 years of experimental heating in situ through open chambers (OTC) in a Phragmites wetland in the Yangtze estuary. The researchers concluded that global warming in this estuary without tidal impacts could increase C reserves, while N and P reserves could increase with moderate warming [106].

Floods are equally important since they cause a distribution of the different mangrove species with their tides. As determined by Zhu et al. [107], many times they are used by the inhabitants to establish aquaculture activities, especially shrimp farming, resulting in variability in mangrove growth due to the nutrients in the drainage of the shrimp ponds. Some species are more sensitive than others.

Researchers Mu et al. [108] showed that Lake Poyang is hydrologically suffering, due to climate change and anthropogenic activities, affecting its water level, and increasingly reducing its vegetation cover as well as its distribution in the dry season as a critical period, thus minimizing its ecological function. Remote sensing images from multiple sources, using the Random Forest technique, digital elevation models, meteorological data, and so on, were used to examine long-term vegetation changes during the dry season.

Similarly, a review article was prepared whose main axis was the carbon budget in Chinese wetlands. The prevailing problem showed uncertainties in the different sequestration data in multiple investigations, so all the pertinent information was synthesized. On another level within the document, it was discussed that a large part of marshes in northeast China, without specifying location, have been lost due to agricultural activities in recent centuries [109].

Now, wetlands in Hangzhou Bay China are in a very economically developed coastal area. Consequently, industrialization and urbanization have had an impact on surrounding ecosystems in recent decades due to changes in land use, being in descending order among the most affected: rice fields, shallow waters, reservoirs and ponds. The total sum of area among all ecosystems in 1990 was approximately 6000 km^2. By 2000 they were reduced by 35.81% and by 2010 by 15.19% [110].

Likewise, Jiaozhou Bay located in the Shandong-China peninsula has an area of approximately 500 km^2, where ponds have been built for aquaculture, becoming the main use of the land of the Dagu estuary, which has generated high alkalinity and salinity in the soil, causing a more complex carbon cycle [111].

China suffers rapid urbanization and industrialization, as estimated by Meng et al. [112], especially Bohai Bay affecting blue carbon ecosystems due to economic development, affecting wetlands from 1979 to 2014 in 1.11×10^5 km^2 of soils for construction and agricultural activities such as aquaculture, represented by 2/3 of the world total.

Indirectly Zhang et al. [113], state that in northeast China in the provinces of Heilongjiang, Jilin, and Liaoning there are different types of wetlands including Ramsar, with a total area of approximately 753.6 km^2. In the last 50 years, these boreal and temperate ecosystems have undergone transformations as a result of climate change due to variation in temperature and rainfall, exerting negative effects on carbon sequestration due to wetland desiccation.

The Jiuduansha Wetland has an area of 423.2 km^2 and is located in the estuary of the Yangtse River. Carbon sequestration and its ecological function, are being affected by the discharge of wastewater by plants located upstream, causing the eutrophication phenomenon and also the increase in tides, high soil respiration and decreasing biomass, thus reducing carbon sequestration [114].

The microbial activity was related to the labile organic carbon of the soil, i.e., dissolved, in the Sanjiang plain in northeast China, with a wetland area of approximately 1.04×10^4 km^2, specifically, in the Honghe International Nature Reserve wetland, a Ramsar site since 2001, but despite its category, agriculture around the reserve has affected the water level in the last 50 years, causing it to drop 4.02 m between 2005 and 2014, affecting its function as a carbon sink [115].

In the Yellow River Delta in China, an attempt was made to reduce carbon loss by improving the connections between wetlands fragmented by anthropogenic activities and land use and cover change (LUCC), moving from crops to urbanization, for which maps were used to observe the variability over time and proposed to protect the wetlands with the regulation of sediments and to observe the dynamic changes of landscapes in some lower reaches and in the delta itself [17].

Bearing in mind Ke et al. [116] who state that the Yancheng National Nature Reserve in China has received all kinds of considerations because it is a place of protection for endangered birds, and despite its international importance, it does not escape anthropogenic activities such as agriculture, aquaculture, and urban expansion, which have degraded and fragmented the landscape due to population growth and economic development. The damage caused was evidenced by remote sensing techniques.

In the north of China, the Baiyandian wetland is the largest freshwater lake in that area, Dong et al. [117] studied its storage capacity because it has been affected by the construction of 150 reservoirs upstream, which has allowed a reduction of its humid area in drought periods, affecting the *Pharagmites australis* as the main biomass plant and primary wetland productivity, reducing its carbon fixation.

On the other hand, in Thailand, the first aspect to highlight is the use of the land around Phang-nga Bay, province of Phuket, tin mining has been present there since 1600. Furthermore, other actions such as palm oil, rubber cultivation, and recently, urbanization. The particulate organic matter in the sediments of the bay was researched, comparing it with seagrasses and mangroves [118].

Similarly, due to its oil wealth, the Arabian Gulf has undergone transformations in its wetlands due to industry, population, and urbanization, achieving mass losses of up to 90% in the last century. Carbon sequestration was found to be higher in mangroves, seagrasses, and marshes, respectively [119].

Now, In Korea, Byun et al. [120], studied the Mud Flat 2 tidal plain, finding that it experiences an anthropogenic impact and pressure for development more than Mud Flat 1, affecting carbon sequestration. Marshes on its shores have been lost by 22% since 1987. Furthermore in the Korean peninsula, Sondak and Chung [121], pointed out that seagrasses have been lost in the last two or three decades due to urbanization, dam construction, and industrialization, reducing the potential for blue carbon sequestration.

In India, in the Gulf of Cambay or Khambhat in the Arabian Sea, there are three districts with approximately 10 million inhabitants. There is an important industrial center, deteriorating estuaries such as marshes, mangroves, and cliffs which also suffer coastal erosion since the last decades. Remote sensing analyses determined that aquaculture and urbanization affect marshes [122].

In West Bengal-India, five districts were chosen to measure water quality, especially inorganic carbon content in algae, which depends on the amount of nutrients since some wetlands receive moderate industrial wastewater. Inorganic carbon and nitrogen may be important for microalgae in their phytocarbonate content [123].

On the coasts of Kerala, mangroves have lost vegetation in the last 50 years due to shrimp farming and coastal erosion as expressed by Harishma et al. [124]. Despite this, results showed that biomass is high and with six varieties of mangroves, highlighting the *Avicenia marina* with the highest amount of biomass and *Sonneratia alba* with the least. It was concluded that they must be conserved due to the high volume of carbon they sequester.

The Chilika lagoon is the largest in Asia with brackish water. It was declared a Ramsar site in 1981. It receives a high load of nitrogen fertilizers in agricultural activities by several freshwater streams. Nutrient concentrations were evaluated showing both spatial and temporal variations regulating chlorophyll [125].

In Kerala, Kannur, among others, the wetlands belong 50% to the state and 50% to individuals. In these, cultivation of coconut trees, excessive exploitation of resources, shrimp farming, urbanization, and roads are developed as the main activities that degrade mangroves. Researchers monitored with remote sensors and geographic information systems to make an assessment of carbon stocks in the ecosystem [126].

The Karuvannur and Chalakkudy rivers discharge a large quantity of nutrient-rich waters to the Kodungallur-Azhikode estuary on the southwest coast of India. The product is from anthropogenic activities, mainly aquaculture, and agriculture because they release inorganic and organic waste that causes quality effects of water and therefore, in the fixation of carbon through phytoplankton [127].

In contrast, an important aspect that Dommain et al. [128] found about carbon sequestration in peat from Southeast Asia, in Kalimantan, Borneo to be exact, is that it decreased with the decrease in sea level. In the Holocene, there was the availability of new lands allowing the formation of mobs, disturbed by anthropic activities such as fires and drainage for agriculture; thus, releasing carbon to a depth of 30 cm.

The Berbak National Park according to Miettinen et al. [129] was subject to changes in its coverage by oil palm plantations despite its state protection. Illegal invasions have grown by cutting down the forest, increasing its vulnerability.

Something important that Bal and Banerjee [130] highlighted is that little is known about mangroves in India, especially their biomass and carbon sequestration, so it was decided to evaluate them in the Bitharkanika Wildlife Sanctuary Wetland and relate them to physicochemical factors to understand the relationship. Carbon per hectare of biomass and that of soil was quantified. They showed that the wetland is affected by aquaculture, agriculture, and urbanization.

In Bangladesh in the Ganges River Delta, agricultural land has progressively changed from use to wetland aquaculture due to salinization, industrialization, population, and so on. Shrimp aquaculture has destroyed 45% of mangroves, reducing their ecosystem services [131].

Now, on the island of Pulau Indah, Malaysia, a multitemporal analysis was carried out with remote sensors using Landsat images and it was identified that the main factor of vegetation loss from the mangroves was the growth of the port infrastructure [132,133].

Due to the very high rates of deforestation in the peatland forests in Borneo, the importance of knowing the impact and magnitude of the disturbance from the past is born, so the researchers used the paleoecological technique of fossil pollen and charcoal to determine the changes, confirming that the last 500 years has been the most critical period of human disturbance with the oil palm cultivation and the logging industry [134].

There is a special type of activity that has been affecting mangroves of Tanjung Piai National Park in the south of Johor, Peninsular Malaysia, and it is the river traffic of boats through the Strait of Malacca towards Singapore, causing waves that generate erosion on the coasts in addition to road construction [135].

In Indonesia, Tareq et al. [136] found that vegetation underwent changes in the Rawa Danau wetland in West Java, in the last 7428 years. It was documented based on forest fires and land uses due to human influences using an elemental analyzer to determine the content of organic carbon and nitrogen. The results allowed us to conclude that these effects are not only related to climate change but also to man.

In another Indonesian wetland, Kusumaningtyas et al. [137], identified that the Segara coastal lagoon is affected by sediment deposition of rivers in the west, achieving reductions in mangrove vegetation. Furthermore, in the center of the lagoon, there is less vegetation than in the east of it. In the lagoon, the carbon reserves were measured and compared with another wetland, Kalimantan, where great variations were evidenced due to the effects on the ecosystem, especially due to activities such as: sedimentation, aquaculture, and overexploitation of resources.

Mangroves and coastal wetlands of Mimika district, Papua province, Indonesia, have undergone modifications due to small indigenous settlements with little impact on them, unlike the mining industry, and palm plantations as the biggest threats [138].

The PT Setia Alam Jaya logging concession has been cutting down the peat forest in the upper basin of the Sebangau river, in Indonesia since before 1998. There, the flow of carbon was studied, especially in the form of greenhouse gases in areas with or without affected vegetation. Due to the water table, more gases were emitted in the dry season [139].

Now, in Sri Lanka, the tourism industry in recent years has specifically focused on four natural parks and due to the high influx of visitors it has led to a degradation of attractions and they should be developed with caution because tourism is already being considered a source of disturbance. Therefore, other forms of tourism must be diversified. A study was made through a survey to estimate the perception of tourists, where they showed their satisfaction with the tour, they evidenced the dirt in the river and other attributes. The study served as input for the necessary corrections to be taken [140].

Likewise, in Sri Lanka Perera and Amarasinghe [141] studied carbon sequestration in micro estuaries and tidal lagoons. The general objective was to assess carbon reserves in the soil in different wetlands. In addition, it was evidenced that the incident factor in the degradation of these, was the population growth that leads to a depletion of the ecosystem and its reserves.

Interestingly, research was carried out in four deltas of two continents on carbon in mangroves, their losses, and gains, in the Asian deltas: Ganges (Bangladesh) and Mekong (Vietnam). This identified that the incident factors in the vegetation loss were coastal erosion and deforestation, respectively. Remote sensing techniques were used to determine the changes in the study [35].

During the Vietnam War between 1964 and 1970, the Can Gio mangrove forest was sprayed with the orange herbicide causing a 57% loss due to defoliation. Currently, the Kien Vang mangroves are being affected by aquaculture, deforestation, and coastal erosion [142].

In this way, as seen in Figure 7, there are nineteen incident factors in Asia. The main factor is aquaculture because the Asian continent is the world's largest producer of aquaculture products [143]. It generates economic growth but also many pollution problems due to discharges of nutrients from aquaculture production units to wetlands. On the other hand, despite the fact that China is one of the main economies, agriculture plays an important role in feeding the most populated continent of the world [144]. Therefore, the pressure exerted by the land and the negative effect on wetlands.

Urbanization and industry are factors related to economic growth and the phenomenon of modernization of cities and their industry that has led China to sustain a growth rate of 6.8% GDP even before the pandemic [145].

On the other hand, the fourteen types of wetlands most affected by these activities are observed in Figure 8, where marine waters and mangroves are those that have the greatest effects. This is mainly due to the lack of wastewater treatment from the activities described in Figure 6.

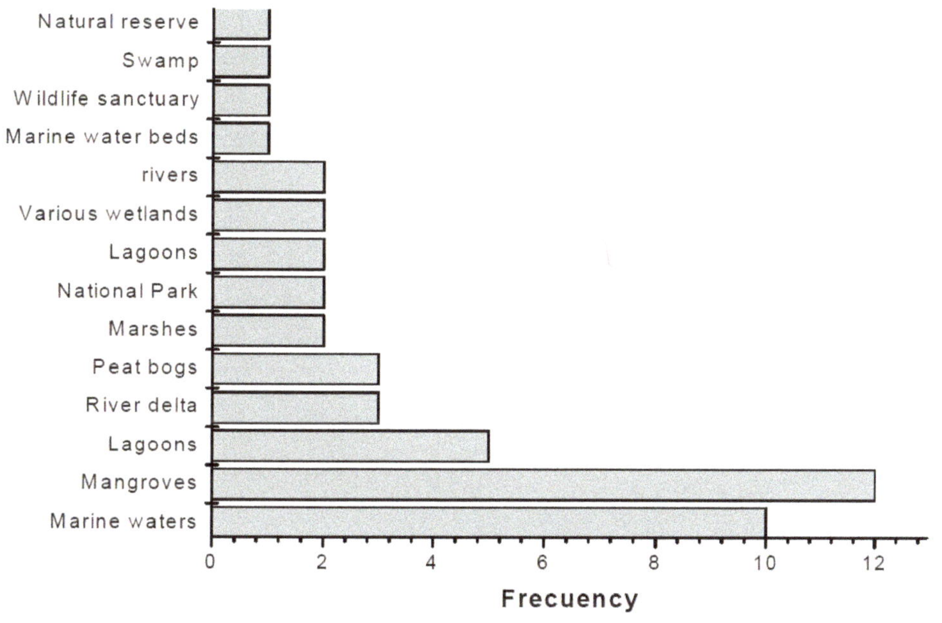

Figure 8. Frequency distribution of the types of wetlands affected in Asia.

3.4. Research on the European Continent

In Table 4, a limited information base on the European continent is described with nine investigations, highlighting Spain with the largest number of researches.

On Schiermonnikoog island in the Netherlands, Elschot et al. [146] analyzed, in different plots of marshlands, the impact degree on carbon sequestration by cattle grazing, determining that in young plots sequestration is limited and in older ones, it increases because the compaction of the cattle trampling decreases oxygen concentration in the soil and therefore reduces carbon decomposition.

In Spain, the remote sensing technique was used with free software to show changes made in 11 wetlands, 7 from the coast and 4 continentals. It was determined that urban development and tourism are the main cause of vegetation loss. This conclusion was reached through digital Landsat image processing [147].

Wetlands in the Ebro River delta are being affected by the construction of dams and the adaptation of rice crops, bringing with it an alteration of the accumulation of allochthonous carbon transported in the sediments. Furthermore, there are alterations in the soil level, showing that where there is greater connectivity, there will be greater results of organic carbon accumulation [148].

Table 4. Data from the European continent on wetlands in coastal, marine, and continental areas (n = 9).

Country	Location and Type of Humidity	Affectation Time (Years)	Incident Factor in Loss of Vegetation and Sequestration	Loss and Sequestration Indicator	Affectation Percentage	Author
New Scotland	Fundi Bay	N.A.	Agriculture	Change of Use	80	[3]
Spain	Lago Fuente de piedra	Last decades	Agriculture and Urbanization	Change of Use	N.A.	[4]
Czech Republic	Biosphere Reserve UNESCO MAB Třeboň	400–600	Aquaculture and Sowing Pastures	Drainage	30	[54]
Holland	Back Barrier Swamp Island of Schiermonnikoog	Last 120	Livestock	Compaction	N.A.	[146].
Spain	11 wetlands in Murcia	Last decades	Agriculture, Urbanization, tourism	Loss of area	N.A.	[147]
Spain	Ebro River Delta	N.A.	Dam Construction and Rice Planting	Reduction of Sediment Flow and Organic Matter	99	[148]
Spain	Wetlands adjacent to the Doñana National Park	N.A.	Aquaculture and Agriculture	Change of Use	N.A.	[149]
Greece	Ten Greek Ramsar sites	30	Agriculture	Change of Use		[150]
United Kingdom	Tadham Moor Somerset Levels and Moors	Historically	Agriculture and Livestock	Drainage	N.A.	[151]

N.A.: Not Available.

On the other hand, Sánchez-Espinosa and Schröder [4], affirm that agriculture expansion has led to changes in land use in recent decades within and around the limits of water resources by modifying water level. The *Fuente de Piedra* saline lake located in the Malaga province, with an area of 1400 ha, a Ramsar site since 1983, does not escape this problem of the effects of olive groves, vineyards, fruit trees, and small urban settlements.

Similarly, Morris et al. [149] assure that the Doñana National Park is the best-protected wetland in Europe by UNESCO and Ramsar. In this, only activities such as forestry are allowed. However, on the surrounding wetlands in private properties, aquaculture and agriculture are the main anthropogenic activities developed. In Doñana, some anthropogenic impacts have reduced the water hydroperiod and when they are extensive, it favors carbon sequestration.

Now, in the Czech Republic from 400 to 600 years ago, some of the wetlands have been used for aquaculture (30%) to this day and others drained for sowing pastures [54].

Greece has 10 Ramsar sites. The wetland area including its catchment area covers 1,858,666 ha and, despite enormous conservation efforts, there have been profit and loss balances as a result of anthropogenic activities from 1986 to 1987 until 2016 to 2017 (30 years). Among the most relevant is the conversion of forested and natural lands to agriculture (22,264 ha), in addition to urban expansion with (14,044 ha) [150].

In Nova Scotia Gallant et al. [3] argue that in the Bay of Fundi, a high percentage of marshes (close to 80%) have been lost due to conversion to agriculture, reducing the wetland social and environmental benefit in carbon sequestration. The Integrated Dynamic Climate and Economy Model (DICE) was used for its economic valuation, estimating an approximate value in the total range from $5105/ha/year to $39,795/ha/year.

In the United Kingdom, as agricultural and livestock production has been historically exploited in the Tadham Moor and Somerset Levels, moors and wetlands have been affected due to their drainage. It became necessary to develop a carbon balance and controls to design conservation management policies. For this, the Eddy Correlation Technique based on flows in the ecosystem was used [151].

In Spain, there are two wetlands dependent on groundwater, the Tablas de Daimiel and the Lagos de Ruidera, some 75 km apart. They have different hydro-geomorphological conditions, both are Ramsar sites and also nature reserves; the first has a humid area of 250 km^2 and only has 20% of its initial area, the Gigüela river provides brackish water and the Guadiana fresher water apart from other sources that feed it; it is affected by potentially toxic trace elements (PTEs) dragged by the river and sediments as stated by Jiménez et al. [152]. In another plane, its water table is 3 m deep because its aquifer has been affected since 1970 with overexploitation, extracting 20,000 million m^3 in the last 40 years, which has caused a drop in the water level of 20 m. Likewise, the second, the Ruidera Lakes, are a series of wetlands interconnected with each other by different water sources, which have had drainage problems since the 1980s due to the pumping of wells for irrigation and have been declared overexploited [153].

Furthermore, in Belgium, the Scheldt estuarine wetland downstream of the city of Ghent is affected by anthropic activities, such as the case of intertidal marshes with heavy metals such as Cd, Cu, Pb, and also Zn in their sediments up to a pro- depth of 1 m, with spatial and temporal variations in its sediments in the short term [154].

In a combined way, Figures 9 and 10 can be explained based on Coles [155], who states that human alteration on wetlands encompasses a wide history of degradation since prehistoric settlements. A first aspect to highlight is that they implemented the cultivation by draining the wetlands, especially when food became scarce in the middle of the 20th century due to the two world wars. Subsequently, since 1950, the drainage process for housing construction, industrialization, and of course, agriculture continued. That explains why agriculture tops the list.

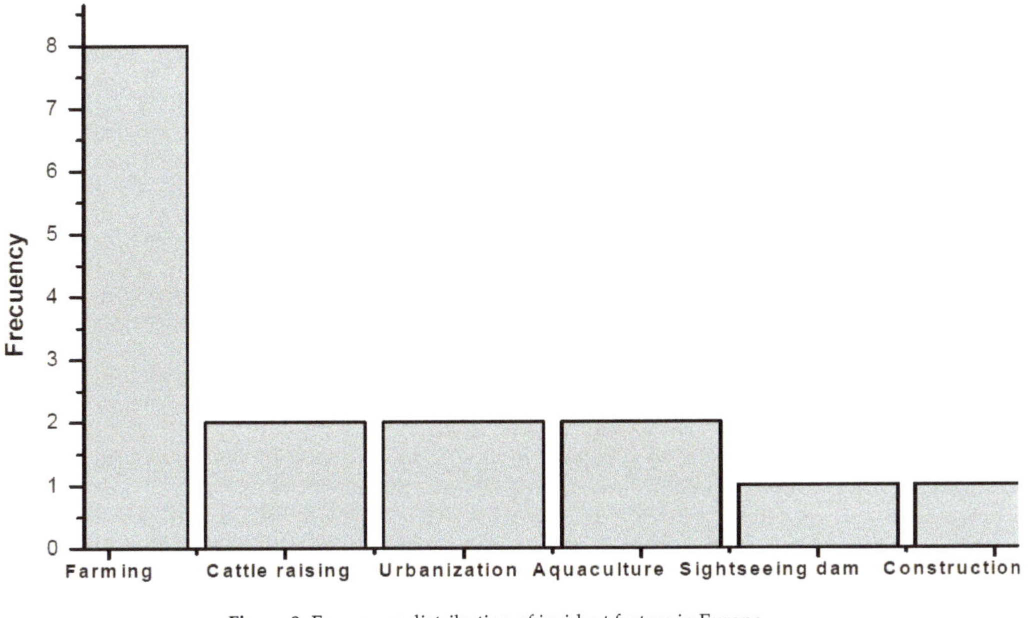

Figure 9. Frequency distribution of incident factors in Europe.

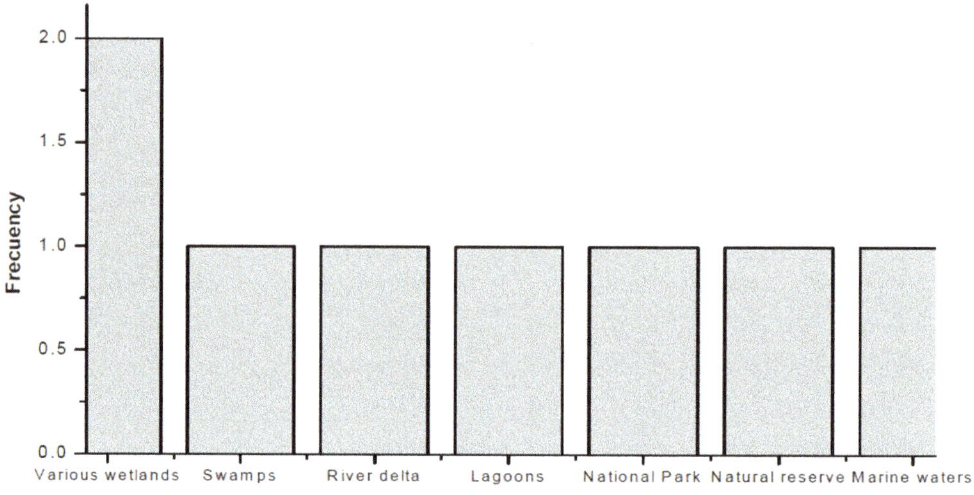

Figure 10. Frequency Distribution of the Types of Wetlands Affected in Europe.

On the other hand, affected wetlands in Europe are varied as can be seen in Figure 9, with seven different types of natural wetlands affected by the six incident factors found in the review. The majority coincide with Asia, with agriculture being the most relevant. Nonetheless, all have their relative importance. Due to lack of information, it is not possible to generalize, but the information gives support in trying to understand that several incident factors are similar in most of the cases analyzed.

3.5. Research in the Oceania/Australia Continent

Oceania and Australia are continents with very important biodiversity for the whole world, the studies on the effects were 14 as can be seen in Table 5, and the majority are focused on Australia with 12 investigations.

Overall, 86% of research studies were developed in Australia, so it is relevant to focus the discussion based on this country. Agriculture as an incident and relevant factor affecting wetlands, and is attributable to the fact of water regulation for agricultural purposes, industry, and urbanization. These externalities are expected to influence climate change to be a preponderant factor in land-use change [156].

In Micronesia, carbon reserves were estimated because they have been little studied by scientists, being an ecosystem service of high value. Two wetlands were studied on two different islands. It was found that the mangroves lost vegetation due to climate change, deriving other effects such as ocean acidification, changes in marine currents, among others that could decrease their productivity and increase mangrove mortality [157].

On Kosrae Island, there are wooded wetlands that are being replaced with agroforestry systems, especially by *Colocasia esculenta* for forestry. Effects of this crop on the carbon cycle within the forest peat were researched [158].

In Australia, Wong et al. [159] found that estuary drainage in the Richmond and Clarence River basins, a product of agricultural activities since 1900, has caused the growth of non-flood-tolerant vegetation in dry or drained areas, and when these occur, organic matter decomposes causing oxygen consumption by microorganisms and providing anoxic conditions within the wetland.

Table 5. Data from the Oceania/Australia continent on wetlands in coastal, marine, and continental zones (n = 14).

Country	Location and Type of Humidity	Affectation Time (Years)	Incident Factor in Loss of Vegetation and Sequestration	Loss and Sequestration Indicator	Affectation Percentage	Author
Micronesia	Mangroves Babeldoab Island and Yap Island	N.A.	Climate Change	Vegetation Loss	N.A.	[156]
Micronesia	Wooded Wetland Kosrae Island	Last 50 years	Agroforestry Crops	Change of Use	N.A.	[157]
Australia	Richmond and Clarence River Basins Estuaries	1900–1970	Agriculture	Drainage	N.A.	[158]
Australia	Queensland wetland	N.A.	Urbanization (Waste Water)	Oxygenation, Water, and Soil Quality.	N.A.	[159]
Australia	Herbert River Queensland	20th Century	Agriculture	Vegetation Loss	variable	[160]
Australia	Halifax bay Wetlands, National Park Insulator Creek among others	Last Century	Deforestation and Degradation	Vegetation Loss	50	[161]
Australia	Victoria State	187	Agricultural Frontier	Drainage	27	[162]
Australia	Fogg Dam Wetland	N.A.	Floods and, natural and arson fires	Gas emission	N.A.	[163]
Australia	Hunter Estuary	1954–1994	Industry, drainage.	Vegetation Loss	30	[164]
Australia	Hunter River Estuary	N.A.	Rising sea level	Possible loss of vegetation	100	[165]
Australia	East Coast Rivers of New South Wales and Queensland	1950–1960	Flood control projects	Vegetation Loss	N.A.	[166]
Australia	Westernport Bay	N.A.	Agriculture, industry and urbanization	Vegetation Loss	N.A.	[167]
Australia	Mangrove forests on western shores Moreton Bay-Queensland	N.A.	Waste water	Eutrophication	N.A.	[168]
Australia	Duck Creek North Freshwater Wetland	European Colonization	Dams, roads	Change in hydrology	N.A.	[169]

N.A.: Not Available.

On the other hand, eutrophication that occurs in the Queensland urbanized wetlands, due to the load of anthropogenic pressures that add organic matter and nutrients, generated a water quality loss due to decrease in dissolved oxygen reaching a saturation <5% exposing the fish to that level daily (Dubuc et al.) [160] which under these conditions are susceptible to dying due to lack of dissolved oxygen in the water.

Wetland vegetation loss in Queensland, since the last century, can be attributed to agriculture especially 56% with stagnant pastures, sugar cane cultivation with 8%, and 4% for other uses. These changes of uses have significantly altered the emission process of greenhouse gases such as CO_2, CH_4, N_2O_2 which varied in emission according to land use [161].

Maleleuca spp. was researched in several wetlands in Australia where there are large areas of forests because it is considered an invasive species and offers antimicrobial properties thanks to the oil extracted from its leaves. More than 1 million ha are found on private land. They suffer high risks of deforestation, preventing their high capacity to retain sediment during floods and sequester carbon [162].

Based on what was researched by Carnell et al. [163], in the State of Victoria a study of carbon sequestration and gas emissions was conducted in different types of wetlands. It was reported that since 1834 (European colonization), 27% of wetlands (147,053 ha) have been lost out of a total of 530,400 ha, which has generated a gas emission between 22.5 and 74.2 million $MgCO_2$. More than 40% of the carbon reserves of these ecosystems are a product of the expansion of the agricultural frontier through the drainage of wetlands.

Exchange of greenhouse gases in the Fogg dam wetland and control exerted by climatological and environmental conditions over these exchanges. It was identified that an influencing factor is floods and also natural fires caused in 1 to 4 years on average, burning extensive areas of alluvial plains. These changes in land use have altered the fragile balance between gas emission and net absorption [164].

Carbon sequestration and area were studied in some undisturbed and disturbed wetlands, mostly by industry development activities coupled with drainage works on Kooragang Island, Australia. It was concluded that rehabilitation has positive benefits and depending on how these wetlands adapt to disturbances, so will their capacity to sequester carbon [165].

Rogers et al. [166], report that the Hunter Natural Park Wetland will possibly lose 100% with a large rise in sea level. Otherwise, if the levels were kept low, it could expand by 35%. This theory resulted from modeling of the management of gates given that this area, despite being a Ramsar site, which with the opening and closing of these, would increase or decrease carbon sequestration, respectively.

In the 1950s and 1960s in Queensland and New South Wales, some hydraulic works were built that affected their rivers, causing losses in wetlands. In another hand, around 4200 structures were also identified preventing the tidal flow towards them [167].

Agriculture, industry and urbanization have degraded melaleuca forests in the Westernport Bay, a site chosen to assess greenhouse gas emissions in sediments, mangrove soils, and marshes. Results showed less carbon content than more tropical wetlands [168].

Assessing the incidence of nutrients in mangroves is important, so in the Moreton Bay-Queensland forests, nutrient-rich pollutants from agricultural runoff caused eutrophication. It was determined that aerial biomass increased more than soil seedlings [169].

Construction of dikes and roads since the European colonization made changes in the hydrology of the Duck Creek North freshwater wetland, because normal flow of water was modified, causing periods of drought and floods that in turn oxygenate or not the soil, varying microbial activity, altering the carbon cycle and emitting greenhouse gases [170].

In the same case as Asia and Europe, agriculture is the incident factor with the highest proportion of the ten identified in this continent. This revelation allows us to clearly observe that the need to produce food is affecting wetlands in these continents, despite the unfavorable climatic conditions in Australia, the Oceania region where the largest number of investigations were found as can be seen in Figures 11 and 12.

Regarding wetland type, seven were identified, being mangroves the most affected, followed by rivers and lagoons. It is very important to notice that agricultural activity in Australia such as the cultivation of *Saccharum officinarum* and *Colocasia esculenta* are highly demanding of water and fertilization, and all nutrient discharges affect wetlands.

In addition, Australia produces wheat, grains (barley, oats, millet, corn, and triticale), rice, oilseeds (rapeseed, sunflower, soybeans, and peanuts), legumes (lupins and chickpeas), cotton, fruits, grapes, tobacco, and vegetables. The main livestock consists of sheep (lamb and wool), beef, pork, poultry, and dairy products. More than 90% of wool and cotton are exported, almost 80% of wheat, more than 50% of barley and rice, more than 40% of meat and grain legumes, more than 30% of dairy products, and almost 20% of fruit production [144].

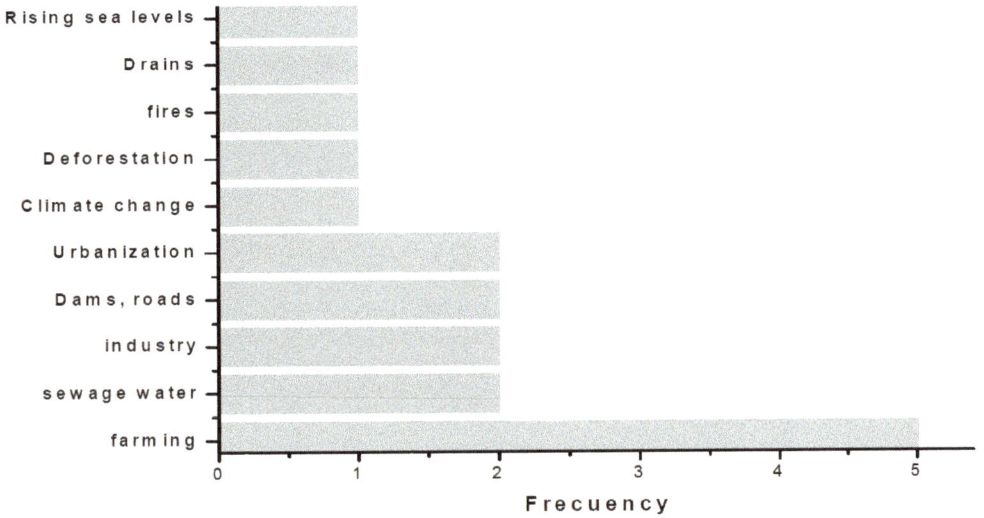

Figure 11. Frequency Distribution of Incident Factors in Oceania/Australia.

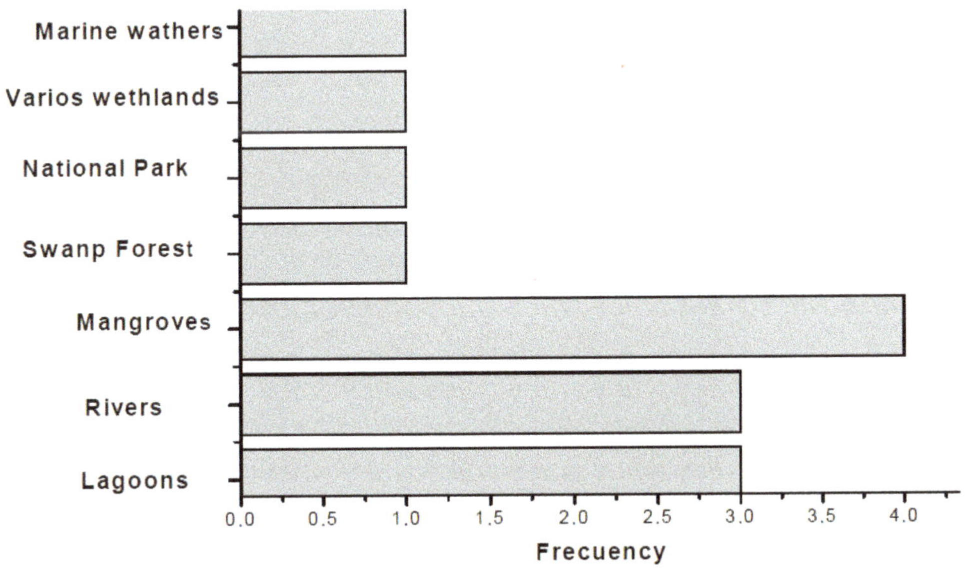

Figure 12. Frequency Distribution of the type of wetlands affected in Oceania/Australia.

3.6. Incident Factors and Effective Losses Globally

Urban and agricultural development of some continents such as Europe, North America, Asia, have deteriorated many wetlands, unlike South America, where populations are widely separated from continental wetlands; but failing that, they deteriorate the coastal areas by urban settlements [171].

On the other hand, in terms of carbon capture, it is diminished every time wetland vegetation is lost, or wastewater is added, causing eutrophication. Consequently, the carbon cycle is broken. First, more greenhouse gases are emitted due to vegetation loss when the soil is exposed, and second, vegetation is affected due to changes in nutrients, especially nitrogen.

Characteristic aspects in all continents are in the loss and sequestering indicators:

- Vegetation loss;
- Eutrophication by wastewater.

Differing aspects are in some incident factors:

- Oil spills;
- Salinity;
- Fires;
- Sea level.

Sixteen factors identified in the different regions are observed in Figure 13, being agriculture farming the most important, followed by urbanization and aquaculture. Two aspects have to do with food and one with industrial development.

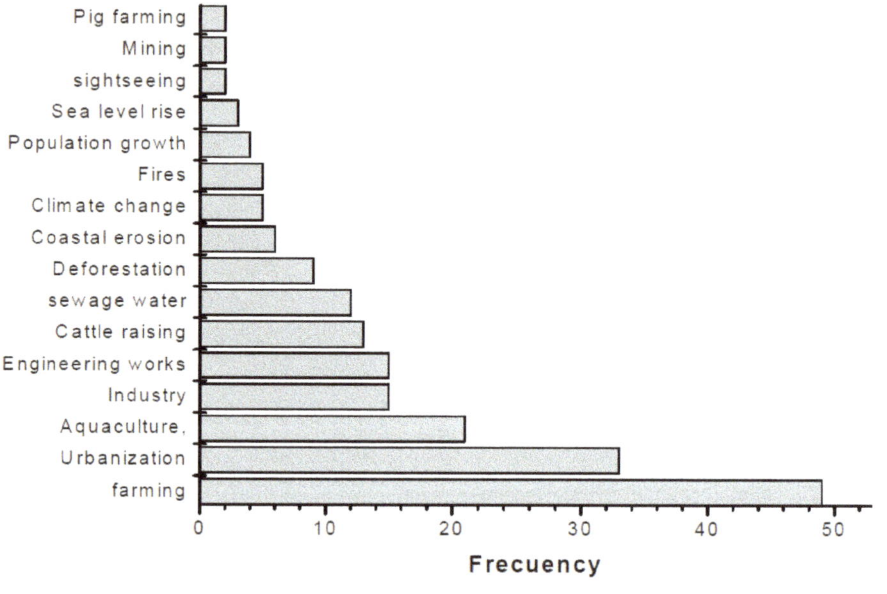

Figure 13. Frequency distribution of incident factors at a global level.

Furthermore, fourteen different types of wetlands have been established on all continents with effective losses as shown in Figure 14. Mangroves, lagoons, marine waters, and several wetlands stand out.

This information is useful because global efforts should be focused on this type of system for future planning, mainly because demand for food will grow together with the population by 2050 [172]. This indicates that wetlands are at risk if measures are not taken to work on their protection and on sustainable agricultural production practices.

By grouping wetlands according to the Ramsar system, they can be classified into marine, coastal, and continental. In Figure 15, results confirm what was stated by [14], "continental wetlands have been more affected than marine and coastal ones". He also debated the theory that wetlands have been lost by 50%, estimating that the value is between 54% and 57% since losses are attributed to economic growth, population, extensive and intensive agriculture, changes in use, and urbanization. The results of this research study identified the most frequent incident factors by continents, establishing a broad panorama of direct and indirect factors that affect them.

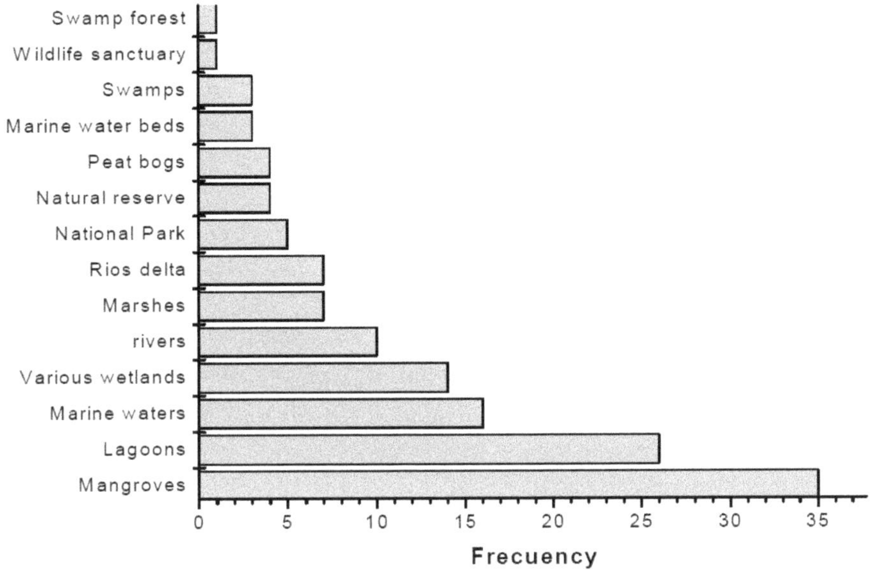

Figure 14. Frequency Distribution of Affected Wetlands at a Global Level.

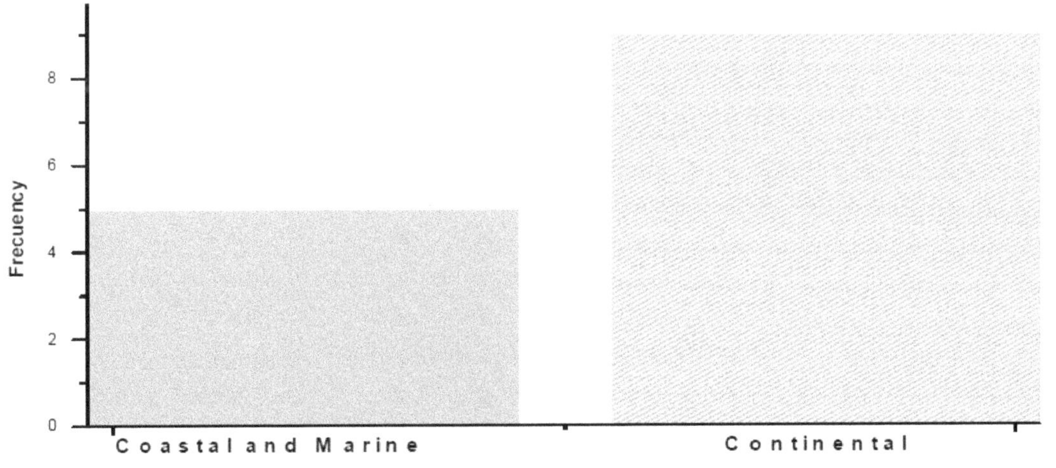

Figure 15. Classification of Wetlands Affected Globally.

In another context, global ineffective public development policies have been having an impact on wetlands, because they uncontrollably allow or promote anthropogenic activities, being urban or economic. Territorial planning goes in a diametrically opposite direction to the ecological function of wetlands because the economic point of view prevails in the first instance. Consequently, it is recommended that the local population participate in decision-making for the conservation of their territories [173].

Proposal to implement the care and environmental management of wetlands in a significant way.

The results, however, will be useful for the entities in charge of monitoring wetlands, especially those that are most susceptible to changes, on which the most incident factors are, in such a way that it serves as an early warning to prevent damage. Unfortunately, those interested in the study of wetlands are very few in Mexico and Colombia. It is an opportunity to make significant contributions. More studies are required. It is important that young people join the research of these environments, from biological, economic and engineering aspects. In Mexico, research is supported in the Laboratory of Wetlands and environmental sustainability of the National Technology of Mexico campus Misantla, as well as in the College of Veracruz, the Institute of Ecology AC, and the Research Institute on Ecosystems and Sustainability (IIES) of the National Autonomous University of Mexico. In the Colombia University of Sucre, its research group GIMAGUAS supports this type of research.

This research work is the starting point that can serve as a guide for wetland studies to consider: biogeochemical factors, geographic location in severe loss, 'vegetation loss', and 'blue carbon loss'.

A comprehensive view of the current situation regarding factors that affect the loss of wetlands at a global level, on the other hand, shows that the impacts are influenced by human activities related to agricultural, industrial and urban development. Therefore, decision-makers are the ones to take measures to manage quality towards healthier wetlands. This study fulfills its objective, and we hope that it will be useful in the future while we continue working on this topic with the same interest.

The sustainable development of wetlands must be linked to the Sustainable Development Goals of the United Nations. Every country with its commitments acquired for its fulfillment must consider actions to maintain them in the future, one of them is the one established by the Ramsar Convention that focuses around three pillars: the wise use of all wetlands, the designation and conservation of Ramsar sites, and the promotion of transboundary management. The Ramsar Strategic Plan for 2016–2024 has four closely related objectives: address wetland loss and degradation, effectively conserve and manage the Ramsar Site network, wisely use wetlands, and improve enforcement. SCR [174]. Furthermore, in the study carried out by the Secretary of the Ramsar Commission in 2018 [174], quantitative data on wetland losses in different continents were reported, however, incident factors were not considered as they were in this research.

This research is pertinent because the level of affectation that the different types of wetlands have suffered at a global level is described in a timely manner and not independently without any type of compilation, considering that their functions are of great importance to counteract the effects of greenhouse gases that produce global warming. The study is original as well because the results of previous investigations were found and served as the basis for the construction of a global map where the countries that do the most research on issues related to wetlands and the factors that affect them are identified, and the little or no research in some regions is also observed.

The contribution of this research is built on the description of factors affecting updated natural wetlands due to population dynamics related to urbanization, and agricultural activities due to the connection they have between the production of housing and food for the population. Nonetheless, despite the fact that the sample studied does not have information from many other countries, it is important to notice that it is very possible that these factors are common factors in other parts of the world. This is mainly because they are factors related to the direct pressures exerted by the human being, also bearing in mind that towards 2050, projections of food and population requirements are alarming and pressure on wetlands may proportionally increase as well [143,175–177].

These damages degrade wetlands that are losing their sequestration functionality and consequently, climate regulation, confirming that global warming will continue to gradually increase in proportion to the damage caused. IPCC [178], unless an international public policy is applied that minimizes consumption of water-soil resources, we will not exceed the capacity for nature self-regulation.

Future Lines of Research

Due to the effects that wetlands suffer from distinct factors, it is advisable to monitor them using remote sensing techniques to identify and quantify the vegetation type, that serves to make comparisons among wetlands of different latitudes using the spectral signatures useful in the identification of coverage and study of the phenology of the wetland vegetation.

It is important to analyze the health status of wetlands using remote sensors through indices that work with the reflectance of the cover and show the degree of health of the vegetation, such as the Normalized Difference Vegetation Index (NDVI), the Normalized Difference Water Index (NDWI), and the Enhanced Vegetation Index (EVI), which will help identify health problems that possibly affect wetlands such as nitrification of their waters. With this information, it is possible to make models of artificial intelligence and large volumes of satellite images from specialized sensors in water and wetlands. A predictive analysis of the future behavior of ecosystems is recommended, considering variables such as population, temperature rise, land-use change, vegetation loss and blue carbon.

Increasing carbon quantification studies to increase global inventory statistics on different continents is suggested.

Rethinking land use planning policies is also advisable because economic interests take precedence over the land ecological component, greatly affecting ecosystems. Thus, sustainability evaluations are necessary with existing methodologies and, failing that, generate new methodologies for their evaluation.

Some indicators shown in the different tables were not analyzed in this document because they were not included within the objective of the research. However, they can be useful for new studies on loss of areas and/or degree of affectation of a quantitative but not qualitative way as it was in this study.

4. Conclusions

After carrying out a macro-analysis of the incident factors and the types of wetlands affected at a global level, of 134 articles, the objective set in the investigation was fulfilled, 'to identify the incident factors in the effective loss of area and carbon sequestration in marine, coastal, and continental wetlands that have had an impact on climate change in the last 14 years at a global level', and it is possible to affirm that this work ratifies the importance of conserving and preserving wetlands due to the incident factors found.

It is confirmed that all over the world, anthropogenic activities that most affected natural wetlands were agriculture (25%), urbanization (16.8%), aquaculture (10.7%), and industry (7.6%). These are direct impacts or pressures exerted by human beings, implying that, increasingly, population growth will be an important factor as a determining agent of damage in the future if this trend is followed.

On the other hand, it was determined that the types of wetlands most affected are: mangroves (25.7%), lagoons (19.11%) and marine waters (11.7%). Nevertheless, after making a summation between marine-coastal and continental wetlands, we find that these systems are affected by 35.7% and 64.3%, respectively. This confirms that more effective environmental management and control measures are urgently needed in order to conserve and preserve them, given the multiple ecological functions that such ecosystems provide.

Finally, this research work is unique and relevant because it exposes the different incident factors at a global level together, evidence was found of little research in other continents on a topic of global interest, the percentages found by continent stand out, they show that there are some with more studies than others, i.e., America (41.04%), Asia (34.32%), Oceania/Australia (10.44%), Africa (7.46%), and Europe (6.74%). Thus, it should be said that the first two have the largest number of coastal wetlands, especially mangroves. Nonetheless, research should be increased in all of them because economic and population development as a result of globalization have been affecting the global climate with GHGs. Therefore, environmental management of conservation and handling of these ecosystems is recommended to reduce their ecological issues by 2050.

Author Contributions: Conceptualization, G.A.B.-D., E.A.B.-T. and L.C.S.H.; methodology, G.F.-L., G.F.-L. and M.C.L.M.; software, E.A.B.-T.; validation, G.F.-L., J.L.M.-M. and M.C.L.M.; formal analysis, G.A.B.-D., E.A.B.-T. and L.C.S.H.; investigation, G.A.B.-D., E.A.B.-T. and L.C.S.H.; resources, J.L.M.-M.; data curation, M.C.L.M.; writing—original draft preparation, G.A.B.-D., L.C.S.H. and G.F.-L.; writing—review and editing, J.L.M.-M., M.C.L.M. and E.A.B.-T.; visualization, G.F.-L., M.C.L.M. and J.L.M.-M.; supervision, G.F.-L., M.C.L.M. and J.L.M.-M.; project administration, G.A.B.-D., L.C.S.H. and E.A.B.-T., funding acquisition, G.A.B.-D., E.A.B.-T. and L.C.S.H. All authors have read and agreed to the published version of the manuscript.

Funding: The study received external funding from the National Council of Science and Technology (CONACYT) with the first author (G.A.B.D.) doctoral fellowship and the correspondence author (E.A.B.-T.) postdoctoral academic stay.

Data Availability Statement: The data supporting the reported results can be requested with the corresponding authors.

Acknowledgments: Thanks to the National Council of Science and Technology (CONACYT) for the scholarship awarded to him to pursue his Doctorate in Engineering Sciences offered by the National Technological Institute of Mexico, Campus Misantla registered in the Quality Postgraduate Register (PNPC), the author also expresses his gratitude to the University of Sucre for all the support provided and received for conducting this research study. Thanks to Luis Carlos Sandoval Herazo, director of the first author's doctoral thesis and advisor to the corresponding author's postdoctoral stay project.

Conflicts of Interest: The authors declare no conflict of interest.

References

1. Costanza, R.; de Groot, R.; Sutton, P.; van der Ploeg, S.; Anderson, S.J.; Kubiszewski, I.; Farber, S.; Turner, R.K. Changes in the global value of ecosystem services. *Glob. Environ. Change* **2014**, *26*, 152–158. [CrossRef]
2. Owers, C.J.; Rogers, K.; Woodroffe, C.D. Spatial variation of above-ground carbon storage in temperate coastal wetlands. *Estuar. Coast. Shelf Sci.* **2018**, *210*, 55–67. [CrossRef]
3. Gallant, K.; Withey, P.; Risk, D.; van Kooten, G.C.; Spafford, L. Measurement and economic valuation of carbon sequestration in Nova Scotian wetlands. *Ecol. Econ.* **2020**, *171*, 106619. [CrossRef]
4. Sánchez-Espinosa, A.; Schröder, C. Land use and land cover mapping in wetlands one step closer to the ground: Sentinel-2 versus landsat 8. *J. Environ. Manag.* **2019**, *247*, 484–498. [CrossRef] [PubMed]
5. Sica, Y.V.; Quintana, R.D.; Radeloff, V.C.; Gavier-Pizarro, G.I. Wetland loss due to land use change in the Lower Paraná River Delta, Argentina. *Sci. Total Environ.* **2016**, *568*, 967–978. [CrossRef]
6. Were, D.; Kansiime, F.; Fetahi, T.; Cooper, A.; Jjuuko, C. Carbon Sequestration by Wetlands: A Critical Review of Enhancement Measures for Climate Change Mitigation. *Earth Syst. Environ.* **2019**, *3*, 327–340. [CrossRef]
7. Mitsch, W.J.; Mander, Ü. Wetlands and carbon revisited. *Ecol. Eng.* **2018**, *114*, 1–6. [CrossRef]
8. Xu, S.; Liu, X.; Li, X.; Tian, C. Soil organic carbon changes following wetland restoration: A global meta-analysis. *Geoderma* **2019**, *353*, 89–96. [CrossRef]
9. Huang, C.; Yuan, C.; Yang, W.; Yang, L. Temporal variations of greenhouse gas emissions and carbon sequestration and stock from a tidal constructed mangrove wetland. *Mar. Pollut. Bull.* **2019**, *149*, 110568. [CrossRef] [PubMed]
10. Boone, J.K.; Bhomia, R.K. Ecosystem carbon stocks of mangroves across broad environmental gradients in West-Central Africa: Global and regional comparisons. *PLoS ONE* **2017**, *12*, e0187749. [CrossRef]
11. Ward, R.D. Carbon sequestration and storage in Norwegian Arctic coastal wetlands: Impacts of climate change. *Sci. Total Environ.* **2020**, *748*, 141343. [CrossRef] [PubMed]
12. Alongi, D.M. Carbon sequestration in mangrove forests. *Carbon Manag.* **2012**, *3*, 313–322. [CrossRef]
13. Sun, X.; Li, Y.; Zhu, X.; Cao, K.; Feng, L. Integrative assessment and management implications on ecosystem services loss of coastal wetlands due to reclamation. *J. Clean. Prod.* **2017**, *163*, S101–S112. [CrossRef]
14. Davidson, N.C. How much wetland has the world lost? Long-term and recent trends in global wetland area. *Mar. Freshw. Res.* **2014**, *65*, 934–941. [CrossRef]
15. Zhao, C.; Liu, S.; Jiang, Z.; Wu, Y.; Cui, L.; Huang, X.; Macreadie, P.I. Nitrogen purification potential limited by nitrite reduction process in coastal eutrophic wetlands. *Sci. Total Environ.* **2019**, *694*, 133702. [CrossRef] [PubMed]
16. Marín-Muñiz, J.L.; Hernández, M.E.; Moreno-Casasola, P. Comparing soil carbon sequestration in coastal freshwater wetlands with various geomorphic features and plant communities in Veracruz, Mexico. *Plant Soil* **2014**, *378*, 189–203. [CrossRef]
17. Liu, H.; Yi, Y.; Yue, Y.; Cui, B. Reducing the likelihood of carbon loss from wetlands by improving the spatial connections between high carbon patches. *J. Clean. Prod.* **2020**, *267*, 121819. [CrossRef]
18. Donato, D.C.; Kauffman, J.B.; Murdiyarso, D.; Kurnianto, S.; Stidham, M.; Kanninen, M. Mangroves among the most carbon-rich forests in the tropics. *Nat. Geosci.* **2011**, *4*, 293–297. [CrossRef]

19. Adame, M.F.; Kauffman, J.B.; Medina, I.; Gamboa, J.N.; Torres, O.; Caamal, J.P.; Reza, M.; Herrera-Silveira, J.A. Carbon Stocks of Tropical Coastal Wetlands within the Karstic Landscape of the Mexican Caribbean. *PLoS ONE* **2013**, *8*, e56569. [CrossRef] [PubMed]
20. Köchy, M.; Hiederer, R.; Freibauer, A. Global distribution of soil organic carbon—Part 1: Masses and frequency distributions of SOC stocks for the tropics, permafrost regions, wetlands, and the world. *Soil* **2015**, *1*, 351–365. [CrossRef]
21. Villa, J.A.; Bernal, B. Carbon sequestration in wetlands, from science to practice: An overview of the biogeochemical process, measurement methods, and policy framework. *Ecol. Eng.* **2018**, *114*, 115–128. [CrossRef]
22. Cui, X.; Liang, J.; Lu, W.; Chen, H.; Liu, F.; Lin, G.; Xu, F.; Luo, Y.; Lin, G. Stronger ecosystem carbon sequestration potential of mangrove wetlands with respect to terrestrial forests in subtropical China. *Agric. For. Meteorol.* **2018**, *249*, 71–80. [CrossRef]
23. Lavery, P.S.; Mateo, M.-Á.; Serrano, O.; Rozaimi, M. Variability in the Carbon Storage of Seagrass Habitats and Its Implications for Global Estimates of Blue Carbon Ecosystem Service. *PLoS ONE* **2013**, *8*, e73748. [CrossRef]
24. Sanderman, J.; Hengl, T.; Fiske, G.; Solvik, K.; Adame, M.F.; Benson, L.; Bukoski, J.J.; Carnell, P.; Cifuentes-Jara, M.; Donato, D.; et al. A global map of mangrove forest soil carbon at 30 m spatial resolution. *Environ. Res. Lett.* **2018**, *13*, 55002. [CrossRef]
25. Bernal, B.; Mitsch, W.J. Carbon sequestration in freshwater wetlands in Costa Rica and Botswana. *Biogeochemistry* **2013**, *115*, 77–93. [CrossRef]
26. Zhang, T.; Cao, G.; Cao, S.; Zhang, X.; Zhang, J.; Han, G. Dynamic assessment of the value of vegetation carbon fixation and oxygen release services in Qinghai Lake basin. *Acta Ecol. Sin.* **2017**, *37*, 79–84. [CrossRef]
27. Von Uexkull, N.; Buhaug, H. Security implications of climate change: A decade of scientific progress. *J. Peace Res.* **2021**, *58*, 3–17. [CrossRef]
28. Lang'at, J.K.S.; Kairo, J.G.; Mencuccini, M.; Bouillon, S.; Skov, M.W.; Waldron, S.; Huxham, M. Rapid losses of surface elevation following tree girdling and cutting in tropical mangroves. *PLoS ONE* **2014**, *9*, e107868. [CrossRef]
29. Githaiga, M.N.; Kairo, J.G.; Gilpin, L.; Huxham, M. Carbon storage in the seagrass meadows of Gazi Bay, Kenya. *PLoS ONE* **2017**, *12*, e0177001. [CrossRef]
30. Juma, G.A.; Magana, A.M.; Michael, G.N.; Kairo, J.G. Variation in Seagrass Carbon Stocks Between Tropical Estuarine and Marine Mangrove-Fringed Creeks. *Front. Mar. Sci.* **2020**, *7*, 696. [CrossRef]
31. Elbasiouny, H.; Abowaly, M.; Gad, A.A.; Abu Alkheir, A.; Elbehiry, F. Restoration and sequestration of carbon and nitrogen in the degraded northern coastal area in Nile Delta, Egypt for climate change mitigation. *J. Coast Conserv.* **2017**, *21*, 105–114. [CrossRef]
32. Eid, E.M.; Shaltout, K.H. Evaluation of carbon sequestration potentiality of Lake Burullus, Egypt to mitigate climate change Egypt. *J. Aquat. Res.* **2013**, *39*, 31–38. [CrossRef]
33. Ekumah, B.; Armah, F.A.; Afrifa, E.K.A.; Aheto, D.W.; Odoi, J.O.; Afitiri, A.R. Geospatial assessment of ecosystem health of coastal urban wetlands in Ghana. *Ocean Coast. Manag.* **2020**, *193*, 105226. [CrossRef]
34. Ligate, E.J.; Chen, C.; Wu, C. Evaluation of tropical coastal land cover and land use changes and their impacts on ecosystem service values. *Ecosyst. Health Sustain.* **2018**, *4*, 188–204. [CrossRef]
35. Lagomasino, D.; Fatoyinbo, T.; Lee, S.; Feliciano, E.; Trettin, C.; Shapiro, A.; Mangora, M.M. Measuring mangrove carbon loss and gain in deltas. *Environ. Res. Lett.* **2019**, *14*, 25002. [CrossRef]
36. Saunders, M.J.; Kansiime, F.; Jones, M.B. Agricultural encroachment: Implications for carbon sequestration in tropical African wetlands. *Glob. Chang. Biol. Bioenergy* **2012**, *18*, 1312–1321. [CrossRef]
37. Were, D.; Kansiime, F.; Fetahi, T.; Hein, T. A natural tropical freshwater wetland is a better climate change mitigation option through soil organic carbon storage compared to a rice paddy wetland. *SN Appl. Sci.* **2020**, *2*, 1–13. [CrossRef]
38. Ellery, W.N.; Grenfell, S.E.; Grenfell, M.C.; Humphries, M.S.; Barnes, K.; Dahlberg, A.; Kindness, A. Peat formation in the context of the development of the Mkuze floodplain on the coastal plain of Maputaland, South Africa. *Geomorphology* **2012**, *141–142*, 11–20. [CrossRef]
39. Torres, J.E. Africa's Economy: 50 Years of Failed Policies]. *Economia* **2008**. Available online: https://go.gale.com/ps/i.do?id=GALE%7CA351436991&sid=googleScholar&v=2.1&it=r&linkaccess=abs&issn=17947634&p=IFME&sw=w&userGroupName=anon~{|b76f7591 (accessed on 12 December 2021).
40. Quimbayo Ruiz, G.A. Territory, sustainability, and beyond: Latin American urbanization through a political ecology. *Environ. Plan E Nat. Space* **2020**, *3*, 786–809. [CrossRef]
41. Delgado, R.; Eguino, H.; Lopes, A. (Eds.) *Fiscal Policy and Climate Change: Recent Experiences of the Finance Ministries of Latin America and the Caribbean*; Inter-American Development Bank (IDB): Washington, DC, USA, 2021. [CrossRef]
42. Ardón, M.; Helton, A.M.; Bernhardt, E.S. Drought and saltwater incursion synergistically reduce dissolved organic carbon export from coastal freshwater wetlands. *Biogeochemistry* **2016**, *127*, 411–426. [CrossRef]
43. Kassakian, J.; Jones, A.; Martinich, J.; Hudgens, D. Managing for No Net Loss of Ecological Services: An Approach for Quantifying Loss of Coastal Wetlands due to Sea Level Rise. *Environ. Manag.* **2017**, *59*, 736–751. [CrossRef] [PubMed]
44. St. Laurent, K.A.; Hribar, D.J.; Carlson, A.J.; Crawford, C.M.; Siok, D. Assessing coastal carbon variability in two Delaware tidal marshes. *J. Coast. Conserv.* **2020**, *24*, 65. [CrossRef]
45. Greiner, J.T.; McGlathery, K.J.; Gunnell, J.; McKee, B.A. Seagrass Restoration Enhances "Blue Carbon" Sequestration in Coastal Waters. *PLoS ONE* **2013**, *8*, e72469. [CrossRef] [PubMed]

46. Riegel, J.B.; Bernhardt, E.; Swenson, J. Estimating Above-Ground Carbon Biomass in a Newly Restored Coastal Plain Wetland Using Remote Sensing. *PLoS ONE* **2013**, *8*, e68251. [CrossRef] [PubMed]
47. Moseman-Valtierra, S.; Gonzalez, R.; Kroeger, K.D.; Tang, J.; Chao, W.C.; Crusius, J.; Bratton, J.; Green, A.; Shelton, J. Short-term nitrogen additions can shift a coastal wetland from a sink to a source of N_2O. *Atmos. Environ.* **2011**, *45*, 4390–4397. [CrossRef]
48. Stralberg, D.; Brennan, M.; Callaway, J.C.; Wood, J.K.; Schile, L.M.; Jongsomjit, D.; Kelly, M.; Parker, V.T.; Crooks, S. Evaluating Tidal Marsh Sustainability in the Face of Sea-Level Rise: A Hybrid Modeling Approach Applied to San Francisco Bay. *PLoS ONE* **2011**, *6*, e27388. [CrossRef] [PubMed]
49. Coverdale, T.C.; Brisson, C.P.; Young, E.W.; Yin, S.F.; Donnelly, J.P.; Bertness, M.D. Indirect human impacts reverse centuries of carbon sequestration and salt marsh accretion. *PLoS ONE* **2014**, *9*, e93296. [CrossRef] [PubMed]
50. Craft, C.B. Tidal freshwater forest accretion does not keep pace with sea level rise. *Glob. Chang. Biol.* **2012**, *18*, 3615–3623. [CrossRef]
51. Siciliano, D.; Wasson, K.; Potts, D.C.; Olsen, R.C. Evaluating hyperspectral imaging of wetland vegetation as a tool for detecting estuarine nutrient enrichment. *Remote Sens. Environ.* **2008**, *112*, 4020–4033. [CrossRef]
52. Hester, M.W.; Willis, J.M.; Baker, M.C. Oil spills in coastal wetlands. In *Encyclopedia of the Anthropocene*; Elsevier Inc.: Amsterdam, The Netherlands, 2017; Volume 5, pp. 67–76. [CrossRef]
53. Yavitt, J.B.; Burtis, J.C.; Smemo, K.A.; Welsch, M. Plot-scale spatial variability of methane, respiration, and net nitrogen mineralization in muck-soil wetlands across a land use gradient. *Geoderma* **2018**, *315*, 11–19. [CrossRef]
54. Craft, C.; Vymazal, J.; Kröpfelová, L. Carbon sequestration and nutrient accumulation in floodplain and depressional wetlands. *Ecol. Eng.* **2018**, *114*, 137–145. [CrossRef]
55. Lane, R.R.; Mack, S.K.; Day, J.W.; Kempka, R.; Brady, L.J. Carbon Sequestration at a Forested Wetland Receiving Treated Municipal Effluent. *Wetlands* **2017**, *37*, 861–873. [CrossRef]
56. Suir, G.M.; Sasser, C.E.; DeLaune, R.D.; Murray, E.O. Comparing carbon accumulation in restored and natural wetland soils of coastal Louisiana. *Int. J. Sediment Res.* **2019**, *34*, 600–607. [CrossRef]
57. Haywood, B.J.; Hayes, M.P.; White, J.R.; Cook, R.L. Potential fate of wetland soil carbon in a deltaic coastal wetland subjected to high relative sea level rise. *Sci. Total Environ.* **2020**, *711*, 135185. [CrossRef] [PubMed]
58. Fennessy, M.S.; Wardrop, D.H.; Moon, J.B.; Wilson, S.; Craft, C. Soil carbon sequestration in freshwater wetlands varies across a gradient of ecological condition and by ecoregion. *Ecol. Eng.* **2018**, *114*, 129–136. [CrossRef]
59. Hemes, K.S.; Chamberlain, S.D.; Eichelmann, E.; Knox, S.H.; Baldocchi, D.D. A Biogeochemical Compromise: The High Methane Cost of Sequestering Carbon in Restored Wetlands. *Geophys. Res. Lett.* **2018**, *45*, 6081–6091. [CrossRef]
60. Newman, S.; Osborne, T.Z.; Hagerthey, S.E.; Saunders, C.; Rutchey, K.; Schall, T.; Reddy, K.R. Drivers of landscape evolution: Multiple regimes and their influence on carbon sequestration in a sub-tropical peatland. *Ecol. Monogr.* **2017**, *87*, 578–599. [CrossRef]
61. Mitsch, W.J.; Hernandez, M.E. Landscape and climate change threats to wetlands of North and Central America. *Aquat. Sci.* **2013**, *75*, 133–149. [CrossRef]
62. Verhoeven, J.T.A.; Laanbroek, H.J.; Rains, M.C.; Whigham, D.F. Effects of increased summer flooding on nitrogen dynamics in impounded mangroves. *J. Environ. Manag.* **2014**, *139*, 217–226. [CrossRef]
63. Dontis, E.E.; Radabaugh, K.R.; Chappel, A.R.; Russo, C.E.; Moyer, R.P. Carbon Storage Increases with Site Age as Created Salt Marshes Transition to Mangrove Forests in Tampa Bay, Florida (USA). *Estuaries Coast* **2020**, *43*, 1470–1488. [CrossRef]
64. Marchio, D.A.; Savarese, M.; Bovard, B.; Mitsch, W.J. Carbon sequestration and sedimentation in mangrove swamps influenced by hydrogeomorphic conditions and urbanization in Southwest Florida. *Forest* **2016**, *7*, 116. [CrossRef]
65. Vences Martínez, J.Á.; Sampedro Rosas, M.L.; Castillo Elías, B.; Olmos Martínez, E.; Juárez López, A.L.; Reyes Umaña, M. Affectation of mangrove by anthropogenic activities at sub-basin nuxco, Guerrero, Mexico. *Mex. J. Agroecosystems* **2016**, *3*, 163–174. Available online: http://ri.uagro.mx/handle/uagro/621 (accessed on 10 December 2021).
66. Cerón, R.M.; Cerón, J.G.; Guerra, J.J.; Zavala, J.C.; Amador, L.E.; Endañu, E.; Moreno, G. Determination of the amount of carbon stored in a disturbed mangrove forest in Campeche, MexicoWIT Trans. *Ecol. Environ.* **2011**, *144*, 327–338. [CrossRef]
67. Vázquez-González, C.; Fermán-Almada, J.L.; Moreno-Casasola, P.; Espejel, I. Scenarios of vulnerability in coastal municipalities of tropical Mexico: An analysis of wetland land use. *Ocean Coast. Manag.* **2014**, *89*, 11–19. [CrossRef]
68. Berlanga-Robles, C.A.; Ruiz-Luna, A.; Bocco, G.; Vekerdy, Z. Spatial analysis of the impact of shrimp culture on the coastal wetlands on the Northern coast of Sinaloa, Mexico. *Ocean Coast. Manag.* **2011**, *54*, 535–543. [CrossRef]
69. Campos, C.A.; Hernández, M.E.; Moreno-Casasola, P.; Cejudo Espinosa, E.; Robledo, R.A.; Infante Mata, D. Soil water retention and carbon pools in tropical forested wetlands and marshes of the Gulf of Mexico. *Hydrol. Sci. J.* **2011**, *56*, 1388–1406. [CrossRef]
70. Bianchi, T.S.; Allison, M.A.; Zhao, J.; Li, X.; Comeaux, R.S.; Feagin, R.A.; Kulawardhana, R.W. Historical reconstruction of mangrove expansion in the Gulf of Mexico: Linking climate change with carbon sequestration in coastal wetlands. *Estuar. Coast. Shelf Sci.* **2013**, *119*, 7–16. [CrossRef]
71. Adame, M.F.; Santini, N.S.; Tovilla, C.; Vázquez-Lule, A.; Castro, L.; Guevara, M. Carbon tocks and soil sequestration rates of tropical riverine wetlands. *Biogeosciences* **2015**, *12*, 3805–3818. [CrossRef]
72. Ochoa-Gómez, J.G.; Lluch-Cota, S.E.; Rivera-Monroy, V.H.; Lluch-Cota, D.B.; Troyo-Diéguez, E.; Oechel, W.; Serviere-Zaragoza, E. Mangrove wetland productivity and carbon stocks in an arid zone of the Gulf of California (La Paz Bay, Mexico). *For. Ecol. Manag.* **2019**, *442*, 135–147. [CrossRef]

73. Beach, T.; Luzzadder-Beach, S.; Terry, R.; Dunning, N.; Houston, S.; Garrison, T. Carbon isotopic ratios of wetland and terrace soil sequences in the Maya Lowlands of Belize and Guatemala. *Catena* **2011**, *85*, 109–118. [CrossRef]
74. Bhomia, R.K.; Kauffman, J.B.; Mcfadden, T.N. Ecosystem carbon stocks of mangrove forests along the Pacific and Caribbean coasts of Honduras. *Wetl. Ecol. Manag.* **2016**, *24*, 187–201. [CrossRef]
75. Rodríguez-Arias, C.E.; Silva Benavides, A.M. Los Humedales de la Quebrada Estero en San Ramón, Costa Rica: Importancia y estado actual. *Posgrado Sociedad Revista Electrónica Sistema Estudios Posgrado* **2017**, *15*, 13–26. [CrossRef]
76. Mitsch, W.J.; Tejada, J.; Nahlik, A.; Kohlmann, B.; Bernal, B.; Hernández, C.E. Tropical wetlands for climate change research, water quality management and conservation education on a university campus in Costa Rica. *Ecol. Eng.* **2008**, *34*, 276–288. [CrossRef]
77. Ramenzoni, V.C.; Besonen, M.R.; Yoskowitz, D.; Sánchez, V.V.; Rivero, A.R.; González-Díaz, P.; Méndez, A.F.; Escuela, D.B.; Ramos, I.H.; Hernández López, N.V.; et al. Transnational research for coastal wetlands conservation in a Cuba–US setting. *Glob. Sustain.* **2020**, *3*, 1–11. [CrossRef]
78. Kauffman, J.B.; Heider, C.; Norfolk, J.; Payton, F. Carbon stocks of intact mangroves and carbon emissions arising from their conversion in the Dominican Republic. *Ecol. Appl.* **2014**, *24*, 518–527. [CrossRef] [PubMed]
79. Pérez-Rojas, J.; Moreno, F.; Quevedo, J.C.; Villa, J. Soil organic carbon stocks in fluvial and isolated tropical wetlands from Colombia. *Catena* **2019**, *179*, 139–148. [CrossRef]
80. Blanco, J.F.; Estrada, E.A.; Ortiz, L.F.; Urrego, L.E. Ecosystem-Wide Impacts of Deforestation in Mangroves: The Urabá Gulf (Colombian Caribbean) Case Study. *ISRN Ecol.* **2012**, *2012*, 1–14. [CrossRef]
81. Ricaurte, L.F.; Olaya-Rodríguez, M.H.; Cepeda-Valencia, J.; Lara, D.; Arroyave-Suárez, J.; Max Finlayson, C.; Palomo, I. Future impacts of drivers of change on wetland ecosystem services in Colombia. *Glob. Environ. Chang.* **2017**, *44*, 158–169. [CrossRef]
82. Palacios Peñaranda, M.L.; Cantera Kintz, J.R.; Peña Salamanca, E.J. Carbon stocks in mangrove forests of the Colombian Pacific. *Estuar. Coast. Shelf Sci.* **2019**, *227*, 106299. [CrossRef]
83. Mateus, F.; Caicedo, Y. *Effect of the Transformation of the Landscape on the Provision of the Ecosystem Service of Provision of Habitat of the Wetland "El Tunjo" (Bogotá-Colombia), from 1940 to 2014*; University of Applied and Environmental Sciences: Bogotá, CO, USA, 2014. Available online: https://core.ac.uk/download/pdf/326431054.pdf (accessed on 15 December 2021).
84. Ayala, K.; Torres, J. *A Socio-Environmental Study of "El Pantanal" Wetland from the Technical University of Machala*; Universidad Técnica de Machala: Machala, Ecuador, 2016. Available online: http://investigacion.utmachala.edu.ec/proceedings/index.php/utmach/article/view/50 (accessed on 27 November 2021).
85. Zamora, A. *Ecosystem Services in Coastal Wetlands from Peru*; University of the South: Sewanee, TN, USA, 2019. Available online: https://repositorio.cientifica.edu.pe/handle/20.500.12805/1383 (accessed on 30 November 2021).
86. Ghermandi, A.; Agard, J.; Nunes, P.A.L.D. Applying Geographic Information Systems to ecosystem services valuation and mapping in Trinidad and Tobago. *Lett. Spat. Resour. Sci.* **2018**, *11*, 289–306. [CrossRef]
87. Ding, Y.; Cawley, K.M.; da Cunha, C.N.; Jaffé, R. Environmental dynamics of dissolved black carbon in wetlands. *Biogeochemistry* **2014**, *119*, 259–273. [CrossRef]
88. Rovai, A.S.; Coelho-Jr, C.; de Almeida, R.; Cunha-Lignon, M.; Menghini, R.P.; Twilley, R.R.; Cintrón-Molero, G.; Schaeffer-Novelli, Y. Ecosystem-level carbon stocks and sequestration rates in mangroves in the Cananéia-Iguape lagoon estuarine system, southeastern Brazil. *For. Ecol. Manag.* **2021**, *479*, 118553. [CrossRef]
89. Finkl, C.W.; Makowski, C. (Eds.) *Coastal Wetlands: Alteration and Remediation*; Springer: Berlin/Heidelberg, Germany, 2017; pp. 159–186. [CrossRef]
90. Lucas, C.M.; Schöngart, J.; Sheikh, P.; Wittmann, F.; Piedade, M.T.F.; McGrath, D.G. Effects of land-use and hydroperiod on aboveground biomass and productivity of secondary Amazonian floodplain forests. *For. Ecol. Manag.* **2014**, *319*, 116–127. [CrossRef]
91. Dos Santos, F.B.; Ferreira, F.C.; Esteves, K.E. Assessing the importance of the riparian zone for stream fish communities in a sugarcane dominated landscape (Piracicaba River Basin, Southeast Brazil). *Environ. Biol. Fishes* **2015**, *98*, 1895–1912. [CrossRef]
92. Ferreira, A.C.; Bezerra, L.E.A.; Matthews-Cascon, H. Aboveground carbon stock in a restored neotropical mangrove: Influence of management and brachyuran crab assemblage. *Wetl. Ecol. Manag.* **2019**, *27*, 223–242. [CrossRef]
93. Portela, A.; Dos Santos, A.; De Lima, J.; Silveira Graudenz, G.; Silva Ruiz, M.; de Mahiques, M.M.; Lopes Figueira, R.C.; de Faria Alvim Wasserman, J.C. Management of Environmental Quality: An International Journal Article Information. 2011. Available online: http://repositorio.ipen.br/bitstream/handle/123456789/31131/26909.pdf?sequence=1&isAllowed=y (accessed on 2 December 2021).
94. Lin, Q.; Yu, S. Losses of natural coastal wetlands by land conversion and ecological degradation in the urbanizing Chinese coast. *Sci. Rep.* **2018**, *8*, 1–10. [CrossRef] [PubMed]
95. Cumings, B.; Deyo, F.C. The Origins and Development of the Northeast Asian Political Economy: Industrial Sectors, Product Cycles, and Political Consequences. In *The Political Economy of the New Asian Industrialism*; Cornell University Press: Ithaca, NY, USA, 2019. [CrossRef]
96. Herbeck, L.S.; Unger, D.; Wu, Y.; Jennerjahn, T.C. Effluent, nutrient and organic matter export from shrimp and fish ponds causing eutrophication in coastal and back-reef waters of NE hainan, tropical China. *Cont. Shelf Res.* **2013**, *57*, 92–104. [CrossRef]

97. Fan, B.; Li, Y.; Pavao-Zuckerman, M. The dynamics of land-sea-scape carbon flow can reveal anthropogenic destruction and restoration of coastal carbon sequestration. *Landsc. Ecol.* **2021**, *36*, 1933–1949. [CrossRef]
98. Jiang, Z.; Liu, S.; Zhang, J.; Wu, Y.; Zhao, C.; Lian, Z.; Huang, X. Eutrophication indirectly reduced carbon sequestration in a tropical seagrass bed. *Plant Soil* **2018**, *426*, 135–152. [CrossRef]
99. Mou, X.; Liu, X.; Sun, Z.; Tong, C.; Huang, J.; Wan, S.; Wang, C.; Wen, B. Effects of Anthropogenic Disturbance on Sediment Organic Carbon Mineralization Under Different Water Conditions in Coastal Wetland of a Subtropical Estuary. *Chin. Geogr. Sci.* **2018**, *28*, 400–410. [CrossRef]
100. Bao, K.; Zhao, H.; Xing, W.; Lu, X.; McLaughlin, N.B.; Wang, G. Carbon Accumulation in Temperate Wetlands of Sanjiang Plain, Northeast China. *Soil Sci. Soc. Am. J.* **2011**, *75*, 2386–2397. [CrossRef]
101. Li, X.; Chen, W.; Song, X.; Wang, M.; Hu, Q.; Deng, C. Effects of reclamation on distribution of soil carbon and nitrogen in saline soil of the yellow river delta. *Acta Pedol. Sin.* **2018**, *55*, 1032–1041. [CrossRef]
102. Zhao, G.; Ye, S.; Laws, E.A.; He, L.; Yuan, H.; Ding, X.; Wang, J. Carbon burial records during the last ~40,000 years in sediments of the Liaohe Delta wetland, China. *Estuar. Coast. Shelf Sci.* **2019**, *226*, 106291. [CrossRef]
103. Huang, Y.; Zhang, T.; Wu, W.; Zhou, Y.; Tian, B. Rapid risk assessment of wetland degradation and loss in low-lying coastal zone of Shanghai, China. *Hum. Ecol. Risk Assess. Int. J.* **2017**, *23*, 82–97. [CrossRef]
104. Duan, J.; Han, J.; Cheung, S.G.; Chong, R.K.Y.; Lo, C.M.; Lee, F.W.F.; Xu, S.J.L.; Yang, Y.; Tam, N.F.; Zhou, H.C. How mangrove plants affect microplastic distribution in sediments of coastal wetlands: Case study in Shenzhen Bay, South China. *Sci. Total Environ.* **2021**, *767*, 144695. [CrossRef] [PubMed]
105. Qu, W.; Han, G.; Eller, F.; Xie, B.; Wang, J.; Wu, H.; Li, J.; Zhao, M. Nitrogen input in different chemical forms and levels stimulates soil organic carbon decomposition in a coastal wetland. *Catena* **2020**, *194*, 104672. [CrossRef]
106. Zhong, Q.; Wang, K.; Nie, M.; Zhang, G.; Zhang, W.; Zhu, Y.; Fu, Y.; Zhang, Q.; Gao, Y. Responses of wetland soil carbon and nutrient pools and microbial activities after 7 years of experimental warming in the Yangtze Estuary. *Ecol. Eng.* **2019**, *136*, 68–78. [CrossRef]
107. Zhu, X.; Hou, Y.; Weng, Q.; Chen, L. Integrating UAV optical imagery and LiDAR data for assessing the spatial relationship between mangrove and inundation across a subtropical estuarine wetland. *ISPRS J. Photogramm. Remote Sens.* **2019**, *149*, 146–156. [CrossRef]
108. Mu, S.; Li, B.; Yao, J.; Yang, G.; Wan, R.; Xu, X. Monitoring the spatio-temporal dynamics of the wetland vegetation in Poyang Lake by Landsat and MODIS observations. *Sci. Total Environ.* **2020**, *725*, 138096. [CrossRef] [PubMed]
109. Xiao, D.; Deng, L.; Kim, D.G.; Huang, C.; Tian, K. Carbon budgets of wetland ecosystems in China. *Glob. Chang. Biol. Bioenergy* **2019**, *25*, 2061–2076. [CrossRef] [PubMed]
110. Lin, W.; Xu, D.; Guo, P.; Wang, D.; Li, L.; Gao, J. Exploring variations of ecosystem service value in Hangzhou Bay Wetland, Eastern China. *Ecosyst. Serv.* **2019**, *37*, 100944. [CrossRef]
111. Wang, X.; Jiang, Z.; Li, Y.; Kong, F.; Xi, M. Inorganic carbon sequestration and its mechanism of coastal saline-alkali wetlands in Jiaozhou Bay, China. *Geoderma* **2019**, *351*, 221–234. [CrossRef]
112. Meng, W.; Feagin, R.A.; Hu, B.; He, M.; Li, H. The spatial distribution of blue carbon in the coastal wetlands of China. *Estuar. Coast. Shelf Sci.* **2019**, *222*, 13–20. [CrossRef]
113. Zhang, Z.; Craft, C.B.; Xue, Z.; Tong, S.; Lu, X. Regulating effects of climate, net primary productivity, and nitrogen on carbon sequestration rates in temperate wetlands, Northeast China. *Ecol. Indic.* **2016**, *70*, 114–124. [CrossRef]
114. Hu, Y.; Wang, L.; Fu, X.; Yan, J.; Wu, J.; Tsang, Y.; Le, Y.; Sun, Y. Salinity and nutrient contents of tidal water affects soil respiration and carbon sequestration of high and low tidal flats of Jiuduansha wetlands in different ways. *Sci. Total Environ.* **2016**, *565*, 637–648. [CrossRef] [PubMed]
115. Xiao, Y.; Huang, Z.; Lu, X. Changes of soil labile organic carbon fractions and their relation to soil microbial characteristics in four typical wetlands of Sanjiang Plain, Northeast China. *Ecol. Eng.* **2015**, *82*, 381–389. [CrossRef]
116. Ke, C.Q.; Zhang, D.; Wang, F.Q.; Chen, S.X.; Schmullius, C.; Boerner, W.M.; Wang, H. Analyzing coastal wetland change in the Yancheng National Nature Reserve, China. *Reg. Environ. Chang.* **2011**, *11*, 161–173. [CrossRef]
117. Dong, W.; Shu, J.; He, P.; Ma, G.; Dong, M. Study on the Carbon Storage and Fixation of Phramites autralis in Baiyangdian Demonstration Area. *Procedia Environ. Sci.* **2012**, *13*, 324–330. [CrossRef]
118. Gillis, L.G.; Ziegler, A.D.; Van Oevelen, D.; Cathalot, C.; Herman, P.M.J.; Wolters, J.W.; Bouma, T.J. Tiny is mighty: Seagrass beds have a large role in the export of organic material in the tropical coastal zone. *PLoS ONE* **2014**, *9*. [CrossRef]
119. Cusack, M.; Saderne, V.; Arias-Ortiz, A.; Masqué, P.; Krishnakumar, P.K.; Rabaoui, L.; Qurban, M.A.; Qasem, A.M.; Prihartato, P.; Loughland, R.A.; et al. Organic carbon sequestration and storage in vegetated coastal habitats along the western coast of the Arabian Gulf. *Environ. Res. Lett.* **2018**, *13*, 74007. [CrossRef]
120. Byun, C.; Lee, S.H.; Kang, H. Estimation of carbon storage in coastal wetlands and comparison of different management schemes in South Korea. *J. Ecol. Environ.* **2019**, *43*, 8. [CrossRef]
121. Sondak, C.F.A.; Chung, I.K. Potential blue carbon from coastal ecosystems in the Republic of Korea. *Ocean Sci.* **2015**, *50*, 1–8. [CrossRef]
122. Misra, A.; Balaji, R. Decadal changes in the land use/land cover and shoreline along the coastal districts of southern Gujarat, India. *Environ. Monit. Assess.* **2015**, *187*, 461. [CrossRef]

123. Jana, B.B.; Nandy, S.K.; Lahiri, S.; Bhakta, J.N.; Biswas, J.K.; Bag, S.K.; Ghosh, P.; Maity, S.M.; Jana, S. Heterogeneity of water quality signature and feedbacks to carbon sequestration in wetlands across some districts of West Bengal, India. *J. Water Clim. Chang.* **2020**, *11*, 434–450. [CrossRef]
124. Harishma, K.M.; Sandeep, S.; Sreekumar, V.B. Biomass and carbon stocks in mangrove ecosystems of Kerala, southwest coast of India. *Ecol. Process.* **2020**, *9*, 31. [CrossRef]
125. Ganguly, D.; Patra, S.; Muduli, P.R.; Vishnu Vardhan, K.; Abhilash, K.R.; Robin, R.S.; Subramanian, B.R. Influence of nutrient input on the trophic state of a tropical brackish water lagoon. *Earth Syst. Sci.* **2015**, *124*, 1005–1017. [CrossRef]
126. Bindu, G.; Rajan, P.; Jishnu, E.S.; Ajith Joseph, K. Carbon stock assessment of mangroves using remote sensing and geographic information system. *Egypt. J. Remote Sens. Space Sci.* **2020**, *23*, 1–9. [CrossRef]
127. Bijoy Nandan, S.; Jayachandran, P.R.; Sreedevi, O.K. Spatio-Temporal Pattern of Primary Production in a Tropical Coastal Wetland (Kodungallur-Azhikode Estuary), South West Coast of India. *J. Coast. Dev.* **2014**, *17*, 392. [CrossRef]
128. Dommain, R.; Couwenberg, J.; Joosten, H. Development and carbon sequestration of tropical peat domes in south-east Asia: Links to post-glacial sea-level changes and Holocene climate variability. *Quat. Sci. Rev.* **2011**, *30*, 999–1010. [CrossRef]
129. Miettinen, J.; Wang, J.; Hooijer, A.; Liew, S. Peatland conversion and degradation processes in insular southeast asia: A case study in Jambi, Indonesia. *Land Degrad. Dev.* **2013**, *24*, 334–341. [CrossRef]
130. Bal, G.; Banerjee, K. Carbon storage potential of tropical wetland forests of South Asia: A case study from Bhitarkanika Wildlife Sanctuary, India. *Environ. Monit. Assess.* **2019**, *191*, 795. [CrossRef] [PubMed]
131. Islam, G.M.T.; Islam, A.K.M.S.; Shopan, A.A.; Rahman, M.M.; Lázár, A.N.; Mukhopadhyay, A. Implications of agricultural land use change to ecosystem services in the Ganges delta. *J. Environ. Manag.* **2015**, *161*, 443–452. [CrossRef] [PubMed]
132. Ahmad, S.; Suratman, M.N. Detection of changes in mangrove forests using Landsat TM in Pulau Indah, Malaysia. 2007. In Proceedings of the Conference of Forestry and Forest Products (CFFPR 2007), Monterey, CA, USA, 26–31 August 2007.
133. Suratman, M.N. Carbon sequestration potential of mangroves in Southeast Asia. In *Managing Forest Ecosystems: The Challenge of Climate Change*; Springer: Dordrecht, The Netherlands, 2008; pp. 297–315. Available online: https://link.springer.com/chapter/10.1007/978-1-4020-8343-3_17 (accessed on 1 January 2022).
134. Cole, L.E.S.; Bhagwat, S.A.; Willis, K.J. Long-term disturbance dynamics and resilience of tropical peat swamp forests. *J. Ecol.* **2015**, *103*, 16–30. [CrossRef] [PubMed]
135. Friess, D.A. Tropical wetlands and REDD+: Three unique scientific challenges for policy. *Int. J. Rural. Law Policy* **2013**, *1*, 1–6. [CrossRef]
136. Tareq, S.M.; Tanoue, E.; Tsuji, H.; Tanaka, N.; Ohta, K. Hydrocarbon and elemental carbon signatures in a tropical wetland: Biogeochemical evidence of forest fire and vegetation changes. *Chemosphere* **2005**, *59*, 1655–1665. [CrossRef]
137. Kusumaningtyas, M.A.; Hutahaean, A.A.; Fischer, H.W.; Pérez-Mayo, M.; Ransby, D.; Jennerjahn, T.C. Variability in the organic carbon stocks, sources, and accumulation rates of Indonesian mangrove ecosystems. *Estuar. Coast. Shelf Sci.* **2019**, *218*, 310–323. [CrossRef]
138. Aslan, A.; Rahman, A.F.; Warren, M.W.; Robeson, S.M. Mapping spatial distribution and biomass of coastal wetland vegetation in Indonesian Papua by combining active and passive remotely sensed data. *Remote Sens. Environ.* **2016**, *183*, 65–81. [CrossRef]
139. Jauhiainen, J.; Takahashi, H.; Heikkinen, J.E.P.; Martikainen, P.J.; Vasander, H. Carbon fluxes from a tropical peat swamp forest floor. *Glob. Chang. Biol.* **2005**, *11*, 1788–1797. [CrossRef]
140. Marasinghe, S.; Perera, P.; Simpson, G.D.; Newsome, D. Nature-based tourism development in coastal wetlands of Sri Lanka: An Importance—Performance analysis at Maduganga Mangrove Estuary. *J. Outdoor Recreat. Tour.* **2021**, *33*, 100345. [CrossRef]
141. Perera, K.A.R.S.; Amarasinghe, M.D. Carbon sequestration capacity of mangrove soils in micro tidal estuaries and lagoons: A case study from Sri Lanka. *Geoderma* **2019**, *347*, 80–89. [CrossRef]
142. Nam, V.N.; Sasmito, S.D.; Murdiyarso, D.; Purbopuspito, J.; MacKenzie, R.A. Carbon stocks in artificially and naturally regenerated mangrove ecosystems in the Mekong Delta. *Wetl. Ecol. Manag.* **2016**, *24*, 231–244. [CrossRef]
143. FAO. *The State of World Fisheries and Aquaculture 2020. Sustainability in Action*; FAO: Rome, Italy, 2022. [CrossRef]
144. FAO. *The State of Food and Agriculture 2021. Make Agri-Food Systems More Resilient to Shocks and Stresses*; FAO: Rome, Italy, 2021. Available online: https://www.fao.org/publications/sofa/sofa-2021/es/ (accessed on 1 January 2022).
145. World Bank Data for China. Upper Middle Income. 2021. Available online: https://data.worldbank.org/?locations=CN-XT (accessed on 1 January 2022).
146. Elschot, K.; Bakker, J.P.; Temmerman, S.; Van De Koppel, J.; Bouma, T.J. Ecosystem engineering by large grazers enhances carbon stocks in a tidal salt marsh. *Mar. Ecol. Prog. Ser.* **2015**, *537*, 9–21. [CrossRef]
147. Martínez-López, J.; Carreño, M.F.; Palazón-Ferrando, J.A.; Martínez-Fernández, J.; Esteve, M.A. Free advanced modeling and remote-sensing techniques for wetland watershed delineation and monitoring. *Int. J. Geogr. Inf. Sci.* **2014**, *28*, 1610–1625. [CrossRef]
148. Fennessy, M.S.; Ibánez, C.; Calvo-Cubero, J.; Sharpe, P.; Rovira, A.; Callaway, J.; Caiola, N. Environmental controls on carbon sequestration, sediment accretion, and elevation change in the Ebro River Delta: Implications for wetland restoration. *Estuar. Coast. Shelf Sci.* **2019**, *222*, 32–42. [CrossRef]
149. Morris, E.P.; Flecha, S.; Figuerola, J.; Costas, E.; Navarro, G.; Ruiz, J.; Rodriguez, P.; Huertas, E. Contribution of Doñana wetlands to carbon sequestration. *PLoS ONE* **2013**, *8*, e71456. [CrossRef] [PubMed]

150. Fitoka, E.; Tompoulidou, M.; Hatziiordanou, L.; Apostolakis, A.; Höfer, R.; Weise, K.; Ververis, C. Water-related ecosystems' mapping and assessment based on remote sensing techniques and geospatial analysis: The SWOS national service case of the Greek Ramsar sites and their catchments. *Remote Sens. Environ.* **2020**, *245*, 111795. [CrossRef]
151. Lloyd, C.R. 2006 Annual carbon balance of a managed wetland meadow in the Somerset Levels, UK. *Agric. For. Meteorol.* **2006**, *138*, 168–179. [CrossRef]
152. Jiménez-Ballesta, R.; García-Navarro, F.J.; Bravo, S.; Amoros, J.A.; Pérez de los Reyes, C.; Mejias, M. Environmental assessment of potential toxic elements contents in the inundated floodplain área of Tablas de Daimiel wetland (Spain). *Environ. Geochem. Health* **2017**, *39*, 1159–1177. [CrossRef] [PubMed]
153. De la Hera, A.; Villarroya, F. Services Evolution of Two Groundwater Dependent Wetland Ecosystems in the "Mancha Húmeda" Biosphere Reserve (Spain). *Resources* **2013**, *2*, 128–150. [CrossRef]
154. Du Laing, G.; De Meyer, B.; Meers, E.; Lesage, E.; Van de Moortel, A.M.K.; Tack, F.M.G.; Verloo, M.G. Metal accumulation in intertidal marshes: Role of sulphide precipitation. *Wetlands* **2008**, *28*, 735–746. [CrossRef]
155. Coles, B. Steps towards the Heritage Management of Wetlands in Europe. *J. Wetl. Archaeol.* **2004**, *4*, 183–198. [CrossRef]
156. Finlayson, C.M.; Davis, J.A.; Gell, P.A.; Kingsford, R.T.; Parton, K.A. The status of wetlands and the predicted effects of global climate change: The situation in Australia. *Aquat. Sci.* **2013**, *75*, 73–93. [CrossRef]
157. Kauffman, J.B.; Heider, C.; Cole, T.G.; Dwire, K.A.; Donato, D.C. Ecosystem carbon stocks of micronesian mangrove forests. *Wetlands* **2011**, *31*, 343–352. [CrossRef]
158. Chimner, R.A.; Ewel, K.C. Differences in carbon fluxes between forested and cultivated micronesian tropical peatlands. *Wetl. Ecol. Manag.* **2004**, *12*, 419–427. [CrossRef]
159. Wong, V.N.L.; Johnston, S.G.; Burton, E.D.; Bush, R.T.; Sullivan, L.A.; Slavich, P.G. Anthropogenic forcing of estuarine hypoxic events in sub-tropical catchments: Landscape drivers and biogeochemical processes. *Sci. Total Environ.* **2011**, *409*, 5368–5375. [CrossRef] [PubMed]
160. Dubuc, A.; Waltham, N.; Malerba, M.; Sheaves, M. Extreme dissolved oxygen variability in urbanised tropical wetlands: The need for detailed monitoring to protect nursery ground values. *Estuar. Coast. Shelf Sci.* **2017**, *198*, 163–171. [CrossRef]
161. Iram, N.; Kavehei, E.; Maher, D.; Bunn, S.; Rashti, M.R.; Farahani, B.S.; Adame, M.F. Greenhouse gas emissions from tropical coastal wetlands and their alternative agricultural lands: Where significant mitigation gains lie. *Biogeosciences Discuss.* **2021**, 1–27. [CrossRef]
162. Adame, M.F.; Reef, R.; Wong, V.N.L.; Balcombe, S.R.; Turschwell, M.P.; Kavehei, E.; Rodríguez, D.C.; Kelleway, J.J.; Masque, P.; Ronan, M. Carbon and Nitrogen Sequestration of Melaleuca Floodplain Wetlands in Tropical Australia. *Ecosystems* **2020**, *23*, 454–466. [CrossRef]
163. Carnell, P.E.; Windecker, S.M.; Brenker, M.; Baldock, J.; Masque, P.; Brunt, K.; Macreadie, P.I. Carbon stocks, sequestration, and emissions of wetlands in south eastern Australia. *Glob. Chang. Biol.* **2018**, *24*, 4173–4184. [CrossRef] [PubMed]
164. Beringer, J.; Livesley, S.J.; Randle, J.; Hutley, L.B. Carbon dioxide fluxes dominate the greenhouse gas exchanges of a seasonal wetland in the wet-dry tropics of Northern Australia. *Agric. For. Meteorol.* **2013**, *182–183*, 239–247. [CrossRef]
165. Howe, A.J.; Rodríguez, J.F.; Saco, P.M. Surface evolution and carbon sequestration in disturbed and undisturbed wetland soils of the Hunter estuary, southeast Australia. *Estuar. Coast. Shelf Sci.* **2009**, *84*, 75–83. [CrossRef]
166. Rogers, K.; Saintilan, N.; Copeland, C. Managed Retreat of Saline Coastal Wetlands: Challenges and Opportunities Identified from the Hunter River Estuary, Australia. *Estuaries Coast* **2014**, *37*, 67–78. [CrossRef]
167. Saintilan, N.; Rogers, K.; Kelleway, J.J.; Ens, E.; Sloane, D.R. Adaptation to Climate Change Impacts on the Coastal Wetlands in the Gulf of Mexico. *Wetlands* **2018**. Available online: http://www.ine.gob.mx/descargas/cclimatico/env_framework_feb09.pdf (accessed on 22 December 2021).
168. Livesley, S.J.; Andrusiak, S.M. Temperate mangrove and salt marsh sediments are a small methane and nitrous oxide source but important carbon store. *Estuar. Coast. Shelf Sci.* **2012**, *97*, 19–27. [CrossRef]
169. Hayes, M.A.; Jesse, A.; Tabet, B.; Reef, R.; Keuskamp, J.A.; Lovelock, C.E. The contrasting effects of nutrient enrichment on growth, biomass allocation and decomposition of plant tissue in coastal wetlands. *Plant Soil* **2017**, *416*, 193–204. [CrossRef]
170. Limpert, K.E.; Carnell, P.E.; Trevathan-Tackett, S.M.; Macreadie, P.I. Reducing Emissions from Degraded Floodplain Wetlands. *Front. Environ. Sci.* **2020**, *8*, 1–18. [CrossRef]
171. Kingsford, R.T.; Basset, A.; Jackson, L. Wetlands: Conservation's poor cousins. *Aquat. Conserv. Mar. Freshw. Ecosyst.* **2016**, *26*, 892–916. [CrossRef]
172. Department of Economic and Social Affairs, Population Division, United Nations. World Population Prospects 2019: Highlights. ST/ESA/SER.A/423. Available online: https://population.un.org/wpp/Publications/Files/WPP2019_Highlights.pdf (accessed on 5 January 2022).
173. Mojica Vélez, J.M.; Barrasa García, S.; Espinoza Tenorio, A. Policies in coastal wetlands: Key challenges. *Environ. Sci. Policy* **2018**, *88*, 72–82. [CrossRef]
174. Convención De Ramsar Sobre Los Humedales. Perspectiva Mundial Sobre Los Humedales: Estado De Los Humedales Del Mundo Y Sus Servicios a Las Personas. Gland (Suiza). 2018. Available online: https://www.ramsar.org/sites/default/files/flipbooks/ramsar_gwo_spanish_web.pdf (accessed on 10 November 2021).

175. FAO. Foro de Expertos de Alto Nivel, del 13-13 de Octubre, Roma 2009 FAO 2009-"Cómo Alimentar al Mundo en 2050". 2009. Available online: https://www.fao.org/fileadmin/templates/wsfs/docs/synthesis_papers/C%C3%B3mo_alimentar_al_mundo_en_2050.pdf (accessed on 15 November 2021).
176. FAO. *The State of Food Insecurity in the World, Addressing Food Insecurity in Protracted Crises*; FAO: Rome, Italy, 2010.
177. FAO. *The State of Food and Agriculture. Climate Change, Agriculture and Food Security*; FAO: Rome, Italy, 2016.
178. IPCC. Global Warming of 1.5 °C: An IPCC Special Report on the Impacts of Global Warming of 1.5 °C above Pre-Industrial Levels and Related Global Greenhouse Gas Emission Pathways, in the Context of Strengthening the Global Response to the Threat of Climate Chang. 2018. Available online: https://www.ipcc.ch/sr15/download/#full (accessed on 20 November 2021).

land

MDPI

Article

The Historical Development of Constructed Wetlands for Wastewater Treatment

Jan Vymazal

Faculty of Environmental Sciences, Czech University of Life Sciences Prague, Kamýcká 129, 165 21 Prague, Czech Republic; vymazal@fzp.czu.cz

Abstract: Constructed wetlands (CWs) for wastewater treatment are engineered systems that are designed and operated in order to use all natural processes involved in the removal of pollutants from wastewaters. CWs are designed to take advantage of many of the same processes that occur in natural wetlands, but do so within a more controlled environment. The basic classification is based on the presence/absence of wastewater on the wetland surface. The subsurface flow of CWs can be classified according to the direction of the flow to horizontal and vertical. The combination of various types of CWs is called hybrid CW. The CWs technology began in the 1950s in Germany, but the major extension across the world occurred during the 1990s and early 2000s. The early CWs in Germany were designed as hybrid CWs; however, during the 1970s and 1980s, horizontal subsurface flow CWs were mostly designed. The stricter limits for nitrogen, and especially ammonia, applied in Europe during the 1990s, brought more attention to vertical subsurface flow and hybrid systems. Constructed wetlands have been used to treat various types of wastewater, including sewage, industrial and agricultural wastewaters, various drainage and runoff waters and landfill leachate. Recently, more attention has also been paid to constructed treatment wetlands as part of a circular economy in the urban environments: it is clear that CWs are a good fit for the new concept of sponge cities.

Keywords: constructed wetlands; macrophytes; pollution; wastewater

Citation: Vymazal, J. The Historical Development of Constructed Wetlands for Wastewater Treatment. *Land* **2022**, *11*, 174. https://doi.org/10.3390/land11020174

Academic Editor: Richard C. Smardon

Received: 2 December 2021
Accepted: 5 January 2022
Published: 21 January 2022

1. Introduction

Wetlands are transitional environments. In a spatial context, they lie between dry land and open water at the coast, around inland lakes and rivers or as mires draped across the landscape. In an ecological context, wetlands are intermediate between terrestrial and aquatic ecosystems. In a temporal context, most wetlands are destined either to evolve into dry land as a result of lowered water tables, sedimentation and plant succession, or to be submerged by rising water tables associated with relative sea-level rise or climatic change. Wetlands often form part of a large continuum of community type, and therefore it is difficult to set boundaries [1].

Wetlands can be defined as areas flooded with shallow water, or soil is saturated with water for long enough to create hydric soils that support specialized macrophytes adapted to life in anaerobic conditions [1,2]. However, Mitsch and Gosselink [1] pointed out that although the concepts of shallow water or saturated conditions, unique wetland soils and vegetation adapted to wet conditions are fairly straightforward, combining these three factors to obtain a precise definition is difficult, due to a number of characteristics that distinguish wetlands from other ecosystems, yet make them less easy to define.

Natural wetlands have been used for wastewater treatment for centuries. In many cases, however, the reasoning behind this use was disposal rather than treatment, and the wetland simply served as a convenient recipient that was closer than the nearest river or other waterways [3]. Such uncontrolled wastewater disposal led to the destruction of many wetlands around the world. The attempts to use natural wetlands for wastewater treatment under controlled conditions continued even in the 1970s and 1980s, especially in the United

States [4–6]. The experiments with natural wetlands revealed the difficulties with system maintenance and, also, the treatment efficiency was quite unpredictable. Therefore, the use of natural wetlands to treat wastewater was replaced with the use of constructed wetlands (CWs), which have been developed since the 1960s.

Constructed wetlands designed to treat wastewater are engineered systems that are built to utilize natural wetland processes involved in the transformation and removal of pollutants, but under more controlled conditions. Constructed wetlands can be built with a greater degree of control than natural systems, thus allowing the construction of treatment facilities with a well-defined composition of substrates, vegetation types and flow patterns. In addition, constructed wetlands offer several additional advantages compared to natural wetlands, including site selection, flexibility in sizing and, most importantly, control over the hydraulic pathways and retention time [7]. Plants are an indispensable part of constructed wetlands; however, their role in pollution removal is rather indirect, such as the insulation of subsurface flow systems, provision of oxygen to otherwise anoxic substrates, provision of surface for attached bacteria, excretion of antibacterial compounds from the roots and the reduction of wind allowing better sedimentation of the suspended solids in surface flow CWs. The direct role is restricted to the uptake of nutrients if the biomass is harvested [8–10].

2. Classification of Constructed Wetlands for Wastewater Treatment

The most comprehensive classification of constructed treatment wetlands was published by Fonder and Headley [11], who pointed out that there are three main characteristics typical of these systems: the presence of macrophytes, the existence of water-logged or saturated substrate conditions for at least part of the time and the inflow of contaminated water with constituents to be removed. Based on the predominant position of the water in the system, two major groups can be recognized: surface flow CWs (sometimes called free water surface CWs) and subsurface flow CWs. Subsurface flow systems can be further classified as horizontal and vertical, according to the direction of the flow.

2.1. *Surface Flow Constructed Wetlands*

Surface flow constructed wetlands usually consist of shallow basins or channels with soil or other suitable mediums, to support the growth of macrophytes if rooted macrophytes are present. One of their primary design purposes is to contact slow-flowing wastewater with reactive biological surfaces [12].

Surface flow CWs can be classified according to the macrophyte type into CWs with (a) free-floating macrophytes, (b) floating-leaved macrophytes, (c) submerged macrophytes, (d) emergent macrophytes and (e) trees [13]. In surface flow constructed wetlands, the organics are removed principally by the bacterial metabolism of both attached and free-living bacteria. Bacteria can be attached to either the roots and rhizomes of free-floating plants, or to the stems and leaves of rooting macrophytes. The removal of suspended solids occurs through gravity sedimentation. The plants minimize the wind-induced turbulence and water stirring, allowing for effective sedimentation [8]. The removal of nitrogen is primarily executed by denitrification, while ammonia volatilization and plant uptake play minor roles. Nitrification occurs in most surface flow constructed wetlands, but this process does not remove nitrogen from wastewater and only transfers ammonia to nitrate. The removal through denitrification can take place in a layer of decomposed plant material at the bottom of the wetland. Phosphorus removal is generally very low because of the limited contact of wastewater with soil particles; therefore, there is a limited precipitation with Fe, Al, Mg or Ca [14].

2.1.1. CWs with Free-Floating Macrophytes

Free-floating macrophytes are highly diverse in form and habitat, ranging from large plants, such as *Eichhornia crassipes* (water hyacinth, Figure 1) or *Pistia stratiotes* (water lettuce) with large leaves and roots, compared to very small plants, such as Lemnaceae (duckweeds, e.g., *Lemna* spp., *Sprodela polyrhiza* or *Wolffia* spp.) with tiny roots [15]. Free-floating plants are highly productive and belong to the fastest growing plants on the planet. *E. crassipes* and *P. stratiotes* are frost sensitive and do not survive in temperate and cold climatic conditions, and are restricted to the tropics and subtropics. On the other hand, Lemnaceae (Figure 2) have a much wider geographic range as they are able to survive even under light frost [16]. Constructed wetlands with free-floating macrophytes were intensively studied in the late 1970s and the early 1980s, but the high operation and maintenance costs were connected with the constant need of plant harvesting, and the subsequent disposal prevented the use of these systems from a wider application [17,18]. It is important to say that duckweed can naturally occur in all types of surface flow constructed wetlands, as these plants can easily be transported by wind or by birds.

Figure 1. Surface flow constructed wetlands with *Eichhornia crassipes* (water hyacinth) in Langtou near Guangzhou, China. Photo Jan Vymazal.

2.1.2. CW with Floating-Leaved Macrophytes

Floating-leaved macrophytes (Figure 3) include plant species that are rooted in the substrate, and their leaves on long peduncles float on the water's surface. Typical examples of this type of macrophyte are water lilies (*Nymphaea* spp.), spatterdock (*Nuphar lutea*) or Indian lotus (*Nelumbo nucifera*). The plants in this group usually have large rhizomes and leaves floating on the water's surface connected to the rhizomes with long peduncles. So far, only several constructed wetlands have used floating-leaved macrophytes.

Figure 2. Constructed wetland with *Lemna* spp. (duckweed) and *Taxodium distichum* (baldcypress) designed to treat stormwater runoff in Orlando, Florida. Photo: Jan Vymazal.

Figure 3. Constructed wetland (Ironbridge, Florida) planted with *Nuphar lutea* (spatterdock), designed for tertiary treatment of 800,000 PE in Orlando, Florida. Photo Jan Vymazal.

2.1.3. CW with Submerged Macrophytes

Submerged macrophytes root in the sediment and the entire plant is submerged in a water column. Submerged plants take up nutrients from the sediments; however, it has been discovered that at least some plants are able to absorb the nutrients directly from the water column [19,20]. The use of submerged macrophytes is restricted to well-oxygenated waters with low concentrations of suspended solids. High concentrations of suspended solids can limit the penetration of PHAR (photosynthetically active radiation) necessary for full photosynthesis. Therefore, it has been recommended to use submerged macrophytes for constructed wetlands designed for tertiary treatment [7]. There is wide variety of species that can be used for constructed wetlands and have been used in laboratory or small-scale systems; however, in full-scale constructed wetlands, *Myriophyllum spicatum* (watermilfoil, Figure 4) has mostly been used [13,21]. In some systems, naturally occurring species were used, such as in the case of the Florida Everglades Stormwater Area constructed wetlands, in which *Najas guadalupensis* (southern naiad) and *Ceratophyllum demersum* (coontail) are present. Submerged plants are naturally covered by periphyton (algal-based assemblage). Periphyton has a beneficial effect on pollutant removal through the release of oxygen necessary for the oxidation of pollutants as well as the uptake of nutrients. On the other hand, the excessive growth of periphyton can seriously limit the photosynthesis of submerged macrophytes by blocking the PHAR [22].

Figure 4. Surface flow CW with submerged macrophytes (mostly *Myriophyllum spicatum*, water milfoil) in Montréal, Canada. Photo Jan Vymazal.

2.1.4. CWs with Emergent Macrophytes

A typical surface flow CW with emergent macrophytes (Figure 5) consists of a shallow basin or sequence of basins, containing 20–30 cm of rooting soil, with a water depth of 10–60 cm and a dense stand of macrophytes. The most common plants used in this type of CW are *Phragmites australis* (common reed), *Typha* spp. (cattails) and *Scirpus/Schoenoplectus* spp. (bulrushes) [23]. The shallow water depth, low flow velocity and presence of the plant stalks and litter regulate the water flow and, especially in long, narrow channels, ensure plug-flow conditions [24].

Typically, surface flow constructed wetlands have aerated zones, especially near the water surface, due to atmospheric diffusion and the production of oxygen by the photosynthetic activity of algae and cyanobacteria. The anoxic and even anaerobic conditions can occur near the bottom and especially within the layer of decaying plant material.

Figure 5. Surface flow CW with emergent plants (*Eleocharis sphacelata*, tall spikerush). Otorohanga, New Zealand. Photo Jan Vymazal.

In Europe, this technology started during the late 1960s. One of the first examples of full-scale FWS CWs with emergent vegetation, are those built in 1967 in Lelystad, in the Netherlands [25] and near Keszthely, Hungary [26].

2.1.5. CWs with Floating Mats of Emergent Macrophytes

Some emergent macrophytes are capable of forming floating mats, even though their individual plants are not capable of such an existence. Under field conditions, the floating islands of emergent macrophytes can naturally occur as a consequence of bottom disturbance [27]. Floating mats are a common phenomenon in wetlands throughout the world, both in temperate [28] and (sub)tropical regions [29]. The floating wetlands (called "plavs"), were first described by Pallis in 1915 from the Danube delta in Romania [30].

Buoyancy in natural systems is supported by the composition of a wetland plant biomass, which contains a large amount of air space (aerenchyma) that makes the biomass less dense than water [31]. Buoyancy can also be promoted by the provision of suspended cables over the water surface, from where the roots of plants that tend to form floating mats can cover the whole surface [32]. Other materials used in constructed floating wetlands (Figure 6) are, for example, Styrofoam, coconut-peat strings or bamboo [33,34].

Figure 6. Floating constructed wetland planted with *Cyperus alternifolius*. Ningbo, China. Photo Jan Vymazal.

2.1.6. CWs with Trees

Constructed wetlands with trees are seldom used for wastewater treatment; however, there are some fine examples of such treatment wetlands. The tree species that were used in constructed wetlands are *Taxodium distichum* (bald cypress) (Figure 1), *Melaleuca quinquenervia* (paper bark tea tree) or mangroves, which can be used to treat saline (waste)waters (Figure 7) or [35,36].

Figure 7. Dapeng Bay, Taiwan: surface flow constructed wetland with mangroves (*Kandelia candel*) for the treatment of mariculture wastewater. Photo Jan Vymazal.

2.2. Subsurface Flow Constructed Wetlands

Constructed wetlands with a subsurface flow can be classified according to the direction of the flow into horizontal (HF CWs) and vertical (VF CWs). The HF CWs are continuously fed, while VF CWs are fed intermittently. The feeding mode creates different redox conditions in the filtration media being anoxic/anaerobic in HF CWs and aerobic in VF CWs. A special category of subsurface CWs is "zero-discharge" CW.

2.2.1. Horizontal Flow Constructed Wetlands

In HF CWs, mechanically pretreated wastewater slowly flows under the surface of the filtration bed filled with porous material planted with emergent macrophytes (Figures 8 and 9). During this passage through the filtration material, the wastewater comes into contact with a network of aerobic, anoxic and anaerobic zones. The aerobic zones are restricted to narrow zones adjacent to the roots and rhizomes that leak oxygen into the substrate [8,37]. The filtration bed is sealed from the surrounding area by an impermeable layer; in most cases, a plastic liner to prevent leakage to the groundwater. The water level in the filtration bed is maintained in the outflow sump using swiveling elbows (Figure 10) or flexible hoses, or plastic pipes that can be held in position by a chain (Figure 10).

Figure 8. Distribution zone filled with large stones and the filtration bed filled with crushed rocks (4–8 mm) before planting. HF CW Čejkovice, Czech Republic. Photo Jan Vymazal.

Figure 9. Constructed wetland with a horizontal subsurface flow. Roseč, Czech Republic. Photo Jan Vymazal.

Figure 10. Water level maintenance. **Left**: outflow pipes hanging on a chain at CW Waikeria, New Zealand. **Right**: swiveling elbow at CW Bouvron, France. Photo Jan Vymazal.

The early experimental HF CWs in Germany, during the 1960s and early 1970s, were filled with coarse sand [38], which was replaced in the late 1970s and early 1980s with soil [39]. However, the soil substrate did not achieve and maintain the necessary hydraulic conductivity and quickly became clogged [40]. The unsuitability of the soil as a filtration material was confirmed in Denmark in the mid-1980s and, following the research conducted in the United Kingdom, proved that gravel was the most suitable filtration medium [41,42]. At present, most constructed wetlands either use washed gravel or crushed rock with a fraction size between 5 and 20 mm, depending on the country.

The macrophytes growing in constructed wetlands help to create suitable conditions for pollution removal. Their role is rather indirect, such as the (1) insulation of the surface during periods of cold weather; the (2) provision of substrates for attached bacteria and the (3) release of the root exudates that can possess antimicrobial properties. The direct role is the sequestration of nutrients from the wastewater in the (aboveground) biomass that can be removed via harvesting [43]. Worldwide, the most frequently used macrophyte is *Phragmites australis* (Common reed), but many other species, such as *Typha* spp. (cattails), *Phalaris arundinacea* (Reed canarygrass, Figure 8), *Scirpus* (*Schoenoplectus*) spp. (Bulrush) and *Iris pseudacorus* (Yellow flag), are used in HF CWs [9].

HF CWs provide a high and steady removal of organics and suspended solids. Organics are degraded by both anaerobic and aerobic microorganisms, but aerobic degradation is mostly restricted to narrow zones adjacent to the roots and rhizomes, in which oxygen can be released [44,45]. Due to the predominant anoxic/anaerobic conditions in the filtration bed, HF CWs provide suitable conditions for denitrification. On the other hand, due to a lack of oxygen, nitrification is very limited and so is volatilization as there is no free water surface. Nitrogen removal through plant uptake and subsequent harvesting is limited and can reach the same values as in natural wetlands, i.e., about 30–60 g N/m^2 on an annual basis for large macrophytes, such as *P. australis* or *T. latifolia* [46]. Phosphorus removal is usually low, provided that the common filtration materials, such as gravel or crushed rock, are used [47]. However, the removal of phosphorus can be enhanced by the use of materials with a high sorption capacity, such as steel slags, shell sand or expanded clay materials [48,49]. In such situations, it necessary to keep in mind that the sorption capacity will be exhausted, and the filter material will have to be replaced in order to keep the high removal. The removal of phosphorus by plant harvesting is limited and usually amounts to 2–5 g P m^{-2} on an annual basis [46].

Horizontal flow constructed wetlands capital costs are higher than those for the surface flow CWs, due to the costs of bed sealing and costs of filtration material, including transportation. The operation and maintenance costs are very low and are derived from the maintenance of pretreatment units. The area required for the filtration beds of HF CWs usually ranges between 5 and 6 m^2 per population equivalent [50–52].

2.2.2. Vertical Flow Constructed Wetlands

Vertical flow constructed wetlands generally consist of a bed of porous material, through which the water moves in a vertical direction. In general, this group of CWs incorporates various hydrologic characteristics. There are three arrangements of vertical subsurface flow constructed wetlands: down flow, up flow and fill and drain [11].

The most common type of vertical flow CWs is the free-drainage down flow unit, in which the outlet is open at the base of the filter bed. The wastewater is intermittently delivered to the surface of the filtration bed in batches. Each new batch is brought, only after the water from the previous batch has percolated through the filter. This allows for air diffusion in the empty bed and, thus, the filtration bed is predominantly aerobic. Wastewater is spread across the filter surface by a network of pipes (Figure 11) with multiple diffusers, to evenly distribute the wastewater to avoid short circuiting. Influent distribution pipes can be positioned on or above the surface of the filtration bed or, mostly in cold climates, can be buried within the coarse material or under the layer of insulating mulch [11]. This concept, developed as early as the mid-1960s by dr. Seidel in Germany, was used to oxidize the anaerobic outflow from a septic tank [53].

Figure 11. Wastewater distribution pipes at Changshu Advanced Materials Industrial Park VF CW, Suzhou, Jiangsu Province, PR China. Photo Jan Vymazal.

The most common filtration material in down flow CWs is sand. Coarse gravel or stones are used in the drainage layer, where perforated drainage pipes are laid. The most frequently used macrophyte in this type of constructed wetland, especially in Europe, is *P. australis* (Figure 12). In Asia, various species, such *Arundo donax* (giant cane), *Miscanthus saccharoflorus* (Amur silvergrass), *Cyperus alternifolius* (umbrella papyrus), *Thalia dealbata* (powdery alligator-flag), *Vetiveria zizanoides* (vetiver grass) or *Canna indica* (Indian shot) are used.

Figure 12. Down flow vertical constructed wetlands planted with *Phragmites australis*. Oberwindhag, Austria. Photo: Jan Vymazal.

Treatment efficiency is high for organics, suspended solids and ammonia, due to the aerobic conditions in the filter bed, and thus results in effective nitrification. As a result of the predominantly aerobic conditions, the down flow vertical CWs do not achieve denitrification. The removal of phosphorus is limited but, similar to the horizontal flow CWS, the removal can be enhanced by filtration material with a high sorption capacity [13].

The surface area required for free drain down flow CWs is smaller than for HF CWs, and it is usually set at 4 m^2 per population equivalent [52,54]. In France, down flow VFs are used to treat raw sewage in a two-step VF system, sometimes called the "French system". In the first stage, sludge treatment, the partial removal of organics and nitrification occur. In the second stage, the further removal of organics and nitrification occur. The system is designed with an area of 1.2 m^2 per PE for the first stage, and 0.8 m^2 per PE for the second stage [55].

The second standard type of system with a vertical subsurface flow is the up flow. In this system, the wastewater is distributed to the bottom of the filter and moves upwards to the filtration bed surface. The outflow can be below or above the bed surface. The up flow vertical constructed wetlands were introduced in the 1980s in Brazil as "filtering soil" (Figure 13). This system is used much less frequently, compared to down flow systems and, in general, provides the same treatment conditions as horizontal flow CWs, due to the saturation of the filtration bed.

Figure 13. Vertical up flow constructed wetlands planted with rice (*Oryza sativa*) called "filtering soil", in Piracicaba, Brazil. Photo Jan Vymazal.

The third type of vertical CW is the system called "fill and drain". The flow typically alternates between an upward and downward flow. The media in these systems has an intermittent saturation level, as it alternates between being saturated and unsaturated as a result of the filling and draining sequences [11]. Due to the alternating aerobic and anaerobic conditions, this system has the potential to remove both ammonia and nitrate [56]. The fill and drain system is also called the "reciprocating" or "tidal flow".

2.3. Zero-Discharge Constructed Wetlands

The zero-discharge constructed wetland (Figure 14) was developed in Denmark in the late 1990s. The function of the system is based on the fact that, during the growing season, the willow evapotranspiration exceeds the wastewater inflow and precipitation [57]. During the winter period without the evapotranspiration, the filtration bed is filled with wastewater and precipitation; however, during the period of evapotranspiration, the water level in the bed decreases. For a single household in Denmark (Figure 13), the area needed typically varies between 150 and 300 m^2. One third of the willow stems is harvested every year to keep the willows in a young and healthy state with high transpiration rates [58].

Figure 14. Zero discharge constructed wetlands planted with *Salix viminalis* (basket willow). Borup, Denmark. Photo Jan Vymazal.

2.4. Hybrid Constructed Wetlands

Various types of constructed wetlands can be combined to complement each other, in order to enhance treatment efficiency, especially for nitrogen. Hybrid constructed wetlands were first introduced in the 1960s in Germany [53]. The original design consisted of several parallel down flow VF CWs (called "filtration beds"), followed by two stages of HF CWs (called "eliminations beds"). In the first VF beds, nitrification was achieved while in the HF stage denitrification took place. Since then, various types of constructed wetlands have been combined [59].

3. Historical Development of Constructed Wetlands as Treatment Technology for Wastewaters

3.1. The Early Stages of Constructed Wetland Treatment Technology Development

Wallace and Knight [60] mentioned that the first documented engineered treatment wetland system was patented as early as 1901 [61]. The treatment system was a typical vertical flow CW; however, the spread of this technology is not documented. It has taken until the early 1950s for the constructed wetland treatment technology to be revived in Germany, by K. Seidel [62–64]. During the early 1960s, Seidel carried out experiments using macrophytes to improve inefficient rural treatment systems (septic tanks and Imhoff tanks). She used highly permeable substrates in modulated basins planted with various macrophytes. The first stage was vertical and aerobic to improve the oxygenation of the septic effluent; the second stage was horizontal [53]. Further cooperation with R. Kickuth in the mid-1960s, resulted in a HF CW commonly known as the "Root Zone Method". The filtration bed was filled with a heavy soil containing clay, and planted with *P. australis* [65]. However, the first full-scale HF CW was put in operation in 1974 in Liebenburg-Othfresen, to treat municipal wastewater.

Despite the fact that most research studies on constructed treatment wetlands, in the 1960s, were aimed at subsurface systems, the full-scale surface (free water surface) constructed wetlands were built in Lelystad, Netherlands [66,67] and in Keszthely, Hungary [68]. On the other hand, in North America, efforts were made in the 1960s to explore the potential of surface flow CWs, both natural and constructed. During the late 1960s, Odum [69] evaluated the use of coastal lagoons for recycling municipal wastewater in North Carolina. During the 1960s, the first experiments with floating plants, especially with *Eichhornia crassipes* (water hyacinth), were carried out. The experiments were restricted to small mesocosms and were performed in locations in which this plant occurs naturally, such as southeast Asia and southern parts of the United States [70,71].

During the 1970s, the research on constructed wetlands for wastewater treatment in Europe was mostly restricted to the Root Zone Method, and mostly only in Germany. In the United States, the research was focused on surface flow constructed wetlands, but subsurface flow technology was also explored. The surface flow large-scale installations included, for example, constructed wetlands for refinery wastewaters in North Dakota [72] or municipal wastewaters in the cold climate in Michigan [73]. During the 1970s, very intensive research was also carried out with the focus on water hyacinth-based constructed wetlands, especially in the Southern States of the United States [74–77]. The first experimental HF mesocosms were built in 1972, and the pilot scale HF CW was built in 1974, in Seymour, Wisconsin [78]. The system could be operated with the water level above as well as below the surface, although the subsurface flow was the preferred option. On the West Coast, the California sub-licensee for the Max Planck Institute (MPI) system, developed by Seidel in Germany, built a hybrid VF-HF CW in Laguna Niguel [79]. Moreover, two large national conferences were organized in the United States with the focus on the use of wetlands for wastewater treatment [80,81].

In Australia, the potential use of aquatic and wetland macrophytes for wastewater treatment was evaluated by Mitchell, during the mid-1970s [82].

3.2. The Rapid Growth of Constructed Wetlands Technology across the World

The 1980s and 1990s can be regarded as the periods that witnessed the rapid extension of the constructed wetlands for wastewater treatment across the world. While until the beginning of the 1980s the technology spread slowly and mostly on a personal exchange of the experience during the late 1980s and during the 1990s, many international conferences with a special focus on this technology were organized in Europe, Asia, Australia and both North and South America. These conferences were mostly organized under the umbrella of the International Water Association (in the 1990s, under the names International Association on Water Pollution Research and Control and International Association on Water Quality).

During the mid-1980s, the international cooperation and exchange on constructed wetland technology in Europe accelerated extensively. In October 1986, the cooperation among ten European countries resulted in the decision to form a European coordinating group, with the two major objectives being the production of design and operation guidelines and to organize a conference in 1990 to bring together experts from around the world [83]. Meanwhile, in the Unites States, two major international conferences on the use of plants and constructed wetlands for wastewater treatment were organized in Orlando, Florida (1986) [84] and in Tallahassee, Tennessee (1988) [85]. These conferences, together with the conference organized in Cambridge in 1990 by the European coordinating group [86], represented a major breakthrough in constructed wetland technology extension around the world.

In Europe during the 1980s, horizontal subsurface flows were the major focus, especially in Germany [87], Denmark [88], Austria [89] and the United Kingdom [90]. In 1983, the soil-based Kickuth-type HF CWs were introduced to Denmark and, shortly after that, about 80 full-scale constructed wetlands were built, mostly for municipal sewage. The evaluation of the performance of these systems revealed that the area originally proposed by Kickuth (m^2/PE) was not enough to provide sufficient treatment, and an area of about 5 m^2/PE was suggested [88]. Moreover, in 1983, a large pilot scale HF CW was built in Mannersdorf near Vienna, Austria. The system was monitored for seven years and the results were summarized by Haberl and Perfler [91]. After a visit to Germany from engineers from U.K. water companies, two full-scale HF constructed wetlands were built (Acle and St. Pauls Walden) in 1985 in UK, and, in the following year, another 21 full-scale systems, such as Freethorpe (Figure 15), were built [90].

Figure 15. Horizontal subsurface flow CW, Freethorpe, United Kingdom, for 900 PE. The early CWs were built as a single bed. Photo Jan Vymazal.

The systems were very intensively monitored, and the major outcome of this research was the recommendation to replace soil with gravel, 5–10 mm in size [92,93]. At the end of the 1980s, the hybrid constructed wetlands of Seidel's configuration (VF-HF) were installed in France [94] and the United Kingdom [95]. At the end of the 1980s, the national guidelines on the design and operation of constructed wetlands for wastewater treatment were issued by ATV in Germany [96], and, shortly after that, the European guidelines were published [97].

In North America, surface flow constructed wetlands were built quite regularly, during the 1980s. Many of these systems served as tertiary treatment units for municipal sewage, and the surface area was usually quite large. Examples of such systems are Incline Village, Nevada (173.28 ha), Ironbridge, Florida (494 ha, Figure 16) or Arcata, California (15.18 ha) [12,98]. In addition to the use of surface flow CWs for municipal sewage, other types of wastewaters were treated in the constructed wetlands, such as urban runoff [99], acid mine drainage [100,101] or livestock feedlot wastewaters [102]. The costly operation and maintenance of water hyacinth-based constructed wetlands, mainly connected with constant harvesting and the disposal of the biomass, caused the gradual cessation of the use of these systems [103].

Figure 16. Surface flow CW at Ironbridge. Tertiary treatment for the city of Orlando, Florida. Photo Jan Vymazal.

The CW consists of 17 cells with a total area of 4.92 km^2. The figure features two of those cells (see also Figure 3).

In North America, subsurface technology was developed slowly, compared to its development in Europe. However, several full-scale installations were put in operation for municipal sewage in California, Louisiana, Alabama, Tennessee and Kentucky [12,104–106], or the paper mill effluent [107]. In the United States, the experience with the design and operation of constructed wetlands was summarized in the manual issued by the U.S. Environmental Protection Agency [108].

The subsurface technology was also developed in Australia during the 1980s, and pilot HF CWs were built to treat piggery waste and abattoir wastewater [109,110]. Pilot scale HF CW experiments with various macrophytes were carried out at the University of Western Sydney at Hawkesbury (Figure 17), where numerous studies were conducted [111]. In Africa, the constructed wetlands have been used since the mid-1980s, especially in South Africa. The constructed wetlands were designed to treat various types of wastewater, including raw and secondary sewage, stormwater runoff and a variety of industrial and mine drainage waters. In these systems, surface flow as well as subsurface flow systems were used [112]. In South America, the constructed wetlands were applied only in Brazil [113]. Research focused on water hyacinth-based systems in combination with up flow vertical CWs, called "filtering soil" (Figure 13). There are no records of full-scale constructed wetlands in Asia during the 1980s.

Figure 17. Experimental HF CW at the University of Western Australia, Hawkesbury. Photo Jan Vymazal.

During the last decade of the 20th century, the constructed wetland technology extended to all continents and all types of constructed wetlands were used. During the 1990s, the technology started in several Asian countries, namely China, India and Nepal. In China, the first full-scale constructed wetland was put in operation in July 1990 at the Longgang Shenzen Special Economic Zone [114]. The constructed wetlands consisted of three stages of HF units with a total surface area of 4589 m^2, and a surface flow cell with a surface area of 1710 m^2. Other hybrid CWs were used to treat pig raising farm wastewater or industrial wastewater [115]. In India, mostly HF constructed wetlands planted with *Phragmites karka* were built to treat municipal sewage [116]. In Nepal, in the 1990s, constructed wetlands drew a lot of attention because of the low costs for the operation and maintenance of them;

however, most systems suffered from a lack of maintenance. The hybrid system in Dhulikel was probably the first constructed wetland built to treat hospital wastewater [117].

The rapid growth of constructed wetland installations, as well as the increased knowledge on processes occurring in constructed wetlands during the treatment of various types of wastewater, resulted in the release of national guidelines in many countries, such as in Austria, [118], Denmark [119], New Zealand [120] and Canada [121], during the late 1990s.

3.3. Constructed Wetlands for Wastewater Treatment in the 21st Century

Constructed wetlands became a "certified" method for wastewater treatment, in many countries across the world in the 21st century. In some countries, such as China, the number of constructed wetlands exceeds one hundred thousand and it is still growing (Figure 18). There is also a growing number of constructed wetlands in South America, especially in Colombia, Argentina and Chile. Unfortunately, the technology has not spread significantly in Africa, where there is great potential for this technology.

Figure 18. Constructed wetlands with horizontal subsurface flow, planted with *Thalia dealbata* (**left**) and *Canna indica* and *Thalia dealbata* (**right**) in Quangdong Province near Guangzhou. Photo Jan Vymazal.

At the beginning of the 21st century, research on the wastewater treatment in constructed wetlands focused on the various design and operation aspects that can lead to the enhanced removal of pollutants [122,123]:

- operation strategies (such as aeration, microbial fuel cells and bioaugmentation);
- supply of electron donors to enhance the removal of selected inorganic anions;
- selection of filter materials for higher sorption capacity and microbial biofilm establishment;
- determination of functions of various bacteria groups on pollution removal;
- selection of macrophytes for enhanced removal of pollutants;
- effect of constructed wetlands on greenhouse gas emissions;
- efficacy of constructed wetlands to remove pharmaceuticals and personal care products.

Constructed wetlands for wastewater treatment are also gaining more attention in relation to sustainable water management in urban settlements, as part of the circular economy and "sponge" cities. Masi et al. [124] pointed out that constructed wetlands can be effectively used in cities to treat sewage, greywater, stormwater overflows and runoff, and can effectively recycle the water within cities. Constructed wetlands can also be a core part of SUDS (Sustainable Urban Drainage Systems). Moreover, Stefanakis [125] stressed that constructed wetlands for wastewater treatment have a great potential to be successfully integrated in urban and peri-urban areas, and fit well within the new concept of sponge cities and the circular economy. The importance of constructed treatment wetlands within urban circularity through the restoration and maintenance of water cycles, water and wastewater treatment, recovery and reuse as well as nutrient recovery and reuse was reported by Atanasova et al. [126].

Unfortunately, most of the research is carried out in small laboratory or greenhouse experiments, with no attempts to include the results of this experiments in a design of full-scale installations. Vymazal [127] pointed out that while, in 1995, 88% of the papers recorded on the Web of Science database were based on full-scale CWs, in 2017 it was only 26%. It seems that the transfer of the laboratory experiment results to the full-scale constructed wetlands will be the major challenge in coming years.

4. The Use of Constructed Wetlands for Various Types of Wastewater

The early use of constructed wetlands was restricted to sewage treatment. The various types of constructed wetlands and their combination enabled the use of CWs for a variety of wastewaters (Tables 1 and 2).

Table 1. The use of constructed wetlands for various types of wastewater.

Type of (Waste) Water	Examples of Use
Sewage	Domestic, municipal, combined sewer overflow
Drainage	Acid/alkaline coal mines, metal ores mines, agricultural tile drainage
Feedlots	Livestock, poultry, pigs, milking parlors
Aquacultures	Freshwater fish, marine fish, shrimp
Food processing	Dairy, cheese, winery, brewery, distillery, sugar, olive mills, fish, soft drinks, abattoir, meat processing
Other industries	Tannery, textile, electroplating, pulp and paper, glass, explosives, refineries, oil drilling water, rubber industry
Runoff waters	Urban, highway, airport, greenhouses, nurseries, golf courses, agricultural fields
Landfill leachate	

Table 2. Examples of the first use of macrophytes and/or constructed wetlands for the treatment of different types of pollution. Experimental (laboratory, mesocosms) and Operational (pilot scale, full scale). Modified and updated from Vymazal and Kröpfelová [13].

	Experimental	
Year	**Wastewater**	**References**
1952	Phenol wastewater	[128]
1956	Sewage and dairy wastewater	[129]
1956	Livestock wastewater	[64]
1965	Sludge dewatering	[130]
1973	Textile wastewater	[131]
1975	Photographic laboratory wastewater	[132]
1978	Acid mine drainage	[133]
1980	Electroplating wastewater	[134]
1980	Removal of cresol	[135]
1980	Piggery effluent	[110]
1980	Abattoir wastewater	[109]
1981	Heavy metals removal	[136]
1981	Tannery wastewater	[137]

Table 2. *Cont.*

Experimental		
Year	**Wastewater**	**References**
1982	Agricultural drainage effluent	[138]
1982	Pesticides	[139]
1982	Sugar refinery wastewater	[140]
1982	Benzene and its derivatives	[141]
1982	Rubber industry effluent	[142]
1983	Pulp/paper mill wastewater	[143,144]
1985	Seafood processing wastewater	[145]
1986	Potato starch industry wastewater	[146]
1986	Cyanides and chlorophenols	[147]
1987	Meat processing wastewater	[148]
1988	Landfill leachate	[149,150]
1989	Chicken farm wastewater	[151]
1991	Fish aquaculture	[152]
1991	Phenanthrene	[153]
1994	Hydrocarbons	[154]
1995	Lignite pyrolysis wastewater	[155]
1997	Winery wastewater	[156]
1998	Coke plant wastewater	[157]
2000	Linear alkylbenzensulfonates (LAS)	[158]
2001	Steel processing industry wastewaters	[159]
2001	Brewery wastewater	[160]
2001	Electric utility wastewater	[161]
2003	Azo dyes removal	[162]
2004	Chlorobenzene removal	[163]
2004	Steel mill effluent	[164]
2006	Sugarcane molasse stillage from ethanol production	[165]
2008	Brine treatment	[166]
2008	Coffee fruit processing wastewater	[167]
2010	Saline aquaculture wastewater	[168]
2010	Brackish shrimp growout system	[169]
2105	Bauxite residue drains	[170]
2017	Rice noodles wastewater	[171]
2018	Glass industry wastewater	[172]
2019	Batik wastewater	[173]
2020	Jewelry industry wastewater	[174]

Table 2. *Cont.*

Year	Operational Wastewater	References
1967	Sewage	[66]
1974	Sludge dewatering	[175]
1975	Oil refinery wastewater	[176]
1978	Textile wastewater	[177]
1979	Fish rearing pond discharge	[178]
1982	Acid mine drainage	[179,180]
1983	Urban stormwater runoff	[99]
1983	Rubber industry effluent	[142]
1985	Dairy wastewater	[41]
1986	Seepage from piled pig muck	[181]
1986	Ash pond seepage	[182]
1987	Thermally affected wastewater	[183]
1988	Livestock wastewater	[102]
1988	Pulp/paper mill wastewater	[107]
1988	Pesticides	[184]
1989	Landfill leachate	[185]
1989	Airport runoff	[186]
1989	Reduction of lake eutrophication	[187]
1990	Lake water	[188]
1991	Woodwaste leachate	[189]
1992	Bakery wastewater	[190]
1992	Channel catfish pond effluent	[191]
1992	Sugar beet processing wastewater	[192]
1992	Combined sewer overflow	[193]
1993	Pesticides in agricultural runoff	[194]
1993	Highway runoff	[195]
1994	Abattoir wastewater	[196]
1994	Airport runoff	[197]
1994	Poultry wastewater	[198]
1995	Greenhouse wastewater	[199]
1995	Nitroaromatic organic compounds	[200]
1995	Potato processing wastewater	[201]
1996	Explosives	[202,203]
1997	Hydrocarbons	[204]
1997	Hospital wastewaters	[117]
1998	Trout farm effluent	[205]
1998	Golf course runoff	[206]
1998	Nylon and ethylene polymers	[207]
1999	Molasses-based distillery effluent	[208]
1999	Winery wastewater	[156]

Table 2. *Cont.*

	Operational	
Year	**Wastewater**	**References**
1999	Distillery wastewater	[208]
2000	Surfactant removal	[209]
2000	Subsurface drainage from grazed dairy pastures	[210]
2001	Tannery wastewater	[211]
2002	Tool factory wastewater	[212]
2002	Pharmaceuticals removal	[213]
2003	Olive mill wastewater	[214]
2004	Sugar factory effluent	[215]
2007	Mountain cheese factory	[216]
2009	Flower farm effluent	[217]
2010	Brewery wastewater	[218]
2010	Brackish shrimp aquaculture wastewater	[168]

Constructed wetlands have been used to treat various types of wastewater, since the 1950s. In the beginning, this treatment technology spread slowly, based only on personal contacts. The worldwide distribution of constructed treatment wetlands occurred in the 1990s, primarily due to several large international conferences. Since the beginning of the 21st century, constructed wetlands for wastewater treatment have become a worldwide phenomenon and an accepted technology in many countries around the world. However, in some countries, the treatment performance of constructed wetlands is still underestimated by water authorities and, therefore, the number of installed constructed wetlands is slow.

Until recently, constructed treatment wetlands have mostly been built and considered with the sole purpose of wastewater treatment. However, in addition to the high treatment efficiency, the constructed treatment wetlands have recently been shown to have a great potential in the new sustainable and circular economy in the urban environment. Constructed treatment wetlands can effectively treat, accumulate and recycle water and nutrients for further use, as suggested in the "sponge city" concept.

Funding: This research received no external funding.

Institutional Review Board Statement: Not applicable.

Informed Consent Statement: Not applicable.

Data Availability Statement: Review paper, no original data.

Conflicts of Interest: The author declares no conflict of interest.

References

1. Mitsch, W.J.; Gosselink, J.G. *Wetlands*, 3rd ed.; John Wiley and Sons: New York, NY, USA, 2000.
2. Cowardin, L.M.; Carter, V.; Golet, F.C.; LaRoe, E.T. *Classification of Wetlands and Deepwater Habitats of the United States. U.S.*; FWS/OBS-79/31; Department of the Interior, Fish and Wildlife Service: Washington, DC, USA, 1992.
3. Wentz, W.A. Ecological/environmental perspectives on the use of wetlands in water treatment. In *Aquatic Plants for Water Treatment and Resource Recovery*; Reddy, K.R., Smith, W.H., Eds.; Magnolia Publishing: Orlando, FL, USA, 1987; pp. 17–25.
4. Kadlec, R.H.; Tilton, D.L. The use of freshwater wetlands as a tertiary wastewater treatment alternative. *CRC Crit. Rev. Environ. Control* **1979**, *9*, 185–212. [CrossRef]
5. Ewel, K.C.; Harwell, M.A.; Kelly, J.R.; Grover, H.D.; Bedford, B.L. Evaluation of the Use of Natural Ecosystems for Wastewater Treatment. In *Ecosystem Research Center Report No. 15*; Cornell University: Ithaca, NY, USA, 1982.

6. Olson, R.K. (Ed.) *Created and Natural Wetlands for Controlling Nonpoint Source Pollution*; U.S. EPA Office of Research and Development, Office of Wetlands, Oceans, and Watersheds: Corvallis, OR, USA, 1993.
7. Brix, H. Wastewater treatment in constructed wetlands: System design, removal processes, and treatment performance. In *Constructed Wetlands for Water Quality Improvement*; CRC Press: Boca Raton, FL, USA, 1993; pp. 9–22.
8. Brix, H. Treatment of wastewater in the rhizosphere of wetland plants–the root zone method. *Water Sci. Technol.* **1987**, *19*, 107–118. [CrossRef]
9. Vymazal, J. Plants used in constructed wetlands with horizontal subsurface flow: A review. *Hydrobiologia* **2011**, *674*, 133–156. [CrossRef]
10. Vymazal, J. Removal of phosphorus in constructed wetlands with sub-surface flow in the Czech Republic. *Water Air Soil Pollut. Focus* **2004**, *4*, 657–670. [CrossRef]
11. Fonder, N.; Headley, T. Systematic classification, nomenclature and reporting for constructed treatment wetlands. In *Water and Nutrient Management in Natural and Constructed Wetlands*; Vymazal, J., Ed.; Springer Science+Business Media B.V.: Dordrecht, The Netherlands, 2010; pp. 191–219.
12. Kadlec, R.H.; Knight, R.L. *Treatment Wetlands*; CRC Pess LLC: Boca Raton, FL, USA, 1996.
13. Vymazal, J.; Kröpfelová, L. *Wastewater Treatment in Constructed Wetlands with Horizontal Sub-Surface Flow*; Springer: Dordrecht, The Netherlands, 2008.
14. Richardson, C.J. Mechanisms controlling phosphorus retention capacity in freshwater wetlands. *Science* **1985**, *228*, 1424–2427. [CrossRef]
15. Brix, H.; Schierup, H.-H. The use of aquatic macrophytes in water pollution control. *Ambio* **1989**, *18*, 100–107.
16. Culley, D.D., Jr.; Epps, E.A. Use of duckweed for waste treatment and animal feed. *J. Water Pollut. Control Fed.* **1973**, *45*, 337–347.
17. Ozimek, T.; Czupryňski, P. Ten years experience of constructed wetlands in Poland. *Publ. Inst. Geogr. Univ. Tartu.* **2003**, *94*, 163–169.
18. Gupta, G. Use of water hyacinths in wastewater treatment (A brief literature review). *J. Environ. Health* **1980**, *43*, 80–82.
19. Denny, P. Solute movement in submerged angiosperms. *Biol. Rev.* **1980**, *55*, 65–92. [CrossRef]
20. Carignan, R.; Kalff, J. Phosphorus sources for aquatic weeds: Water or sediments? *Science* **1980**, *207*, 987–989. [CrossRef] [PubMed]
21. Toet, S.; Van Logtestijn, R.S.P.; Schreier, M.; Kampf, R.; Verhoeven, J.T.A. The functioning of a wetland system used for polishing effluent from a sewage treatment plant. *Ecol. Eng.* **2005**, *25*, 101–124. [CrossRef]
22. Twilley, R.R.; Kemp, W.M.; Staver, K.W.; Stevenson, J.C.; Boynton, W.R. Nutrient enriccommunitihment of estuarine submersed vascular plant communities. 1. Algal growth and effects on production of plants and associated es. *Mar. Ecol. Prog. Ser.* **1985**, *23*, 179–191. [CrossRef]
23. Vymazal, J. Emergent plants used in free water surface constructed wetlands: A review. *Ecol. Eng.* **2013**, *61P*, 582–592. [CrossRef]
24. Reed, S.C.; Middlebrooks, E.J.; Crites, R.W. *Natural Systems for Waste Management and Treatment*, 2nd ed.; McGraw-Hill Book Company: New York, NY, USA, 1995.
25. De Jong, J. The purification of wastewater with the aid of rush or reed ponds. In *Biological Control of Water Purification*; Pierson, R.W., Ed.; Pennsylvania University Press: Philadelphia, PA, USA, 1976; pp. 133–139.
26. Lakatos, G. Hungary. In *Constructed Wetlands for Wastewater Treatment in Europe*; Vymazal, J., Brix, H., Cooper, P.F., Green, M.B., Haberl, R., Eds.; Backhuys Publishers: Leiden, The Netherlands, 1998; pp. 191–206.
27. Vymazal, J. Wetlands constructed for wastewater treatment and control. In *Encyclopedia of Sustainable Science and Technology*; Meyers, R.E., Ed.; Springer: Dordrecht, The Netherlands, 2012; pp. 11891–11910.
28. Hogg, E.H.; Wein, R.W. The contribution of *Typha* components to floating mat buoyancy. *Ecology* **1988**, *69*, 1025–1031. [CrossRef]
29. Thompson, K. Emergent plants of permanent and seasonally flooded wetlands. In *The Ecology and Management of African Wetlands*; Denny, P., Ed.; Dr. W. Junk: The Hague, The Netherlands, 1985; pp. 43–107.
30. Pallis, M. The structure and history of plavs: The floating fen of the delta of the Danube. *Linn. Soc. J. Bot.* **1915**, *43*, 133–290.
31. Lynch, J.; Fox, L.J.; Owen, J.S., Jr.; Sample, D.J. Evaluation of commercial floating treatment wetland technologies for nutrient remediation of stormwater. *Ecol. Eng.* **2015**, *75*, 61–69. [CrossRef]
32. Headley, T.R.; Tanner, C.C. Constructed wetlands with floating emergent macrophytes: An innovative stormwater treatment technology. *Crit. Rev. Environ. Sci. Technol.* **2012**, *42*, 2261–2310. [CrossRef]
33. Borne, K.E.; Fassman, E.A.; Tanner, C.C. Floating treatment wetlands retrofit to improve stormwater pond performance for suspended solids, copper and zinc. *Ecol. Eng.* **2013**, *54*, 173–182. [CrossRef]
34. Tanner, C.C.; Headley, T.R. Components of floating emergent macrophyte treatment wetlands influencing removal of stormwater pollutants. *Ecol. Eng.* **2011**, *37*, 474–486. [CrossRef]
35. Bolton, K.G.E.; Greenway, M. Pollutant removal capacity of a constructed *Melaleuca* wetland receiving primary settled sewage. *Water Sci. Technol.* **1999**, *39*, 199–206. [CrossRef]
36. Huang, C.M.; Yuan, C.S.; Yang, W.B.; Yang, L. Temporal variations of greenhouse gas emissions and carbon sequestration and stock from a tidal constructed mangrove wetland. *Mar. Pollut. Bull.* **2019**, *149*, 110568. [CrossRef]
37. Vymazal, J. Types of constructed wetlands for wastewater treatment: Their potential for nutrient removal. In *Transformations of Nutrients in Natural and Constructed Wetlands*; Vymazal, J., Ed.; Backhuys Publishers: Leiden, The Netherlands, 2001; pp. 1–93.
38. Seidel, K. Reinigung von Gewässern durch höhere Pflanzen. *Deutsche Naturwissenschaft* **1966**, *12*, 289–297.
39. Kickuth, R. A low-cost process for purification of municipal and industrial waste water. *Der Trop.* **1982**, *83*, 141–154.

40. Haberl, R.; Perfler, R. Nutrient removal in a reed bed system. *Wat. Sci. Technol.* **1991**, *23*, 729–737. [CrossRef]
41. Brix, H.; Schierup, H.-H. Sewage treatment in constructed wetlands–Danish experience. *Water Sci. Technol.* **1989**, *21*, 1665–1668. [CrossRef]
42. Findlater, B.C.; Hobson, J.A.; Cooper, P.F. Reed bed treatment systems: Performance evaluation. In *Constructed Wetlands for Water Pollution Control*; Cooper, P.F., Findlater, B.C., Eds.; Pergamon Press: Oxford, UK, 1990; pp. 193–204.
43. Brix, H. Do macrophytes play a role in constructed treatment wetlands? *Water Sci. Technol.* **1997**, *35*, 11–17. [CrossRef]
44. Vymazal, J.; Kröpfelová, L. Removal of organics in constructed wetlands with horizontal sub-surface flow: A review of field experience. *Sci. Total Environ.* **2009**, *407*, 3911–3922. [CrossRef]
45. Drew, M.C. Plant responses to anaerobic conditions in soil and solution culture. *Curr. Adv. Plant Sci.* **1979**, *11*, 179–199.
46. Vymazal, J. Removal of nutrients in constructed wetlands for wastewater treatment through plant harvesting–Biomass and load matter the most. *Ecol. Eng.* **2020**, *155*, 105962. [CrossRef]
47. Vymazal, J. Removal of nutrients in various types of constructed wetlands. *Sci. Total Environ.* **2007**, *380*, 48–65. [CrossRef]
48. Vohla, C.; Põldvere, C.; Noorvee, A.; Kuusemets, V.; Mander, Ü. Alternative filter media for phosphorus removal in a horizontal subsurface flow constructed wetlands. *J. Environ. Sci. Health* **2005**, *40*, 1251–1264. [CrossRef]
49. Vohla, C.; Koiv, M.; Bavor, H.J.; Chazarenc, F.; Mander, Ü. Filter materials for phosphorus removal from wastewater in treatment wetlands—A review. *Ecol. Eng.* **2011**, *37*, 70–89. [CrossRef]
50. Tanner, C.C.; Headley, T.; Dakers, A. *Guidelines for the Use of Horizontal Subsurface-Flow Constructed Wetlands in On-Site Treatment of Household Wastewaters*; National Institute of Water & Atmospheric Research: Hamilton, New Zealand, 2011.
51. U.S. EPA. *Constructed Wetlands Treatment of Municipal Wastewater*; EPA/625/R-99/010; Office of Research and Development: Cincinnati, OH, USA, 2000.
52. *Standard DWA-A 262A*; Principles for Dimensioning, Construction and Operation of Wastewater Treatment Plants with Planted and Unplanted Filters for Treatment of Domestic and Municipal Wastewater. The German Association for Water—Wastewater and Waste: Hannef, Germany, 2018.
53. Seidel, K. Neue Wege zur Grundwasseranreicherung in Krefeld. In Hydrobotanische Reinigungsmethode vol. II. *GWF Wasser/Abwasser* **1965**, *30*, 831–833.
54. *ÖNORM B 2505*; Wastewater Treatment Plants–Intermittently Loaded Effluent Filtration Systems (Constructed Wetlands)–Application, Dimensioning, Installation, Operation, Service and Inspection. Austrian Standards Institute: Vienna, Austria, 2009.
55. Molle, P.; Boutin, C.; Merlin, G.; Iwema, A. How to treat raw sewage with constructed wetlands: An overview of the French systems. *Water Sci. Technol.* **2005**, *51*, 11–21. [CrossRef] [PubMed]
56. Behrends, L.L.; Houke, L.; Bailey, E.; Jansen, P.; Brown, D. Reciprocating constructed wetlands for treating industrial, municipal and agricultural wastewater. *Water Sci. Technol.* **2001**, *44*, 399–405. [CrossRef]
57. Gregersen, P.; Brix, H. Zero-discharge of nutrients and water in a willow dominated constructed wetland. *Water Sci. Technol.* **2001**, *44*, 407–412. [CrossRef]
58. Gregersen, P.; Gabriel, S.; Brix, H.; Faldager, I. *Guidelines for Willow Systems up to 30 PE*; Økologisk Byfornyelse og Spildevandsrensning No. 27; Ministry of Environment and Energy: Copenhagen, Denmark, 2003. (In Danish)
59. Vymazal, J. The use of hybrid constructed wetlands for wastewater treatment with special attention to nitrogen removal: A review of a recent development. *Water Res.* **2013**, *47*, 4795–4811. [CrossRef]
60. Wallace, S.D.; Knight, R.L. *Small-Scale Constructed Wetland Treatment Systems. Feasibility, Design Criteria, and O&M Requirements*; Water Environmental Research Foundation: Alexandria, Virginia, 2006.
61. Monjeau, C. Purifying Water. U.S. Patent 681,884, 18 December 1901.
62. Seidel, K. Pflanzungen zwischen Gewässern und Land. *Mitt. Max-Planck Gesselschaft* **1953**, 17–20.
63. Seidel, K. Abbau von Bacterium coli durch höhere Pflanzen. *Naturwissenschaften* **1964**, *51*, 395. [CrossRef]
64. Seidel, K. Zur Problematik der keim-und Pflanzengewasser. *Verh. Des Int. Ver. Limnol.* **1961**, *14*, 1035–1039.
65. Kickuth, R. Abwasserreinigung in Wurzelraumverfahren. *Wasser Luft Betr.* **1980**, *11*, 21–24.
66. De Jong, J.; Kok, T.; Koridon, A.H. *The Purification of Sewage with the Aid of Ponds Containing Bulrushes and Reeds in the Netherlands*; Rapport 1977–7 Bbw; Rijskdienst voor de IJsselmeerpolders: Lelystad, The Netherlands, 1977.
67. Greiner, R.W.; De Jong, J. *The Use of Marsh Plants for the Treatment of Waste Water in Areas Designated for Recreation and Tourism*; Report No. 225; RIJP: Lelystad, The Netherlands, 1984.
68. Hatvani, I.G.; Clement, A.; Kovács, J.; Kovács, I.S.; Korponai, J. Assessing water-quality data: The relationship between the water quality amelioration of Lake Balaton and the construction of its mitigation wetland. *J. Great Lakes Res.* **2014**, *40*, 115–125. [CrossRef]
69. Odum, H.T. *Self-Organization of Estuarine Ecosystems in Marine Ponds Receiving Treated Sewage. Data from Experimental Pond Studies at Morehead City, North Carolina, 1968–1972. A Data Report*; UNC-SG-85-04; University of North Carolina Sea Grant Publications: Chapel Hill, NC, USA, 1985.
70. Sinha, S.N.; Sinha, L.P. Studies on the use of water hyacinth culture in oxidation ponds treating digested sugar wastes and effluents of septic tank. *Environ. Health* **1969**, *11*, 197–207.
71. Sheffield, C.W. Water hyacinth for nutrient removal. *Hyacinth Control J.* **1967**, *6*, 27–30.
72. Litchfield, D.K. Constructed wetlands for wastewater treatment at Amoco Oil Company's Mandan, North Dakota Refinery. In *Constructed Wetlands for Water Quality Improvement*; Moshiri, G.A., Ed.; Lewis Publishers: Boca Raton, FL USA, 1993; pp. 485–488.

73. Kadlec, R.H. Wastewater treatment at the Houghton Lake wetland: Hydrology and water quality. *Ecol. Eng.* **2009**, *35*, 1287–1311. [CrossRef]
74. Dinges, R. Upgrading stabilization pond effluent by water hyacinth culture. *J. Water Pollut. Control Fed.* **1978**, *50*, 833–845.
75. Ornes, W.H.; Sutton, D.L. Removal of phosphorus from static sewage effluent by water hyacinth. *Hyacinth Control J.* **1975**, *13*, 56–58.
76. Wolverton, B.C.; McKown, M.M. Water hyacinth for removal of phenol from polluted waters. *Aquat. Bot.* **1976**, *30*, 29–37. [CrossRef]
77. Stewart, E.A., III. Utilization of water hyacinths for control of nutrients in domestic wastewater–Lakeland, Florida. In *Aquaculture Systems for Wastewater Treatment: Seminar Proceedings and Engineering Assessment*; EPA430/9-80-006; Bastian, R.K., Reed, S.C., Eds.; U.S. EPA: Washington, DC, USA, 1979; pp. 273–293.
78. Fetter, C.W.; Sloey, W.E.; Spangler, F.L. Potential replacement of septic tank drain fields by artificial marsh wastewater treatment systems. *Ground Water* **1976**, *16*, 1–7.
79. Pope, P.R. *Wastewater Treatment by Rooted Aquatic Plants in Sand and Gravel Trenches*; U.S. EPA-600/2-81-091; U.S. Environmental Protection Agency: Cincinnati, OH, USA, 1981.
80. Tilton, D.L.; Kadlec, R.H.; Richardson, C.J. (Eds.) *Freshwater Wetlands and Sewage Effluent Disposal*; NSF/RANN Conference: Ann Arbor, MI, USA, 1796.
81. Greeson, P.E.; Clark, J.R.; Clark, J.E. (Eds.) *Wetland Functions and Values: The State of our Understanding*; American Water Resources Association: Minneapolis, MN, USA, 1979.
82. Mitchell, D.S. The potential for wastewater treatment by aquatic plants in Australia. *Water Aust.* **1976**, *5*, 15–17.
83. Cooper, P.F. UK experience with reed beds and constructed wetland systems 1985 to 2003. In Proceedings of the International Seminar The Use of Aquatic Macrophytes for Wastewater Treatment in Constructed Wetlands, Lisbon, Portugal, 8–10 May 2003; Dias, V., Vymazal, J., Eds.; ICN and INAG: Lisbon, Portugal, 2003; pp. 403–421.
84. Reddy, K.R.; Smith, W.H. Preface. In *Aquatic Plants for Water Treatment and Resources Recovery*; Reddy, K.R., Smith, W.H., Eds.; Magnolia Publishing: Orlando, FL, USA, 1987; pp. 1–3.
85. Hammer, D.A.; Bastian, R.K. Wetland ecosystems: Natural water purifiers? In *Constructed Wetlands for Wastewater Treatment*; Hammer, D.A., Ed.; Lewis Publishers: Chelsea, MI, USA, 1989; pp. 5–19.
86. Cooper, P.F. The use of reed bed systems to treat domestic sewage. The European design and operation guidelines for reed bed treatmentsystems. In *Constructed Wetlands for Water Quality Improvement*; Moshiri, G.A., Ed.; CRC Press/Lewis Publishers: Boca Raton, FL, USA, 1993; pp. 203–217.
87. Bucksteeg, K. Sewage treatment in helophyte beds–first experience with a new treatment process. *Water Sci. Technol.* **1987**, *19*, 1–10. [CrossRef]
88. Schierup, H.-H.; Brix, H.; Lorenzen, B. Wastewater treatment in constructed reed beds in Denamrk–state of the art. In *Constructed Wetlands for Water Pollution Control*; Cooper, P.F., Findlater, B.C., Eds.; Pergamon Press: Oxford, UK, 1990; pp. 495–504.
89. Haberl, R.; Perfler, R. Root-zone system: Mannersdorf—New results. In *Constructed Wetlands for Wastewater Treatment*; Hammer, D.A., Ed.; Lewis Publishers: Chelsea, MI, USA, 1989; pp. 606–621.
90. Cooper, P.F.; Boon, A.G. The use of Phragmites for wastewater treatment by the Root Zone Method: The UK approach. In *Aquatic Plants for Water Treatment and Resource Recovery*; Reddy, K.R., Smith, W.H., Eds.; Magnolia Publishing: Orlando, FL, USA, 1987; pp. 153–174.
91. Haberl, R.; Perfler, R. Seven years of research work and experiences with wastewater treatment by a reed bed system. In *Constructed Wetlands for Water Pollution Control*; Cooper, P.F., Findlater, B.C., Eds.; Pergamon Press: Oxford, UK, 1990; pp. 205–214.
92. Coombes, C. Reed bed treatment systems in Anglian Water. In *Constructed Wetlands for Water Pollution Control*; Cooper, P.F., Findlater, B.C., Eds.; Pergamon Press: Oxford, UK, 1990; pp. 223–234.
93. Cooper, P.F.; Hobson, J.A. Sewage treatment by reed bed systems: The present situation in the United Kingdom. In *Constructed Wetlands for Wastewater Treatment*; Hammer, D.A., Ed.; Lewis Publishers: Chelsea, MI, USA, 1989; pp. 153–171.
94. Liénard, A.; Boutin, C.; Esser, D. France. In *Constructed Wetlands for Wastewater Treatment in Europe*; Vymazal, J., Brix, H., Cooper, P.F., Green, M.B., Haberl, R., Eds.; Backhuys Publishers: Leiden, The Netherlands, 1998; pp. 153–167.
95. Burka, U.; Lawrence, P. A new community approach to wastewater treatment with higher plants. In *Constructed Wetlands for Water Pollution Control*; Cooper, P.F., Findlater, B.C., Eds.; Pergamon Press: Oxford, UK, 1990; pp. 359–371.
96. *ATV-Hinweis H 262*; Behandlung von häuslichen Abwasser in Pflanzenbeeten. Regelwerk Abwasser-Abfall, Gesellschaft zur Förderung der Abwassertechnik: Hennef, Germany, 1989.
97. Cooper, P.F. (Ed.) *European Design and Operation Guidelines for Reed Bed Treatment Systems. Prepared for the European Community/European Water Pollution Control Association Emergent Hydrophyte Treatment System Expert Contact Group*; Water Research Center Report UI 17; Water Research Center Report: Swindon, UK, 1990.
98. Knight, R.L.; Kadlec, R.H.; Wilhelm, M.; Demgen, F.C.; Gearheart, R.A.; Dyer, J.C.; Jackson, J.; Shearer, J.S.; Richwine, D.; Newberry, L.; et al. *Constructed Wetlands for Wastewater Treatment and Wildlife Habitat. 17 Case Studies*; EPA832-R-93-005; U.S. EPA: Washington, DC, USA, 1993.
99. Silverman, G.S. Development of an urban runoff treatment wetlands in Fremont, California. In *Constructed Wetlands for Wastewater Treatment*; Hammer, D.A., Ed.; Lewis Publishers: Chelsea, MI, USA, 1989; pp. 669–676.

100. Wieder, R.K. A survey of constructed wetlands for acid coal mine drainage treatment in the eastern United States. *Wetlands* **1989**, *9*, 299–315. [CrossRef]
101. Brodie, G.A.; Hammer, D.A.; Tomljanovich, D.A. Man-made wetlands for acid drainage control. In Proceedings of the 8th Annual National Abandoned Mine Lands Conference, Billings, MT, USA, 10–15 August 1986; pp. 87–105.
102. Hammer, D.A. Designing constructed wetlands systems to treat agricultural nonpoint source pollution. *Ecol. Eng.* **1992**, *1*, 49–82. [CrossRef]
103. Stewart, E.A., III; Haselow, D.L.; Wyse, N.M. Review of operations and performance data of five water hyacinth based treatment systems in Florida. In *Aquatic Plants for Water Treatment and Resource Recovery*; Reddy, K.R., Smith, W.H., Eds.; Magnolia Publishing: Orlando, FL, USA, 1987; pp. 279–288.
104. Watson, J.T.; Reed, S.C.; Kadlec, R.H.; Knight, R.L.; Whitehouse, A.E. Performance expectations and loading rates for constructed wetlands. In *Constructed Wetlands for Wastewater Treatment*; Hammer, D.A., Ed.; Lewis Publishers: Chelsea, MI, USA, 1989; pp. 319–358.
105. Gersberg, R.M.; Elkins, B.V.; Goldman, C.R. Use of artificial wetlands to remove nitrogen from wastewater. *J. Water Pollut. Control Fed.* **1984**, *56*, 152–156.
106. Gersberg, R.M.; Lyons, S.R.; Brenner, R.; Elkins, B.V. Fate of viruses in artificial wetlands. *J. Appl. Microbiol.* **1987**, *53*, 731–736. [CrossRef] [PubMed]
107. Thut, R.N. Feasibility of treating pulp mill effluent with a constructed wetland. In *Constructed Wetlands for Water Quality Improvement*; Moshiri, G.A., Ed.; Lewis Publishers: Boca Raton, FL, USA, 1993; pp. 441–447.
108. U.S. EPA. *Design Manual: Constructed Wetlands and Aquatic Plant System for Municipal Wastewater Treatment*; EPA 625/1-88/022; U.S. EPA Office of Water: Cincinnati, OH, USA, 1988.
109. Finlayson, M.; Chick, A. Testing the potential of aquatic plants to treat abbatoir effluent. *Water Res.* **1983**, *17*, 415–422. [CrossRef]
110. Finlayson, M.; Chick, A.; von Oertzen, I.; Mitchell, D. Treatment of piggery effluent by an aquatic plant filter. *Biol. Wastes* **1987**, *19*, 179–196. [CrossRef]
111. Bavor, H.J.; Roser, D.J.; McKersie, S. Nutrient removal using shallow lagon-solid matrix macrophyte systems. In *Aquatic Plants for Water Treatment and Resource Recovery*; Reddy, K.R., Smith, W.H., Eds.; Magnolia Publishing: Orlando, FL, USA, 1987; pp. 227–235.
112. Wood, A. The application of artificial wetlands in South Africa. In *Constructed Wetlands for Water Pollution Control*; Cooper, P.F., Findlater, B.C., Eds.; Pergamon Pres: Oxford, UK, 1990; pp. 235–244.
113. Salati, E., Jr.; Salati, E.; Salati, E. Wetland projects developed in Brazil. *Water Sci. Technol.* **1999**, *43*, 19–25. [CrossRef]
114. Yang, Y.; Zhencheng, X.; Kangping, H.; Junsan, W.; Guizhi, W. Removal efficiency of the constructed wetland wastewater treatment system at Bainikeng, Shenzhen. In Proceedings of the 4th International Conference Wetland Systems for Water Pollution Control, Guangzhou, China, 6–10 November 1994; ICWS Secretariat: Guangzhou, China, 1994; pp. 94–103.
115. Wang, J.; Cai, X.; Chen, Y.; Yang, Y.; Liang, M.; Zhang, Y.; Wang, Z.; Li, Q.; Liao, X. Analysis of the configuration and the treatment effect of constructed wetland wastewater treatment system for different wastewaters in South China. In Proceedings of the 4th International Conference Wetland Systems for Water Pollution Control, Guangzhou, China, 6–10 November 1994; ICWS Secretariat: Guangzhou, China, 1994; pp. 114–120.
116. Billore, S.K.; Singh, N.; Sharma, J.K.; Dass, P.; Nelson, R.M. Horizontal subsurface flow gravel bed constructed wetland with *Phragmites karka* in central India. *Water Sci. Technol.* **1999**, *40*, 163–171. [CrossRef]
117. Laber, J.; Haberl, R.; Sherstha, R. Two-stage constructed wetland for treating hospital wastewater in Nepal. *Water Sci. Technol.* **1999**, *40*, 317–324. [CrossRef]
118. *ÖNORM B 2505*; Sub-surface flow Constructed Wetlands–Application, Dimensioning, Installation and Operation. Österreichisches Normungsinstitut: Vienna, Austria, 1997. (In German)
119. Danish Environmental Protection Agency. *Environmental Guidelines for Root Zone Systems up to 30 PE*; Ministry of Environment and Energy: Copenhagen, Denmark, 1999. (In Danish)
120. Tanner, C.C.; Kloosterman, V.C. *Guidelines for Constructed Wetland Treatment of Farm Dairy Wastewaters in New Zealand*; National Institute of Water and Atmospheric Research Science and Technology Series No. 48; National Institute of Water and Atmospheric Research Science and Technology: Hamilton, New Zealand, 1997.
121. Tousignant, E.; Fankhauser, O.; Hurd, S. *Guidance Manual for the Design, Construction and Operations of Constructed Wetlands for Rural Applications in Ontario*; Stantec Consulting, University of Guelph and South Nation Conservation: Guelph, ON, Canada, 1999.
122. Wu, S.; Kuschk, P.; Brix, H.; Vymazal, J.; Dong, R. Development of constructed wetlands in performance intensifications for wastewater treatment: A nitrogen and organic matter targeted review. *Water Res.* **2014**, *57*, 40–55. [CrossRef]
123. Vymazal, J.; Zhao, Y.; Mander, Ü. Recent research challenges in constructed wetlands for wastewater treatment: A review. *Ecol. Eng.* **2021**, *169*, 106318. [CrossRef]
124. Masi, F.; Rizzo, A.; Regelsberger, M. The role of constructed wetlands in a new circular economy, resource oriented, and ecosystem services paradigm. *J. Environ. Manag.* **2018**, *216*, 275–284. [CrossRef]
125. Stefanakis, A.I. The role of constructed wetlands as green infrastructure for sustainable urban water management. *Sustainability* **2019**, *11*, 6981. [CrossRef]
126. Atanasova, N.; Castellar, J.A.C.; Pineda-Martos, R.; Nika, C.E.; Katsou, E.; Istenič, D.; Pucher, B.; Andreucci, M.B.; Langergraber, G. Nature-based solutions and circularity in cities. *Circ. Econ. Sustain.* **2021**, *1*, 319–332. [CrossRef]

127. Vymazal, J. Do laboratory scale experiments improve constructed wetland treatment technology? *Environ. Sci. Technol.* **2018**, *52*, 12956–12957. [CrossRef] [PubMed]
128. Seidel, K. Phenol-Abbau in Wasser durch *Scirpus lacustris* L. wehrend einer versuchsdauer von 31 Monaten. *Naturwissenschaften* **1965**, *52*, 398–406. [CrossRef]
129. Seidel, K. Macrophytes and water purification. In *Biological Control of Water Pollution*; Tourbier, J., Pierson, R.W., Eds.; Pennsylvania University Press: Philadelphia, PA, USA, 1976; pp. 109–122.
130. Bittmann, M.; Seidel, K. Entwässerung und Aufbereitung von Chemieschlamm mit Hilfe von Pflanzen. *GWF* **1967**, *10*, 488–491.
131. Widyanto, L.S. The effect of industrial pollutants on the growth of water hyacinth (*Eichhornia crassipes*) tropical pest biology program BIOTROP. *Proc. Indones. Weed Sci. Conf.* **1975**, *3*, 328–329.
132. Wolverton, B.C.; McDonald, R.C. *Water Hyacinths (Eichhornia Crassipes) for Removing Chemical and Photographic Pollutants from Laboratory Wastewaters*; NASA Tech. Memorandum TM-X-72731; National Space Technology Laboratories: Bay St. Louis, MS, USA, 1976.
133. Huntsman, B.E.; Solch, J.G.; Porter, M.D. Utilization of *Sphagnum* sp. dominated bog for coal acid mine drainage abatement. In *Geological Society of America 91th Annual Meeting Book of Abstracts*; Geological Society of America: Boulder, CO, USA, 1978; p. 322.
134. Shroff, K.C. Reuse of water and sludge for cultivation of variety of value added botanical species. In *ENPC-IIT Joint Workshop on Strategy and Technology for Water Quality Management*; Indian Institute of Technology: Bombay, India, 1982; pp. 379–425.
135. Wolverton, B.C.; McDonald, R.C. Natural processes for treatment of organic chemical waste. *Environ. Prof.* **1981**, *3*, 99–104.
136. Gersberg, R.M.; Lyon, S.R.; Elkins, B.V.; Goldman, C.R. The removal of heavy metals by artificial wetlands. In *Proceddings Conference Future of Water Use*; AWWA Research Foundation: Denver, CO, USA, 1984; pp. 639–648.
137. Prasad, B.G.S.; Madhavakrishna, W.; Nayudamma, Y. Utilization of water hyacinth in the treatment and disposal of tannery wastewater. In *International Conference on Water Hyacinth*; Synopsis of Papers: Hyderabad, India, 1983; p. 56.
138. Reddy, K.R.; Campbell, K.L.; Graetz, D.A.; Portier, K.M. Use of biological filters for treating agricultural drainage effluents. *J. Environ. Qual.* **1982**, *11*, 591–595. [CrossRef]
139. Gudekar, V.R.; Borkar, L.P.; Kavadia, K.M.; Trivedy, R.K. Studies on the feasibility of removal of sodium pentachlorophenate (SPCP) with water hyacinth. *Pollut. Res.* **1984**, *3*, 71–75.
140. Yeoh, B.G. Use of Water Hyacinth in Wastewater Treatment. In *Terminal Report*; Standard and Industrial Research Institute of Malaysia: Shah Alam, Malaysia, 1983.
141. Wolverton, B.C.; McDonald, R.C.; Marble, L.K. Removal of benzene and its derivatives from polluted water using the reed/microbial filter technique. *J. Miss. Acad. Sci.* **1984**, *29*, 119–127.
142. John, C.K. Use of water hyacinth in the treatment of effluents from rubber industry. In *Proceedings Conference on Water Hyacinth*; UNEP Nairobi: Hyderabad, India, 1983; pp. 699–712.
143. Allender, B.M. Water quality improvement of pulp and paper mill effluents by aquatic plants. *Appita* **1984**, *3*, 303–306.
144. Thut, R.N. Utilization of artificial marshes for treatment of pulp mill effluents. *Tappi J.* **1990**, *73*, 93–96.
145. Guida, V.G.; Kugelman, I.J. Experiments in wastewater polishing in constructed tidal marshes: Does it work? Are the results predictable? In *Constructed Wetlands for Wastewater Treatment*; Hammer, D.A., Ed.; Lewis Publishers: Chelsea, MI, USA, 1989; pp. 727–734.
146. De Zeeuw, W.; Heijnen, G.; De Vries, J. Reed bed treatment as a wastewater (post) treatment alternative in the potato starch industry. In *Constructed Wetlands in Water Pollution Control*; Cooper, P.F., Findlater, B.C., Eds.; Pergamon Press: Oxford, UK, 1990; pp. 551–553.
147. Wolverton, B.C.; Bounds, B.K. Aquatic plants for pH adjustment and removal of toxic chemicals and dissolved minerals from waste supplies. *J. Miss. Acad. Sci.* **1988**, *33*, 71–80.
148. Van Oostrom, A.J.; Cooper, R.N. Meat processing effluent treatment in surface-flow and gravel-bed constructed wastewater wetlands. In *Constructed Wetlands in Water Pollution Control*; Cooper, P.F., Findlater, B.C., Eds.; Pergamon Press: Oxford, UK, 1990; pp. 321–332.
149. Staubitz, W.W.; Surface, J.M.; Steenhuis, T.S.; Peverly, J.H.; Lavine, M.J.; Weeks, N.C.; Sanford, W.E.; Kopka, R.J. Potential use of constructed wetlands to treat landfill leachate. In *Constructed Wetlands for Wastewater Treatment*; Hammer, D.A., Ed.; Lewis Publishers: Chelsea, MI, USA, 1989; pp. 735–742.
150. Birkbeck, A.E.; Reil, D.; Hunter, R. Application of natural and engineered wetlands for treatment of low-strength leachate. In *Constructed Wetlands in Water Pollution Control*; Cooper, P.F., Findlater, B.C., Eds.; Pergamon Press: Oxford, UK, 1990; pp. 411–418.
151. Vymazal, J. Use of reed-bed system for the treatment of concentrated wastes from agriculture. In *Constructed Wetlands in Water Pollution Control*; Cooper, P.F., Findlater, B.C., Eds.; Pergamon Press: Oxford, UK, 1990; pp. 347–358.
152. Zachritz, W.H., II.; Jacquez, R.B. Treating intensive aquaculture recycled water with a constructed wetlands filter system. In *Constructed Wetlands for Water Quality Improvement*; Moshiri, G.A., Ed.; CRC Press/Lewis Publishers: Boca Raton, FL, USA, 1993; pp. 609–614.
153. Machate, T.; Noll, H.; Behrens, H.; Kettrup, A. Degradation of phenanthrene and hydraulic characteristics in a constructed wetland. *Water Res.* **1997**, *31*, 554–560. [CrossRef]
154. Salmon, C.; Crabos, J.L.; Sambuco, J.P.; Bessiere, J.M.; Basseres, A.; Caumette, P.; Baccou, J.C. Artificial wetland performances in the purification efficiency of hydrocarbon wastewater. *Water Air Soil Pollut.* **1998**, *104*, 313–329. [CrossRef]

155. Wiessner, A.; Kuschk, P.; Sottmeister, U.; Struckmann, D.; Jank, M. Treating a lignite pyrolysis wastewater in a constructed subsurface flow wetland. *Water Res.* **1999**, *33*, 1296–1302. [CrossRef]
156. Grismer, M.E.; Carr, M.A.; Shepherd, H.L. Evaluation of constructed wetland treatment performance for winery wastewater. *Water Environ. Res.* **2003**, *75*, 412–421. [CrossRef] [PubMed]
157. Jardinier, N.; Blake, G.; Mauchamp, A.; Merlin, G. Design and performance of experimental constructed wetlands treating coke plant effluents. *Water Sci. Technol.* **2001**, *44*, 485–491. [CrossRef]
158. Del Bubba, M.; Lepri, L.; Cincinelli, A.; Griffini, O.; Tabani, F. Linear alkylbenzensulfonates (LAS) removal in a pilot submerged horizontal flow constructed wetland. In Proceedings of the 7th International Conference Wetland Systems for Water Pollution Control, Lake Buena Vista, FL, USA, 11–16 November 2000; University of Florida: Gainesville, FL, USA, 2000; pp. 919–925.
159. Yang, L.; Lin, H.Y.; Shih, P.Y. Feasibility of constructed wetlands applied to industrial wastewater recirculating treatment systems. In Proceedings of the 8th International Conference Wetland Systems for Water Pollution Control, Arusha, Tanzania, 16–19 September 2002; University of Dar es Salaam and IWA: Dar es Salaam, Tanzania, 2002; pp. 460–471.
160. Kalibbala, H.M.; Nalubega, M.; Kulabako, R.N. Challenges in the use of constructed wetland in the treatment of industrial wastewater. In Proceedings of the 8th International Conference Wetland Systems for Water Pollution Control, Arusha, Tanzania, 16–19 September 2002; University of Dar es Salaam and IWA: Dar es Salaam, Tanzania, 2002; pp. 504–513.
161. Ye, Z.H.; Lin, Z.-Q.; Whiting, S.N.; de Souza, M.P.; Terry, N. Possible use of constructed wetland to remove selenocyanata, arsenic, and boron from electric utility wastewater. *Chemosphere* **2003**, *52*, 1571–1579. [CrossRef]
162. Davies, L.C.; Carias, C.C.; Novais, J.M.; Martins-Dias, S. Phytoremediation of textile effluents containing azo dye by using *Phragmites australis* in a vertical flow intermittent feeding constructed wetland. *Ecol. Eng.* **2005**, *25*, 594–605. [CrossRef]
163. Braeckevelt, M.; Rokadia, H.; Mirschel, G.; Weber, S.; Imfeld, G.; Stelzer, N.; Kuschk, P.; Kastner, M.; Richnow, H.H. Biodegradation of chlorobenzene in a constructed wetland treating contaminated groundwater. In Proceedings of the 10th International Conference Wetland Systems in Water Pollution Control, Lisbon, Portugal, 23–29 September 2006; Dias, V., Vymazal, J., Eds.; MAOTDR: Lisbon, Portugal, 2006; pp. 1927–1935.
164. Yang, L.; Hu, C.C. Treatment of oil-refinery and steel-mill wastewaters by mesocosm constructed wetland system. *Water Sci. Technol.* **2005**, *51*, 157–164. [CrossRef]
165. Olguín, E.J.; Sánchez-Galván, G.; Gonzáles-Portela, R.E.; López-Vela, M. Constructed wetland mesocosms for the treatment of diluted sugarcane molasses stillage from ethanol production using *Pontederia sagittata*. *Water Res.* **2008**, *42*, 3659–3666. [CrossRef]
166. Chakraborti, R.K.; Bays, J.S. Natural treatment of high-strength reverse osmosis concentrate by constructed wetlands for reclaimed water use. *Water* **2020**, *12*, 158. [CrossRef]
167. Fia, R.; De Matos, A.T.; De Matos, M.P.; Abreu, E.C.; Fia, F.R.L. Treatment of the wastewater of coffee fruit processing in anaerobic filter system followed by constructed wetland system: I—Removal of organic material. *Eng. Agrícola* **2010**, *30*, 1191–1202. [CrossRef]
168. Gao, F.; Li, C.; Jin, W.H. Study on saline aquaculture wastewater treatment by constructed wetland. In Proceedings of the International Conference on Electric Communications and Control, Ningbo, China, 9–11 September 2011; pp. 3938–3941.
169. Shi, Y.; Zhang, G.; Liu, J.; Zhu, Y.; Xu, J. Performance of a constructed wetland in treating brackish wastewater from commercial recirculating and super-intensive shrimp growout systems. *Bioresour. Technol.* **2011**, *102*, 9416–9424. [CrossRef]
170. Hua, T.; Haynes, R.J.; Zhou, Y.F. Removal of Al, Ga, As, V and Mo from alkaline wastewater using pilot-scale constructed wetlands. *Environ. Sci. Pollut. Res.* **2019**, *26*, 35121–35130. [CrossRef] [PubMed]
171. Nguyen, X.C.; Chang, S.W.; Tran, T.C.P.; Nguyen, T.T.N.; Hoang, T.Q.; Banu, J.R.; Al-Muhtaseb, A.H.; La, D.C.; Guo, W.; Ngo, H.H.; et al. Comparative study about the performance of three types of modified natural treatment systems for rice noodle wastewater. *Bioresour. Technol.* **2019**, *282*, 163–170. [CrossRef] [PubMed]
172. Gholipour, A.; Zahabi, H.; Stefanakis, A.I. A novel pilot and full-scale constructed wetland study for glass industry wastewater treatment. *Chemosphere* **2020**, *247*, 125966. [CrossRef] [PubMed]
173. Rahmadyanti, E.; Audina, O. The performance of hybrid constructed wetland system for treating the batik wastewater. *J. Ecol. Eng.* **2020**, *21*, 94–103. [CrossRef]
174. Pratiwi, N.I.; Mukimin, A.; Zen, N.; Septarina, I. Integration of electrocoagulation, adsorption and wetland technology for jewelry industry wastewater treatment. *Sep. Purif. Technol.* **2021**, *279*, 119690. [CrossRef]
175. Neurohr, G.A. Use of aquatic macrophytes for sludge treatment. In Proceedings of the 6th Symposium on Wastewater Treatment, Montréal, QC, Canada, 16–17 November 1983; pp. 262–284.
176. Litchfield, D.K.; Schatz, D. Constructed wetlands for wastewater treatment at Amoco Oil Company's Mandan, North Dakota refinery. In *Constructed Wetlands for Wastewater Treatment*; Hammer, D.A., Ed.; Lewis Publishers: Chelsea, MI, USA, 1989; pp. 233–237.
177. Kickuth, R. Das Wurzelraumverfahren-ein kosten-günstiges Klärverfahren für den dezentralen Einsatz in Kommunen und Gewerbe. *Der Trop.* **1982**, *83*, 141–154.
178. Hammer, D.A.; Rogers, P. *Treating Fish Rearing Pond Discharge Waters with Artificial Wetlands*; Tennessee Valley Authority Report: Knoxville, TN, USA, 1980.
179. Stone, R.W. The presence of iron and manganese-oxidizing bacteria in natural and simulated bogs. In *Treatment of Mine Drainage by Wetlands*; Burris, J.E., Ed.; The Pennsylvania University: University Park, PA, USA, 1984; pp. 30–36.

180. Pesavento, B.G. Factors to be considered when constructing wetlands for utilization as biomass filters to remove minerals from solution. In *Treatment of Mine Drainage by Wetlands*; Burris, J.E., Ed.; The Pennsylvania University: University Park, PA, USA, 1984; pp. 45–49.
181. Gray, K.R.; Biddlestone, A.J.; Job, G.; Galanos, E. The use of reed beds for the treatment of agricultural effluents. In *Constructed Wetlands in Water Pollution Control*; Cooper, P.F., Findlater, B.C., Eds.; Pergamon Press: Oxford, UK, 1990; pp. 333–346.
182. Brodie, G.A.; Hammer, D.A.; Tomljanovich, D.A. Constructed wetlands for treatment of ash pond seepage. In *Constructed Wetlands for Wastewater Treatment*; Hammer, D.A., Ed.; Lewis Publishers: Chelsea, MI, USA, 1989; pp. 211–219.
183. Ailstock, M.S. Utilization and treatment of thermal discharge by establishment of a wetlands plant nursery. In *Constructed Wetlands for Wastewater Treatment*; Hammer, D.A., Ed.; Lewis Publishers: Chelsea, MI, USA, 1989; pp. 719–726.
184. Rodgers, J.H.; Dunn, A. Developing design guidelines for constructed wetlands to remove pesticides from agricultural runoff. *Ecol. Eng.* **1992**, *1*, 83–95. [CrossRef]
185. Surface, M.J.; Peverly, J.H.; Steenhuis, T.S.; Sanford, W.E. Effect of season, substrate composition, and plant growth on landfill leachate treatment in a constructed wetland. In *Constructed Wetlands for Water Quality Improvement*; Moshiri, G.A., Ed.; Lewis Publishers: Boca Raton, FL, USA, 1993; pp. 461–472.
186. Higgins, J.; Dechaine, L. The use of large sub-surface flow wetlands to treat glycol-contaminated stormwater from aircraft deicing operations. In *Proceedings 10th International Conference Wetland Systems for Water Pollution Control, Lisbon, Portugal, 23–29 September 2006*; MAOTDR: Lisbon, Portugal, 2006; pp. 1879–1889.
187. Szilagyi, F.; Somlyody, L.; Koncsos, L. Operation of the Kis-Balaton reservoir: Evaluation of nutrient removal rates. *Hydrobiologia* **1990**, *19*, 297–305. [CrossRef]
188. Vincent, G. Use of artificial wetlands for the treatment of recreational wastewater. In Proceedings of the 3rd International Conf. Wetland Systems in Water Pollution Control, Sydney, Australia, 11–16 November 1992; pp. 28.1–28.4.
189. Hunter, R.; Birkbeck, A.E.; Coombs, G. Innovative marsh treatment systems for control of leachate and fish hatchery wastewater. In *Constructed Wetlands for Water Pollution Improvement*; Moshiri, G.A., Ed.; CRC Press/Lewis Publishers: Boca Raton, FL, USA, 1993; pp. 477–484.
190. Vymazal, J. Constructed wetlands for wastewater treatment in the Czech Republic-state of the art. In Proceedings of the 4th International Conferences on Wetland Systems for Water Pollution Control, Guangzhou, China, 6–10 November 1994; pp. 129–137.
191. Schwartz, M.F.; Boyd, C.E. Constructed wetlands for treatment of channel catfish pond effluent. *Progress. Fish-Cult.* **1995**, *57*, 255–266. [CrossRef]
192. Anderson, P. Constructed wetland treatment of sugarbeet process wastewater-1993 update-American Crystal Sugar Company. *Spec. Group Use Macrophytes Water Pollut. Control Newsl.* **1993**, *8*, 6–7.
193. Cooper, P.F.; Job, G.D.; Green, M.B.; Shutes, R.B.E. *Reed Beds and Constructed Wetlands for Wastewater Treatment*; WRc Publications: Medmenham, Marlow, UK, 1996.
194. Braskerud, B.C.; Haarstad, K. Screening and retention of thirteen pesticides in a small constructed wetland. *Water Sci. Technol.* **2003**, *48*, 267–274. [CrossRef] [PubMed]
195. Swift, J.; Landsdown, R.V. Design of a vegetative system for motorway run-off treatment: Constraints and design criteria. In Proceedings of the 4th International Conference Constructed Wetland Systems for Water Pollution Control, Guangzhou, China, 6–10 November 1994; pp. 697–703.
196. Vymazal, J. Czech Republic. In *Constructed Wetlands for Wastewater Treatment in Europe*; Vymazal, J., Brix, H., Cooper, P.F., Green, M.B., Haberl, R., Eds.; Backhuys Publishers: Leiden, The Netherlands, 1998; pp. 95–121.
197. Worrall, P. Reed bed for glycol treatment. *Spec. Group Use Macrophytes Water Pollut. Control. Newsl.* **1995**, *12*, 13–14.
198. Hill, D.T.; Rogers, J.W. Auburn University constructed wetland for the treatment of poultry lagoon effluent—A case study. In *Constructed Wetlands for Animal Waste Treatment. A Manual on Performance, Design and Operation with Case Histories*; Payne, V.W.E., Knight, R.L., Eds.; Gulf of Mexico Program; Stennis Space Center: Hancock County, MS, USA, 1997; pp. II-34–II-40.
199. Prystay, W.; Lo, K.V. Assessment of constructed wetlands for the reduction of nitrogen and phosphorus from greenhouse wastewaters. In *Proceedings 6th International Conference on Wetlands Systems for Water Pollution Control*; Tauk-Tornisielo, S.M., Salati Filho, E., Eds.; Universidade Estadual Paulista: Sao Paulo, Brazil, 1998; pp. 101–114.
200. Novais, J.M.; Martins-Dias, S. Constructed wetlands for industrial wastewater treatment contaminated with nitroaromatic organic compounds and nitrate at very high concentrations. In *Proceedings Conference The Use of Aquatic Macrophytes for Wastewater Treatment in Constructed Wetlands*; Dias, V., Vymazal, J., Eds.; ICN and INAG: Lisbon, Portugal, 2003; pp. 277–288.
201. Burgoon, P.S.; Kadlec, R.H.; Henderson, M. Treatment of potato processing wastewater with engineered natural systems. *Water Sci. Technol.* **1999**, *40*, 211–215. [CrossRef]
202. Best, E.P.H.; Miller, J.L.; Larson, S.L. Explosives removal from groundwater at the Volunteer Army Ammunition Plant, TN, in small-scale wetland modules. In *Wetlands and Remediation*; Means, J.L., Hinchee, R.E., Eds.; Battelle Press: Columbus, OH, USA, 2000; pp. 365–373.
203. Behrends, L.L.; Sikora, F.J.; Bader, D.F. Phytoremediation of explosives-contaminated groundwater using constructed wetlands. In *Wetlands and Remediation*; Means, J.L., Hinchee, R.E., Eds.; Battelle Press: Columbus, OH, USA, 2000; pp. 375–381.
204. Moore, B.J.; Ross, S.D.; Gibson, D.; Callow, L. Constructed wetlands for treatment of dissolved phase hydrocarbons in cold climates. In *Wetlands and Remediation*; Means, J.L., Hinchee, R.E., Eds.; Battelle Press: Columbus, OH, USA, 2000; pp. 333–340.

205. Comeau, Y.; Brisson, J.; Réville, J.-P.; Forget, C.; Drizo, A. Phosphorus removal from trout farm effluents by constructed wetlands. *Water Sci. Technol.* **2001**, *44*, 55–60. [CrossRef] [PubMed]
206. Kohler, E.A.; Poole, V.L.; Reicher, Z.J.; Turco, R.F. Nutrient, metal and pesticide removal during storm and nonstorm events by a constructed wetland on an urban golf course. *Ecol. Eng.* **2004**, *23*, 285–298. [CrossRef]
207. Snyder, J.A.; Mokry, L.E. Chemical industry´s constructed wetland—An environmental education success story. In Proceedings of the 7th Internat. Conference Wetland Systems for Water Pollution Control, Lake Buena Vista, FL, USA, 11–16 November 2000; Univeristy of Florida: Gainesville, FL, USA, 2000; pp. 1349–1355.
208. Billore, S.K.; Singh, N.; Ram, H.K.; Sharma, J.K.; Singh, V.P.; Nelson, R.M.; Das, P. Treatment of a molasses based distillery effluent in a constructed wetland in central India. *Water Sci. Technol.* **2001**, *44*, 441–448. [CrossRef]
209. Billore, S.K.; Ram, H.; Singh, N.; Thomas, R.; Nelson, R.M.; Pare, B. Treatment performance evaluation of surfactant removal from domestic wastewater in a tropical horizontal subsurface constructed wetland. In Proceedings of the 8th Internat. Conf. Wetland Systems for Water Pollution Control, Arusha, Tanzania, 16–19 September 2002; University of Dar es Salaam: Dar es Salaam, Tanzania, 2002; pp. 393–399.
210. Tanner, C.C.; Long Nguyen, M.; Sukias, J.P.S. Using constructed wetlands to treat subsurface drainage from intensively grazed dairy pastures in New Zealand. *Water Sci. Technol.* **2003**, *48*, 207–213. [CrossRef] [PubMed]
211. Kucuk, O.S.; Sengul, F.; Kapdan, L.K. Removal of ammonia from tannery effluents in a reed bed constructed wetland. *Water Sci. Technol.* **2003**, *48*, 176–186. [CrossRef]
212. Maine, M.A.; Hadad, H.; Sánchez, G.; Caffaratti, S.; Bonetto, C. Removal efficiency in a constructed wetland for wastewater treatment from a tool factory. In Proceedings of the 10th International Conference Wetland Systems in Water Pollution Control, Lisbon, Portugal, 23–29 September 2006; Dias, V., Vymazal, J., Eds.; MAOTDR: Lisbon, Portugal, 2006; pp. 1753–1761.
213. Gross, B.; Montgomery-Brown, J.; Naumann, A.; Reinhardt, M. Occurrence and fate of pharmaceuticals and alkylphenol ethoxylate metabolites in an-effluent-dominated river and wetland. *Environ. Toxicol. Chem.* **2004**, *23*, 2074–2083. [CrossRef] [PubMed]
214. Kapellakis, I.E.; Tsagarakis, K.P.; Angelakis, A.N. Performance of free water surface constructed wetlands for olive mill wastewater treatment. In Proceedings of the 9th International Conference Wetland Systems for Water Pollution Control, Avignon, France, 26–30 September 2004; ASTEE: Nanterre, France, 2004; pp. 113–120.
215. Bojcevska, H.; Raburu, P.O.; Tonderski, K.S. Free water surface wetlands for polishing sugar factory effluent in western Kenya-macrophyte nutrient recovery and treatment results. In Proceedings of the 10th International Conference Wetland Systems for Water Pollution Control, Lisbon, Portugal, 23–29 September 2006; Dias, V., Vymazal, J., Eds.; MAOTDR: Lisbon, Portugal, 2006; pp. 709–718.
216. Comino, E.; Riggio, V.; Rosso, M. Mountain cheese factory wastewater treatment with the use of a hybrid constructed wetland. *Ecol. Eng.* **2011**, *37*, 1673–1680. [CrossRef]
217. Kimani, R.W.; Mwangi, B.M.; Gichuki, C.M. Treatment of flower farm wastewater effluents using constructed wetlands in lake Naivasha Kenya. *Indian J. Sci. Technol.* **2012**, *5*, 1870–1878. [CrossRef]
218. Crous, L.; Britz, P. The use of constructed wetland technology in the treatment and beneficiation of brewery effluent for aquaculture. In Proceedings of the 12th International Conference Wetland Systems for Water Pollution Control, Venice, Italy, 4–8 October 2010; Masi, F., Nivala, J., Eds.; IWA: London, UK; IRIDRA Srl and Pan Srl: Padova, Italy, 2010; pp. 1255–1259.

land

MDPI

Article

The Effects of Tidal Flat Reclamation on the Stability of the Coastal Area in the Jiangsu Province, China, from the Perspective of Landscape Structure

Yanhui Chen [1,2], Guosheng Li [1,*], Linlin Cui [1,3], Lijuan Li [1,2], Lei He [4] and Peipei Ma [1,2]

1 Institute of Geographic Sciences and Natural Resources Research, Chinese Academy of Sciences, Beijing 100101, China; chenyh.18b@igsnrr.ac.cn (Y.C.); cuilinlin@cuit.edu.cn (L.C.); lilj.17b@igsnrr.ac.cn (L.L.); mapp.19b@igsnrr.ac.cn (P.M.)
2 University of Chinese Academy of Sciences, Beijing 100049, China
3 College of Resources and Environment, Chengdu University of Information Technology, Chengdu 610225, China
4 School of Tourism and Urban Management, Jiangxi University of Finance and Economics, Nanchang 330013, China; helei@jxufe.edu.cn
* Correspondence: ligs@igsnrr.ac.cn

Abstract: As one of the most important wetland systems, coastal wetlands play an important role in conserving water, regulating the climate and protecting biodiversity. However, due to large-scale and long-term tidal flat reclamations, the landscape structure and function of the coastal wetlands have been greatly affected. Therefore, it is necessary to understand the spatio-temporal characteristics of the impact of tidal flat reclamation on regional ecology and to quantitatively assess the relationships between them. In this study based on long-term, multiperiod remote sensing data, the main spatio-temporal variation characteristics of stability, and the relationship between stability and tidal flat reclamation were analyzed with regard to the influence scope of tidal flat reclamation. The results showed that a substantial decrease in natural wetlands in 1980, mainly caused by tidal flat reclamation, was discovered in the Jiangsu coastal area, and the influence scope of tidal flat reclamation on regional landscape ecology was roughly 30 km. In the affected area, the overall stability had a tendency to improve, but the stability change characteristics between reclamation area and non-reclamation area varied greatly. Especially in the reclamation area, the stability of construction wetlands and non-wetlands deteriorated. Spatially, the stability outside the reclamation area had the characteristics of first deteriorating and then improving as the distance from the reclamation area increased. Under the influence of tidal flat reclamation, the influence of different use types of TFR on stability was not completely consistent, and the influence of the same uses type of tidal flat reclamation on different landscapes was also different.

Keywords: coastal wetlands; tidal flat reclamation; stability; impact; Jiangsu coastal area

Citation: Chen, Y.; Li, G.; Cui, L.; Li, L.; He, L.; Ma, P. The Effects of Tidal Flat Reclamation on the Stability of the Coastal Area in the Jiangsu Province, China, from the Perspective of Landscape Structure. *Land* **2022**, *11*, 421. https://doi.org/10.3390/land11030421

Academic Editor: Richard C. Smardon

Received: 30 January 2022
Accepted: 11 March 2022
Published: 14 March 2022

1. Introduction

Coastal wetland, as an important landscape cover type and unique wetland ecosystem in coastal areas, plays an extremely important role in maintaining water, regulating climate and protecting biodiversity [1,2]. However, a great loss of coastal wetlands has been caused by excessive anthropogenic activities at different spatial scales [3,4], which will inevitably lead to the structural degradation of wetland systems and ultimately seriously affect the service supply capacity of regional ecosystems [5,6]. Therefore, to reduce the negative impact of human activities on the ecological environment, a better understanding of the ecological effect of human activities is urgent.

Tidal flat reclamation (TFR), as one of the most important human activities in the coastal area, has a long history in China [7] with a long coastline and abundant coastal wetlands [8,9]. According to statistics, a large number of coastal wetlands have been converted into agricultural land, industrial land, urban land, etc., to meet the needs of various industries [10]. From 1984 to 2018 alone, the coastal wetlands in China shrank from 10,263 km^2 to 7400 km^2, a decrease of approximately 27.9% [11]. In recent decades, TFR has provided a large number of reserve land resources for the development of various undertakings in coastal areas, which has greatly promoted economic development [12]; however, it has also caused a series of ecological and environmental problems [13,14]. A growing body of research suggests that rapid and large-scale TFR can lead to a sharp decline in coastal wetland resources [15], loss of diversity [16], decrease in coastal ecosystem functions and services [17], etc. Finally, these negative impacts, in turn, can have an impact on the regional environment and are detrimental to the sustainable development of coastal areas [13]. Therefore, facing irreversible ecological degradation caused by TFR, the evaluation of the ecosystem state has become particularly necessary for explicating the impact on ecosystems, which is significant to regional ecological conservation.

In recent years, the ecological impact of TFR has received an increasing amount of attention, and many scholars have conducted relevant research and achieved a series of results [18,19]. As the understanding of the ecological impact of TFR has increased, managers have taken a more scientific approach to formulating land planning and resource conservation policies [20]. However, most of the research has focused on very specific areas of aquatic or terrestrial ecosystems such as biodiversity loss [21], carbon flux [22], heavy metal contamination [23] and the bacterial community [24] and the comprehensive impact of TFR on the ecosystem is still relatively lacking [25].

Stability is an important feature in the structure and function of ecosystems that determines the rise and fall of ecosystems. As an important comprehensive indicator of the ecosystem state, stability has been the focus of ecological researchers to evaluate the ecological conditions of terrestrial ecosystems [26,27]. As ecological stability is a multidimensional concept that covers the different aspects of the dynamics of the system and its response to perturbations, the concept of stability has not yet been defined exactly [27,28]. Due to the differences in the professional backgrounds and research angles of researchers, different scholars often assign different connotations to ecosystem stability according to actual research needs [26,27]. Although there are many different concepts of ecological stability, they all contain the following implications: the ability of a system to remain in the status quo after a disturbance and the ability of a system to return to its original state after being disturbed.

To evaluate the ecological stability of wetlands, many scholars have performed excellent work and constructed many meaningful evaluation indicators. The evaluation of stability has mainly been based on the research objective of selecting a single indicator to characterize stability [29] or based on the structural relationship between multiple indicators by constructing a composite stability evaluation index to comprehensively evaluate stability [30,31]. Commonly used indicators can be roughly divided into three categories: structural indicators, functional indicators and external environmental factor indicators. Structural indicators include the components of animals, plants, microorganisms, soil [32,33], structural characteristic indices [31], etc., in the ecosystem. Functional indicators include the productivity level of the ecosystem [32], carbon absorption capacity [31], surface water availability [34], etc. External environmental factor indicators mainly include the external effects exerted by nature or humans on wetlands [35,36]. However, many functional indicators are difficult to measure, or raw data are difficult to obtain, making it impossible to conduct large-scale, long-term studies. Moreover, external environmental factor indicators are easily subject to subjective cognition and environmental influences and are not easy to quantify. There are also some indicators that are limited to theoretical research and cannot be applied to practice [27].

In view of the fact that most coastal areas tend to display long-term and large-scale TFR, it is necessary to construct an appropriate evaluation indicator when comprehensively evaluating the impact of TFR. The realization of landscape functions requires the support of the landscape structure [37], and the landscape structure can be expressed through the rich information contained in the landscape types and its spatial distribution forms and combinatorial relationships such as the diversity of landscape types, the area of landscape types and the landscape pattern. The change in this kind of information characterizes some of the most critical implications of the complex interactions between natural environment changes and human activities [5,38]. Remote sensing technology is widely used in landscape ecology due to its wide spatial range, long time series and easy access and can obtain rich landscape ecological information [39]. Therefore, a stability evaluation index can be constructed from the perspective of landscape structure based on remote sensing data.

In addition, previous research on the ecological impact of TFR was mostly limited to the impact of a single or several projects [40,41], and the change in the ecological environment before and after the project(s) was usually compared in the whole study area [42]. However, many studies have shown that the impact of human activities on regional ecology has obvious spatio-temporal characteristics [43], indicating that the characteristics of the impact of TFR inside and outside the reclamation area are not necessarily the same. Even within the non-reclamation area, the influence in different locations may be different. At the same time, the ecological impact of TFR has obvious spatial-temporal accumulation characteristics, and the impact at different times and different locations is constantly superimposed. Therefore, combined with long-term and multiperiod data, research on the spatio-temporal characteristics of the ecological impact of TFR is still lacking.

In this paper, we aim to make a modest step toward understanding the ecological impact of TFR from the perspective of landscape ecological stability by choosing the Jiangsu coastal area as a case study. Combined with multiperiod remote sensing data from 1980 to 2018, this study evaluated the impact of TFR on landscape pattern and determined the impact range of TFR. On this basis, the spatio-temporal variation characteristics of stability were analyzed, and the relationship between TFR and stability was quantitatively analyzed, both of which are expected to provide a meaningful reference for the future conservation and management of coastal wetland resources.

2. Materials and Methods

2.1. Study Area

The Jiangsu coastal area, as one of the regions with the richest wetland resources, is located in eastern China and adjacent to the Yellow Sea (Figure 1). The coverage is between 119°28′–121°59′ E and 31°39′–34°31′ N. In this region, there are a total of 10 cities and counties, as follows: Xiangshui, Binhai, Sheyang, Dafeng, Dongtai, Haian, Rudong, Tongzhou, Haimen, and Qidong. Due to the differences in the natural geographical environment, the northern part of the coastal wetlands is mainly dominated by rocky sand and substrate, and the central and southern parts are dominated by muddy coast, which is conducive to the formation of tidal flats. According to previous surveys, the coastline is approximately 954 km, accounting for approximately 5% of the total length of China's coastline [44,45]. Abundant natural wetlands provide important habitats for endangered animal species and maintain regional biodiversity and at the same time provide sufficient reserve land resources to meet different land needs such as agricultural land and construction land. However, with the development of the economy, the short-term impact of human activities on the regional landscape ecology and its cumulative effect have gradually emerged. Therefore, the Jiangsu coastal area, due to its unique natural conditions and intense TFR, is a typical area for studying the ecological impact of TFR on the coastal area.

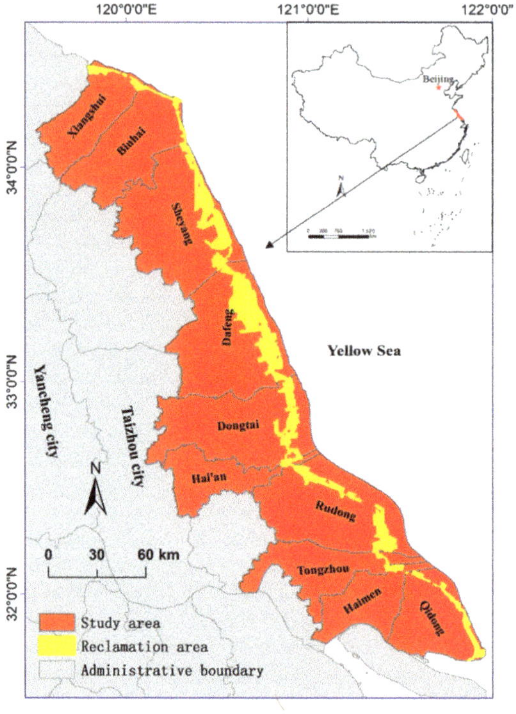

Figure 1. Location of the study area and geographic distribution of reclamation area since 1980.

2.2. Data Sources and Preprocessing

The data we collected and used in this study involved both primary and secondary data. The primary data included Landsat MSS/TM/ETM+/OLI data for almost 40 years, as follows: 1980, 1983, 1986, 1992, 1995, 2000, 2005, 2008, 2011, 2014 and 2018 [45]. To fully cover the entire study area, three scenes of images were used for each target year. All the remote sensing data involved in this study were mainly cloud-free and acquired from August to October, with path/row numbers of 120/036, 119/037 and 118/038 (the path/row numbers for 1980 are 129/036), and with a spatial accuracy of 30m (the accuracy of remote sensing data in 1980 is 80 m). Considering that the size of the landscape patches in the study area was between 1.83 and 3.06 km, the 30m or 80m spatial accuracy was sufficient [5].

To meet the requirements of this study, 17 landscape types (Table 1), including natural wetlands, artificial wetlands, and non-wetlands (for the sake of discussion, non-wetlands were considered a special type of wetlands in this study), were extracted based on an object-oriented remote sensing interpretation method supplemented by visual interpretation. The interpretation results showed that the overall accuracy was more than 90%, which met our research requirements. In addition, vector data, including the provincial, municipal and county administrative boundaries of the administrative divisions of Jiangsu province, were used as secondary data.

Table 1. Landscape types in Jiangsu coastal wetlands.

Categories	Landscape Types
Natural wetlands	River (RR), grassy marshland (GM), Phragmites Australis (PA), Suaeda glauca (SG), Spartina alterniflora (SA), mudflat (MT)
Artificial wetlands	Paddy field (PF), pool (PL), salt field (SF), mariculture farm (MF)
Non-wetlands	Dryland (DL), forest (FT), bareland (BD), levee (LE), urban land (UL), rural residential land (RL), construction land (CL)

2.3. *The Identify of Use Types of Reclamation*

Based on the remote sensing interpretation results and combined with the main use types of reclamation in the Jiangsu coastal area in the past 38 years (regarding 2018, the same as below), five main use types of reclamation were identified by overlay analysis.

To identify the transformation of TFR from the change in landscape types in each time period, sequential spatial overlay analysis was run on the landscape vector data layers for 1980, 1983, 1986, 1992, 1995, 2000, 2005, 2008, 2011, 2014 and 2018 using the overlay module of ArcGIS version 10.2. The results showed great continuous transformation of landscape change over the past 38 years. As the transformation processes information during a specific time period (e.g., 1980–1983, 1983–1986 and 1986–1992) can be identified by comparing the start-point landscape types with the end-point landscape types (e.g., a transformation from a tidal flat into a port area indicates a typical reclamation) [46], five major use types of reclamation were identified (Table 2). These types included aquaculture land (AQL), arable land (ARL), port construction land (POL), salt industry land (SAL) and hydraulic engineering land (HYL). In addition, as the area from tidal flats to woodland and bare land was minimal, neither transition was taken into account.

Table 2. The identification of use types of reclamation according to their start-and end-point landscape types.

Use Types	Start-Point Landscape Type	End-Point Landscape Type
AQL	Nature wetlands: mudflat, Suaeda glauca, Phragmites Australis, grasslands, rivers, and Spartina alterniflora	Mariculture farm
ARL		Paddy land, dryland
POL		Rural land, urban land, construction land
SAL		Salt pond
HYL		Pool, levee

2.4. *Buffer Analysis*

Buffer analysis, as an important spatial analysis tool, is often used to identify changes in ecosystems, which are affected by human activities or other disturbance [47]. To analyze the spatial characteristics of landscape pattern and stability under the influence of TFR, one-sided buffer zones were constructed along the boundary of the reclamation area based on the Geographic Information System (GIS). A total of 12 bands with a width of 4.5 km were distributed on the landward side of the reclamation area (the southwest side). The width of levees was not accounted for in this study. To facilitate the further discussion, each buffer zone was numbered 1-12 from nearest to farthest according to its distance to the reclamation area.

2.5. *Evaluation of the Ecological Stability in the Jiangsu Coastal Area*

Based on the background structure (the optimal area proportion of the landscape types) of the Jiangsu coastal area, the stability index, as an indicator of local ecological stability, was used to evaluate the impact of TFR on ecological stability in the Jiangsu coastal area. The ecological stability index equation is as follows:

$$I = \sqrt[2]{\frac{\sum_{i=0}^{m} (\frac{A_i}{A_s} - B_i)^2}{m}} \quad (1)$$

where I is the stability index of the target area; A_i is the area of landscape type i in the study area; A_s is the total area of the study area; B_i is the proportion of type i's area to the area of the study area in the background structure; m is the number of landscape types in the study area. Stability index I can be understood as the standard deviation of the landscape structure over a certain time period and the background structure, so the smaller the I, the better the stability in a region.

The calculation of the background structure was mainly based on the theory of competition/coexistence. First, the ecological service game/competition model was constructed by using the reference point-based non-dominated sorting (NSGA-III) algorithm. The optimal solution as the optimal balance of ecosystem services was obtained using this model [45]. Finally, the corresponding landscape type area and area ratio (background structure; Table 3) were obtained based on the cascading relationship of "process-function-service". (For a specific calculation method, see reference [45]).

Table 3. Optimal area ratio of various landscapes in Jiangsu coastal area (unit: %).

Types	GM	MT	DL	RR	FT	PA	BD	SA	PL	PF	SG	SF	MF
Optimal	9.94	0.07	0.87	1.81	35.81	7.27	2.53	4.24	0.44	29.53	0.15	1.30	6.00

2.6. Statistical Analyses

To analyze the stability trend with distance from the reclamation area, the ordinary least squares (OLS) method was used in this study. In addition, the Pearson correlation coefficients between the stability index and the cumulative area of reclamation in the affected area, the reclamation area and non-reclamation area were calculated to quantitatively analyze the effects of TFR on stability. All statistical analyses were conducted in SPSS 23. Statistical significance was at the 0.05 level.

2.7. Landscape Indices

To determine the spatial extent of the impact of TFR on the regional ecology, based on previous research [5,43,48,49], and taking into account the ecological significance of landscape patter indices, the fragmentation index (fragmentation), cohesion index (COHESION) and Shannon's Diversity Index (SHDI) were selected in this study. The landscape pattern indices of each buffer zone were calculated according to the formulas as follows:

Fragmentation characterizes the degree of fragmentation of the landscape, reflects the complexity of the spatial structure of the landscape, and to a certain extent reflects the degree to which human activities affect the landscape. The formula is as follows:

$$C_i = \frac{N_i}{A_i} \quad (2)$$

where C_i is the fragmentation of landscape type i ; N_i is the number of patches of landscape type i ; A_i is the total area of landscape type i. To understand the degree of fragmentation of all types as a whole, we divided the total number of patches in a target area by the total area of this area.

COHESION reflects the aggregation and dispersion of patches in the landscape,

$-1<$ COHESION < 1. When the value is -1, the patches are completely dispersed; when the value is 0, the patches are randomly distributed; when the value is 1, the patches are clustered.

$$\text{COHESION} = \left[1 - \frac{\sum_{j=1}^{m} P_{ij}}{\sum_{j=1}^{m} P_{ij}\sqrt{a_{ij}}}\right] * [1 - \frac{1}{\sqrt{A}}]^{-1} * 100 \tag{3}$$

where a_{ij} is the area(m^2) of the j th patch in the landscape type i ; p_{ij} presents the circumference (m) of the j th patch in the landscape type i; A is the total area of the landscape type i (hm^2).

SHDI is used to describe the diversity and complexity of landscape patches, SHDI ≥ 0. When SHDI = 0, the landscape contains only 1 landscape. When SHDI is large, the proportional distribution of area among landscape types becomes more equitable, and the complexity of the ecosystem composition usually tends to increase.

$$\text{SHDI} = -\sum_{i=1}^{m} p_i \ln(p_i) \tag{4}$$

where P_i is the area ratio of patch type i to a target area; m is the number of all patch types.

In addition, the overall transfer probability (P) of the target area was also calculated in this study. The probability of landscape transfer indicates the likelihood of a landscape transformation within an area. It is calculated as follows:

$$P = \sum_{i=1}^{n} \frac{l_i}{S} \tag{5}$$

where P is the overall transfer probability of the target area; l_i is the area of landscape type i that is converted to other landscape types in a certain period; S is the area of the target area; n is the number of landscape types in the start year of the target period;

3. Results

3.1. Major Landscape Transformation Features

3.1.1. Analysis of the Landscape Change Process

In the past 38 years, the main landscape change throughout the study area was characterized by a significant decrease in natural wetlands and a significant increase in artificial wetlands and non-wetlands (Figure 2). Specifically, natural wetlands decreased by 2477.64 km^2 (41.63%), and artificial wetlands and non-wetlands increased by 1987.32 km^2 (21.77%) and 378.89 km^2 (2.96%), respectively.

Among natural wetlands, the mudflat and coastal marshes decreased by 1600.42 km^2 and 756.59 km^2, respectively and accounted for 60.11% and 28.26% of the total natural wetlands outflow area, respectively. Mariculture farms and paddy fields were the main contributors to the increase in artificial wetlands, increasing by 1736.73 km^2 and 657.34 km^2, respectively, and accounting for 72.51% and 27.44% of the total inflow area of artificial wetlands, respectively. The increase in non-wetlands was mainly due to the increase in residential land (including rural residential land and urban land) and construction land, and the area of residential land and construction land increased by 945.69 km^2 and 296.54 km^2, respectively, accounting for 73.67% and 23.10% of the total non-wetlands inflow area, respectively.

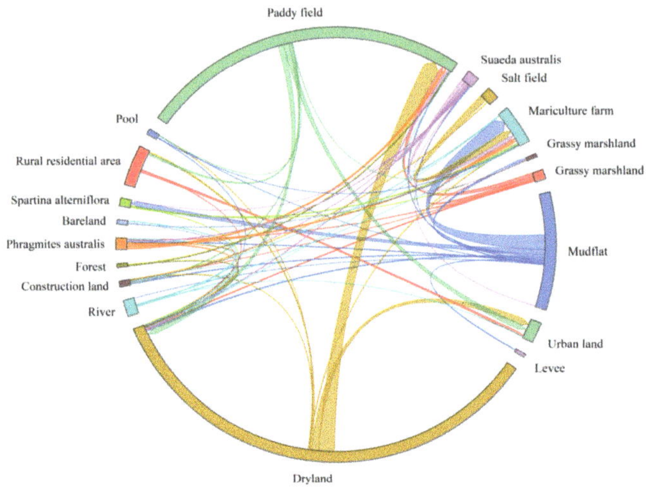

Figure 2. The area of the landscape transformation in the Jiangsu coastal area during the period 1980–2018.

3.1.2. Analysis of TFR process

Due to the natural conditions and policy-oriented reclamation planning, landscape transformation in the Jiangsu coastal area was spatially distinct. The reduced natural wetlands were mainly concentrated in the reclamation area in the tidal flats. Overall, from 1980 to 2018, a total of 1969.86 km^2 of coastal wetlands was reclaimed across the tidal flat area in Jiangsu province, and a consistently increasing annual rate of 51.84 km^2 was observed. From the perspective of time course (Figure 3), the intensity of reclamation showed a generally fluctuating upward trend during the period 1980–2011, reaching a maximum value in 2008–2011, and then showing a rapid downward trend.

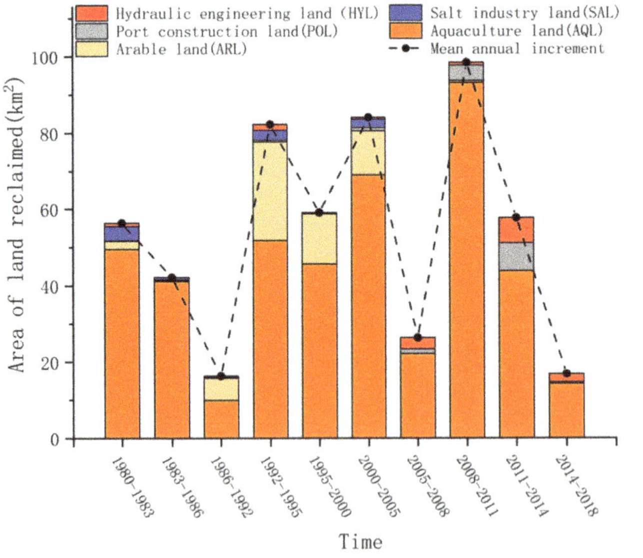

Figure 3. The reclamation process from 1980 to 2018.

The majority of reclamation occurred in the middle of the coastal wetlands. Specifically, over 85.1% of the reclamation occurred in Dongtai, Rudong, Sheyang and Dafeng. Especially, approximately 35% of the reclamation occurred in Dafeng, where the average annual reclamation area between 1990 and 2005 (36.3 km^2) was two times higher than that over the past 38 years. For administrative reasons, there has been only 12.6 km^2 wetlands reclaimed in Tongzhou in the past 38 years, with a rate of less than 1 km^2 per year.

In terms of use types, nature wetlands were mainly occupied by aquaculture land (APL) and arable land (ARL), followed by hydraulic engineering land (HYL), port construction land (POL) and, salt industry land (SAL), accounting for the total amount of reclamation 81.94%, 12.40%, 2.60%, 2.42%, and 1.64%, respectively. Aquaculture land was an absolute advantage at all time. When aquaculture land was not considered, the main use types were salt industry land (SAL) and arable land (ARL) in 1980–2005. After 2005, the main use types were port construction land (POL) and hydraulic engineering land (HYL), indicating that the use types of TFR shifted from agriculture to industry.

In non-reclamation areas, landscape transformation mainly occurred near tidal flats and at the junction of paddy field and dry land. Except for the large-scale transition between paddy field and dryland, a large amount of construction land and urban and rural settlements have been transferred from paddy field and dryland. The transformation from rural residential land and construction land to urban land also accounted for a certain amount.

3.2. Change in Cological Stability in the Coastal Area

3.2.1. The Impact Scope of TFR in the Jiangsu Coastal Area

At the landscape level, the general impact of human activities is the transformation of the landscape in the target area, the results of which are changing the landscape structure, affecting the system functions [48,49], and ultimately affecting the stability of the landscape ecology [50]. Therefore, the stability change in regional ecology is fundamentally a change in the landscape structure. In this regard, to identify the spatial extent of TFR in the Jiangsu coastal area, the overall transfer probability and three landscape pattern indices (fragmentation, COHESION and SHDI) within each buffer zone were calculated, as shown in Figure 4.

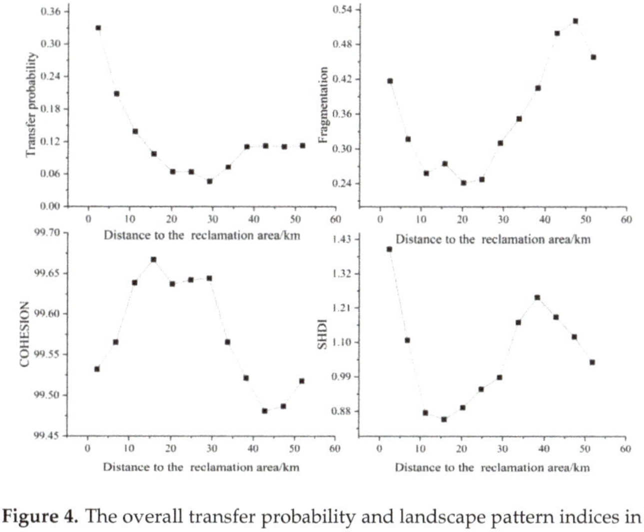

Figure 4. The overall transfer probability and landscape pattern indices in buffer zones.

In Figure 4, it can be seen that the overall transfer probability generally showed a downward trend. Specifically, the overall transfer probability gradually decreased with the distance from the reclamation area in the range of 0–29.25 km (buffer zones 1–7) and then increased to the max at 38.25 km. After 38.25 km, the overall transfer probability tended to stabilize. As the landscape transformed, the landscape pattern also changed at the same time. Specifically, within 24.75 km from the reclamation area, fragmentation showed a general downward trend and reached a minimum value at 24.75 km from the reclamation area, after which the overall trend of fragmentation was upward. COHESION was roughly negatively correlated with the fragmentation. Within 15.75 km from the reclamation area, COHESION rose significantly with the increase in distance and reached a maximum value at 15.75 km. Then, COHESION was basically stable within 20.25–29.25 km. After 29.25 km, COHESION decreased rapidly and reached a minimum at 42.75 km. Due to the reduction in landscape types, SHDI descended rapidly within 15.75 km from the reclamation area. After 15.75 km, SHDI gradually rose and then began to accelerate at 29.25 km, reaching a maximum at 38.25 km. After 38.25 km, SHDI gradually declined.

In summary, in the range of 24.95–29.25 km from the reclamation area, a significant turning point could be identified for the transfer probability and all three landscape pattern indices. Therefore, we believe that the maximum impact extent of TFR is approximately 29.25 km.

3.2.2. Temporal and Spatial Change in Stability

Based on the stability index calculation method, the stability index of the reclamation area, non-reclamation area (including the area all the buffer zones covered within the influence scope of TFR), and the affected area (including the reclamation area and non-reclamation area) were obtained for 11 target years (Figure 5).

In the affected area, the stability index was basically on a downward trend before 2000 and fluctuated after 2000 (including 2000), while the stability index of natural wetlands maintained a slight downward trend with small fluctuations. The change characteristics of the stability index of constructed wetlands and non-wetlands were basically consistent with those of the wetlands, but the stability index of construction wetlands had a slight upward trend after 2000. The above shows that the stability of construction wetlands and non-wetlands had a significant impact on the stability of wetlands in the entire affected area.

In the non-reclamation area, the change characteristics of wetlands stability were generally consistent with those of the affected area. The stability index of natural wetlands remained basically unchanged, and the stability index of construction wetlands showed small fluctuations after a significant decline in 1983–1986. The stability index change process of the non-wetlands was basically the same as that of the wetlands. The above shows that the stability of non-wetlands played a leading role in the stability of the non-reclamation area.

In the reclamation area, the stability index of wetlands maintained a downward trend after a short period of increase in 1980–1983, and generally remained unchanged after 2011. The variation characteristics of the stability index of natural wetlands were basically the same as those of wetlands, but the former declined faster. The stability index of construction wetlands gradually fluctuated and rose, while the stability index of the non-wetlands maintained a slight upward trend with small fluctuations. Overall, the stability of the reclamation area was mainly dominated by the stability of natural wetlands.

On average, the average stability index for the past 38 years was 0.17, 0.14 and 0.17 in the affected area, reclamation area and non-reclamation area, respectively (Figure 6), indicating that the stability level of the reclamation area was better than that of the non-reclamation area. From the perspective of different wetland types (Figure 6), in the reclamation area, the stability index of different wetland types was not much different, but the stability index of various types of wetlands in the non-reclamation area had a significant gap, which was basically consistent with that of the affected area. The stability index of natural wetlands and constructed wetlands in the reclamation area was greater than that in the non-reclamation area, and the stability index of non-wetlands in the reclamation area was smaller than that in non-reclamation area.

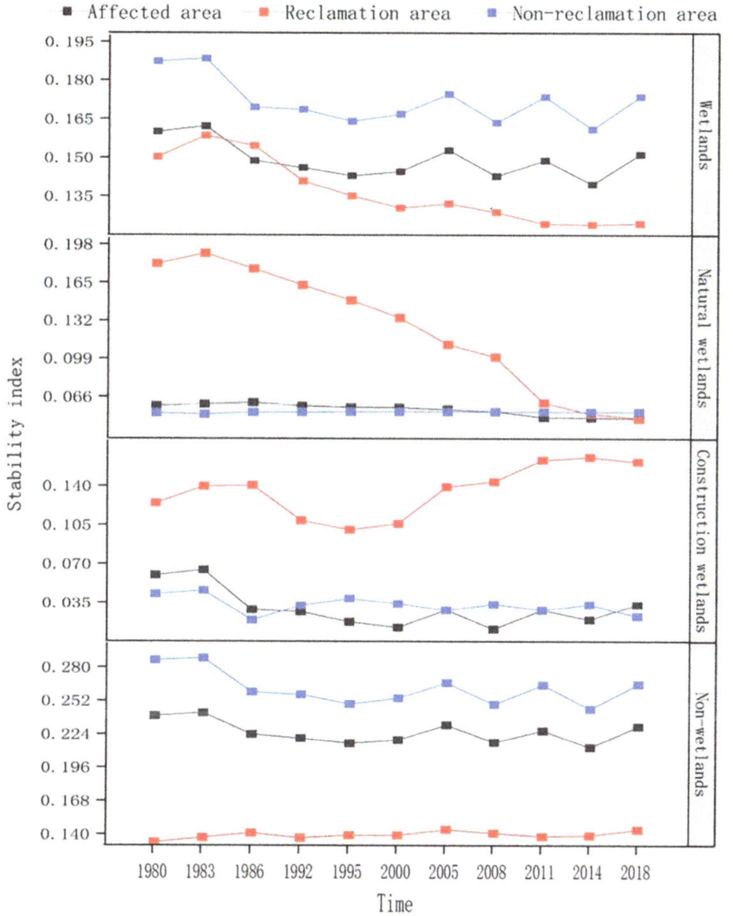

Figure 5. Temporal variation of different wetland types in different regions.

To further explore the spatial distribution characteristics of stability, the stability index of each buffer zone in the non-reclamation area was calculated. The general results of stability change in different buffer zones (Figure 7) showed that TFR can significantly affect the spatial distribution of stability. Figure 7 shows that the effect of LFTF on the stability index of the non-reclamation area was not linear but showed a cubic function relationship ($R^2 = 0.98$, $p < 0.01$). Specifically, 11.25 km was the threshold distance from the reclamation area for stability index change, within which the closer to the reclamation area, the smaller the stability index (or the better the stability level), and outside which, the closer to the reclamation area, the greater the stability index (or the worse the stability level). In addition, the standard deviation of the stability index in each buffer zone over the past 38 years also exhibited obvious spatial characteristics. Similarly, approximately 11.25 km was the threshold for the change of standard deviation of the stability index. Within 11.25 km from the reclamation area, the overall stability index fluctuation was larger, and with the increase in distance, the stability index fluctuation increased. After 12.5 km from the reclamation area, the overall stability fluctuation was smaller, and as the distance increased, the fluctuation of the stability index gradually decreased until it was almost close to zero. The above shows that within the affected range the impact of TFR on stability was mainly concentrated within a 15.75 km range from the reclamation area (buffer zones 1–4).

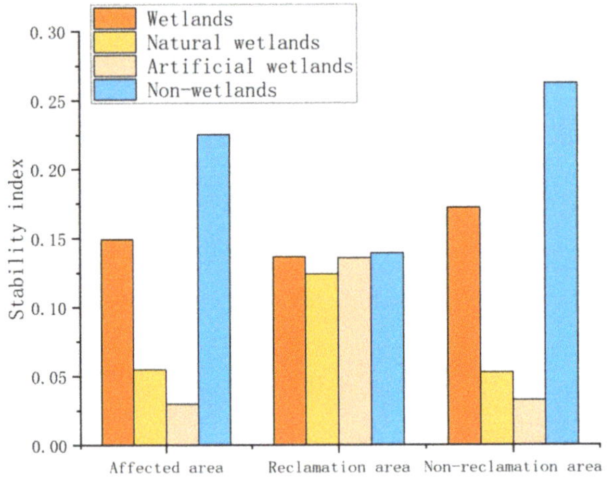

Figure 6. Stability index of different wetland types in different regions.

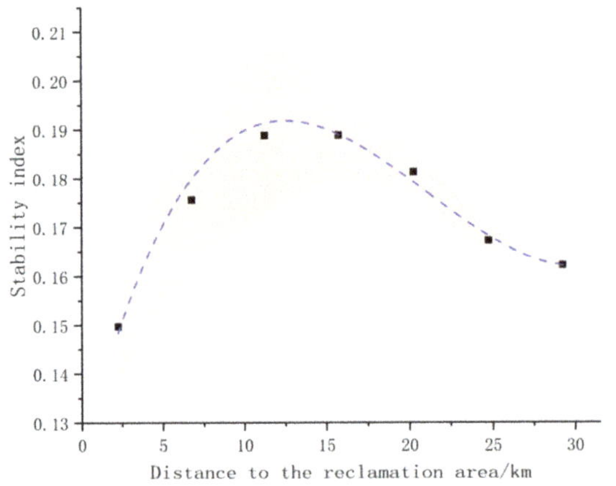

Figure 7. Stability index and deviation of stability in the different buffer zones.

3.3. Quantitative Relationship between TFR and Ecological Stability

Overall, the cumulative area of reclamation was negatively correlated with the stability index of natural wetlands in the affected area and natural wetlands, artificial wetlands, and wetlands in the reclamation area, and it was positively correlated with the stability index of natural wetland in the non-reclamation area (Table 4).

For specific use types of reclamation, in the affected area, all five use types were negatively correlated with the stability index of natural wetlands. Type ARL and SAL were negatively correlated with the stability index of artificial wetlands and coastal wetlands. In the reclamation area, all five types were negatively correlated with the stability index of natural wetlands and coastal wetlands. Type POL and HYL were positively correlated with the stability index of artificial wetlands, and type SAL was positively correlated with the stability index of non-wetlands. In the non-reclamation area, type AQL, ARL, and SAL were positively correlated with the stability index of natural wetlands, while type

ARL and SAL were negatively associated with the stability index of non-wetlands and coastal wetlands.

Table 4. Correlation matrix between stability index and area of reclamation.

Use Types	Affected Area				Reclamation Area				Non-Reclamation Area			
	SW	SNW	SCW	SOW	SW	SNW	SCW	SOW	SW	SNW	SCW	SOW
CA	−0.488	−0.936 **	−0.525	−0.385	−0.909 **	−0.987 **	0.667 *	0.596	−0.486	0.686 *	−0.474	−0.476
AQL	−0.570	−0.899 **	−0.618 *	−0.474	−0.932 **	−0.962 **	0.581	0.582	−0.563	0.712 *	−0.441	−0.556
ARL	−0.675 *	−0.749 **	−0.777 **	−0.589	−0.962 **	−0.843 **	0.267	0.593	−0.675 *	0.821 **	−0.397	−0.670 *
POL	−0.376	−0.977**	−0.321	−0.281	−0.794 **	−0.965 **	0.793 **	0.364	−0.343	0.522	−0.418	−0.333
SAL	−0.631 *	−0.707 *	−0.737 **	−0.547	−0.880 **	−0.817 **	0.353	0.666*	−0.644 *	0.735 *	−0.454	−0.637 *
HYL	−0.405	−0.977 **	−0.357	−0.308	−0.824 **	−0.979 **	0.755 **	0.444	−0.384	0.565	−0.444	−0.374

"**" Indicates significant correlation at the 0.01 level (two tails); "*" indicates significant correlation at the 0.05 level (two tails); SW, SNW, SCW, and SOW represent the stability index of coastal wetlands, nature wetlands, construction wetlands, and non-wetlands, respectively. CA represents the cumulative area of five use types of reclamation.

4. Discussion

4.1. The Impact Scope of TFR

Tourism development, such as port construction, urban expansion and road networks, can have an impact on the surrounding environment. For TFR, the various use types of reclamation represent the comprehensive impact of human activities. According to the first law of geography, "everything is related to everything else, but near things are more related than distant things [51]", so this effect of TFR diminishes with distance and accumulates over a certain time period and space.

The ecosystem structure is the basis of the system function, and a certain structure supports a certain function [37]. Changes in landscape structure are the basis for functional change. The landscape pattern indices condense a variety of rich information about the landscape pattern. According to previous research, human activities can significantly affect the fragmentation, COHESION, and SHDI; the AREA_MN (mean patch area); the FRAC_MN (mean fractal dimension index); the AI (aggregation index) [5,49,52]; etc. Combined with the ecological implications of landscape indices, we chose three indices (fragmentation, COHESION, and SHDI) to depict the landscape structure. According to the buffer analysis, the selected landscape pattern indices had obvious spatial distribution characteristics. Based on this, it was concluded that 29.25 km away from the reclamation area was the abrupt distance of the landscape pattern change and 29.25 km was used as the maximum influence range of TFR in the Jiangsu coastal area.

There are only a few studies on the impact extent of TFR on landscape ecology in China. Based on buffer analysis, Di et al. [53] found that the intensity of human activity had a significant gradient within 30 km from the coastline, indicating that the effects of TFR were within 30 km. Considering that the average distance between the western boundary of the reclamation area and the coastline was 9.32 km in this study, if the coastline was taken as the starting point, the influence range of TFR in this study was roughly 39 km, which was obviously greater than 30 km. Possible reasons for this are as follows: TFR has obvious spatio-temporal accumulation characteristics, so the impact of the TFR on the landscape ecology is the result of the combination of long-term and multi-regional reclamation. In Di's study, the time span was 10 years, while our study spanned almost 40 years, so the impact scope of TFR was greater in our study.

4.2. The Rationality of the Index Construction

At present, there are no fixed indicators for stability evaluation due to the complexity of ecological stability itself and the inconsistency of the concepts of ecological stability [26,31]. Most studies have selected stability indicators based on specific research objectives. This study built a local stability index by combining the composition ratio of landscape types of optimal ecological backgrounds and realistic landscapes.

The selection of the index needs to be based on the characteristics of the study area [31,54]. In the Jiangsu coastal area, due to the long-term and large-scale TFR, while the landscape ecology in the reclamation area changed, it could also significantly affect the surrounding landscape pattern through natural or socioeconomic factors and ultimately affect the service supply and ecological stability of the ecosystem. At the landscape level, this is manifested by the transformation of landscape types before and after reclamation as well as the difference between the transformation of landscape types or the landscape pattern of the reclamation area and non-reclamation area. The result of the change or difference led to the changes in the areas of different landscape types over a certain time period. As a result, the impact of human activities can be reflected by proportional relationships between the areas of the various landscape types.

To evaluate the proportional relationships between different periods or different regions, a reasonable reference is necessary. According to Li et al. [45], the optimal ecological background structure depicted an ecological competition result in the case of artificial participation with minimal human intervention. By this they meant that the various functions of the ecosystem were not only coordinated and unified but also achieved approximately the best-reachable, natural stable condition before humans carried out large-scale production and life transformation in the area. Therefore, this situation can be regarded as the background quantitative structure of the Jiangsu coastal area.

According to the definition of stability we constructed, the stability index essentially reflects the closeness of various service supplies of the study area between reality scenarios and the best equilibrium state in a certain period. Therefore, the closer the landscape type proportion of the ecosystem in the study area is to the background structure, the better the ecology stability in this region. When the stability changes, e.g., becomes smaller, it can not only indicate that the landscape type proportion of the system is closer to the background structure but also that the various services of the ecosystem have been maximized after ecological competition by all parties.

In addition, Figure 4 shows that in the region close to the reclamation area (buffer zones 1–2), the fragmentation was larger and the COHESION was lower, meaning that the region had a higher degree of fragmentation and scattered patches. However, the SHDI was relatively large; the reason for this was that due to the proximity to the tidal flats, there were many landscape types, and patch distribution was also more balanced, which enabled the system to resist external disturbances [55,56]. In the zones that were slightly farther away (buffer zones 3–4), the fragmentation and SHDI were low, and the COHESION was relatively high. Combined with the results of remote sensing surveys, there were very few landscape types here with a large area of arable land and construction land, which was not conducive to its resistance to external disturbances. In the more distant zones (buffer zones 5—7), the fragmentation and COHESION remained at a low and high level, respectively, but the SHDI increased significantly as the landscape types in this region increased, and the distribution was relatively balanced. Therefore, the resistance to the external disturbances in the most distant zones (buffer zones 5–7) was enhanced relative to the slightly farther away zones (buffer zones 3–4).

Combined with the spatial variation characteristics of the standard deviation of the landscape pattern indices, it can also be seen that with the increase in the distance from the reclamation area, the standard deviation (especially for the SHDI and fragmentation) basically showed the characteristics of first increasing and then decreasing (Figure 8), indicating that the ability of the system to maintain its own state under human disturbance had this spatial change feature. The above shows that within the influence scope stability showed the characteristics of first deteriorating and then improving with the increase in distance from the reclamation area. This was basically consistent with the stability change characteristics indicated by the stability index in this study (t7). Furthermore, through correlation analysis, it was found that the stability index had a significant positive correlation with the SHDI ($p < 0.05$), a marginally positive correlation with the fragmentation ($p = 0.059$), and a nonsignificant negative correlation with COHESION. The standard deviation of the landscape pattern indices had a positive correlation with stability index, but only SHDI passed the significance test ($p < 0.05$). Therefore, we believe that the stability index constructed in this study is reasonable.

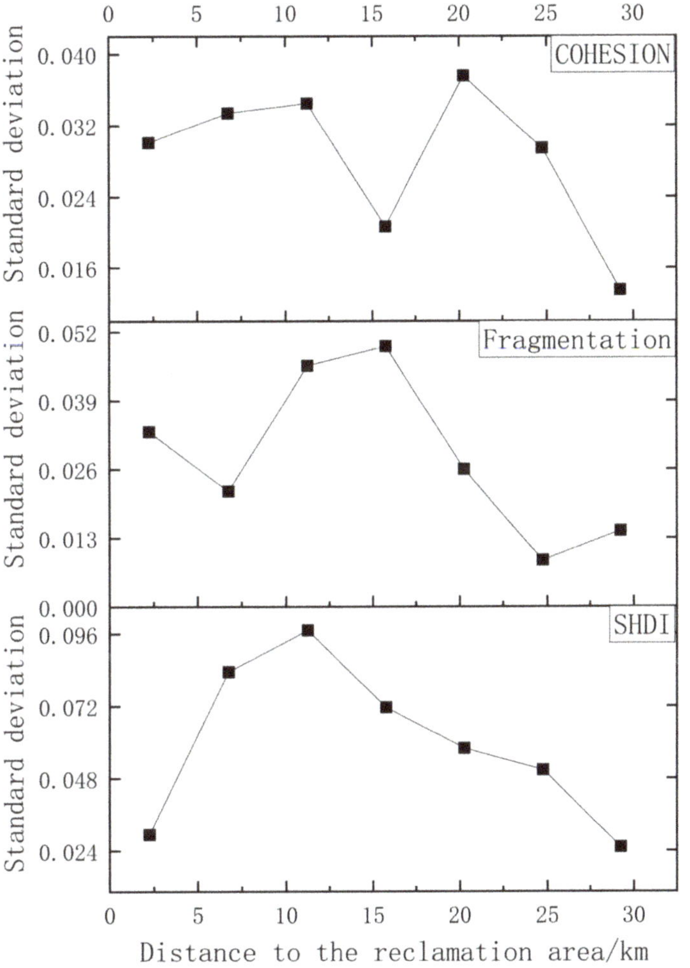

Figure 8. Standard deviation of the landscape pattern indices in different buffer zones.

4.3. Landscape Transformation and Stability

In the past 38 years, the landscape structure in the Jiangsu coastal area has undergone significant changes due to the influence of human activities. Intense human activities, especially TFR, can not only directly change the landscape pattern of the reclamation area [57] but also have spillovers within a certain range. The spillovers can affect the surrounding landscape ecology through a variety of factors such as the development and construction of ports and tourist resort area, which can drive the surrounding economy and gradually change and reshape the surrounding landscape pattern.

In this study, we constructed a local stability index to characterize the stability of the coastal area in Jiangsu province and to study the impact of TFR on regional ecology. As can be seen from Figures 5 and 6, the stability of the reclamation and non-reclamation areas was very different. In the reclamation area, the stability index of the wetlands and natural wetlands showed a downward trend, but the stability of natural wetlands has remained almost unchanged in recent years. The stability index of constructed wetlands and non-wetlands has shown an upward trend. In the non-reclamation area, the stability index change characteristics were generally consistent with those of the entire affected area.

In the reclamation area (mainly distributed in the tidal flat area), a large number of natural wetlands were converted to artificial wetlands, such as mariculture farm and salt fields in the early stage (1983–2000) [43]. The result of this was that the proportion of natural wetlands and artificial wetlands gradually decreased and increased, respectively, and both tended to be close to the background structure. At the same time, although the urban land and construction land increased, the area proportion of non-wetlandss remained relatively stable, so the stability level of the reclamation area was determined by the stability level of natural wetlands and construction wetlands, which was manifested as a tendency to improve. In the later period (2000–2018), in addition to the conversion of natural wetlands into aquaculture land, they were mainly converted to non-wetlands such as dry land and construction land [43]. After 2008, although the natural wetlands were in a relatively stable state, the stability of non-wetlands gradually deteriorated, and the construction wetlands were in a very unstable state. After 2015, the government and Jiangsu province strengthened the protection of coastal wetlands, and the proportion of construction wetlands decreased, so the overall stability of wetlands showed no significant fluctuations or deviations.

In the non-reclamation area, paddy fields, dryland, construction land, urban land, and rural residential land were mainly distributed, of which the area proportion of non-wetlands was significantly greater than that in the background structure at any time. Therefore, in the non-reclamation area, the stability change was mainly affected by the area change in non-wetlands. In the early stage (1983–2000), the landscape transformation was mainly from arable land to construction land and residential land, and the area proportion of non-wetlands increased, but it was closer to the background structure, so the stability tended to be slightly better. In the later period (2005–2018), in addition to the early main landscape transformation, the mutual conversion between paddy fields and dryland increased at the junction of paddy fields and dryland near the reclamation area, resulting in obvious fluctuations in the proportion of non-wetlands, which were eventually manifested as obvious fluctuations in stability.

In addition, spatially, both the stability index and the volatility of the stability index in the non-reclamation area increased first and then decreased with the distance from the reclamation area. The main reasons for this were as follows: In the area closer to the reclamation area (buffer zones 1 and 2), there were mainly natural wetlands and artificial wetlands such as paddy fields and mariculture farms, and the landscape types were diverse. Compared with other more distant areas, the landscape structure composition of this area was closer to the tidal flat area with a higher stability level, so the stability index of this area was relatively smaller. In the past 38 years, the main transformation was the continuous transformation of large areas of natural wetlands to artificial wetlands or non-wetlands. As a result, the proportion of artificial wetlands and non-wetlands changed dramatically, which was manifested by significant fluctuations in the wetland stability index. In the slightly further areas (buffer zones 3 and 4), paddy field, dryland, urban land, and rural residential land were the main landscapes, but the non-wetland types were dominant, so the landscape structure deviated from the background structure more. Because the non-wetlands such as residential land and construction land were more susceptible to economic factors, the area of these landscape types (non-wetlands) has had a significant increase with the huge losses in paddy fields and dryland in the past 38 years, resulting in a dramatic change in the proportion of non-wetlandss and a significant increase in the volatility of the wetland stability index. In the more distant areas (buffer zones 5–7), although paddy fields and dryland were also the main landscapes, there were fewer construction land areas than in slightly further areas, so the contribution of non-wetlands to the stability of this region was weakened, and the wetland stability index was relatively small. In the past 38 years, the area of various landscape types has not changed significantly, and the stability index has not fluctuated much.

4.4. Implications for Protection and Management of Wetland Resources

Regarding the contradiction that the demand for land resources is large and resource protection is imminent, it is of great significance to deal with a series of ecological and environmental problems caused by the destruction of coastal wetlands such as reduced diversity, poorer water quality, and reduced carbon storage to achieve the sustainable development of tidal flat resources [11,58].

Although the average annual reclamation intensity accounted for a small proportion of the coastal area, the impact of TFR was continuous and cumulative, and a long period of large-scale TFR led to landscape degradation within the reclamation area as well as non-reclamation area in the Jiangsu coastal area [5]. The results of this study further showed that TFR had a significant impact on the stability of the non-reclamation area, the size of which was much larger than that of the reclamation area. Although the stability of the entire affected area had a slight tendency to improve, the stability of artificial and natural wetlands in the reclamation area was deteriorating, and the stability of non-wetlands in non-reclamation areas has also deteriorated in the past 10 years, which requires attention. In addition, different use types of TFR had different effects on wetland stability, and the same use type of TFR had different effects on different landscape types. All of the above indicate that future reclamation plans need to carefully consider the scale and use types of reclamation.

In addition, the concept of the compact city, whose basic principle is to form a better ecological environment for human living by increasing the density of development in a relatively compact area [59] can be borrowed and applied to TFR to solve the problem of unreasonable reclamation planning, by increasing the TFR in fewer areas and strengthening the efficiency of land use in tidal flat areas.

4.5. Limitations

This study has several major limitations that need to be addressed through future research.

First, we constructed 12 one-sided buffer zones with a width of 4.5 km outside the reclamation area to analyze the influence scope of TFR and the spatial variation of stability in this study. However, the spatial scale is an important factor that significantly influences the identification of the landscape structure and functional features. As the width of the buffer zone gradually increases or decreases, the results obtained may vary. Therefore, determining the optimal analytical scale is a problem that needs to be solved.

In addition, as the impact of TFR on regional landscape ecology is not immediate, the time when different use types of TFR have an impact on the ecological environment may also be different. When the interval time between adjacent target years is too short, it may not accurately reflect the impact of TFR in this period. In this study, although the time span was relatively long (between 3 and 6 years), it may have still omitted or overcalculated the impact of TFR in the target period to varying degrees.

Finally, when calculating the stability index in this study, we uniformly used the optimal background structure of the entire study area, but theoretically, each specific area should have its own optimal background structure. Therefore, it may be more appropriate to calculate the best background structure of the corresponding region when calculating the stability index of a certain area. For example, different buffer zones can have their own optimal background structure. However, when the optimal structure is obtained for the departmental areas, the entire study area will not necessarily achieve the best structure. Therefore, balancing local optimality with global optimality is a new problem that needs to be solved.

5. Conclusions

The rapid and large-scale TFR strongly changed the landscape pattern of the coastal area of Jiangsu province and also affected the structure and function of the landscape. The results showed that in the past 38 years, the main transformation features were the decrease in natural wetlands and the increase in artificial wetlands and non-wetlands. Among them, the reduction of natural wetlands was mainly caused by reclamation, which has reached 1969.86 km^2 in the past 38 years. Except for the huge conversion to aquaculture land in each period, the use types of the reclaimed natural wetlands have gradually changed from agricultural land to industrial land since 2008.

According to the spatial analysis of landscape transfer probability and landscape pattern indices, the impact range of TFR on regional landscape ecology was roughly 30 km. Within the influence scope, overall, the stability of the affected area had a tendency to improve, but the volatility has increased in the past 10 years. In the reclamation area, the overall stability of the reclamation area had a tendency to improve, but the stability of construction wetlands and non-wetlands deteriorated. In the non-reclamation area, the stability change characteristics were generally consistent with the entire affected area. In addition, spatially, both stability and its volatility had the characteristics of first deteriorating and then improving as the distance from the reclamation area increased. Through correlation analysis, the relationship between the cumulative area of TFR and the stability index was quantitatively analyzed. It was found that under the influence of TFR, the influence of different use types of TFR on stability was not completely consistent, and the influence of the same use type of TFR on different landscape types was also different.

In summary, based on multitime remote sensing data and combined with the spatial analysis of the landscape structure indices, the spatial scope of the ecological impact of TFR was determined. On this basis, the spatio-temporal characteristics of stability were analyzed through buffer analysis by constructing a local stability index, and the impact of TFR on regional stability was quantitatively analyzed through correlation analysis. However, this study also has some drawbacks, as described in the previous discussion. All the shortcomings should be explored in future studies to better understand the impact of TFR on regional ecology and to better support the future conservation and management of wetland resources in the coastal area of Jiangsu province, China.

Author Contributions: Conceptualization, G.L.; methodology, G.L., L.C. and Y.C.; software, Y.C.; formal analysis, G.L.; investigation, L.C., Y.C. and L.L.; data curation, L.L., P.M. and L.H.; writing—original draft preparation, Y.C.; writing—review and editing, G.L. and Y.C.; visualization, Y.C. and L.L.; funding acquisition, G.L. and L.C. All authors have read and agreed to the published version of the manuscript.

Funding: This research was funded by the National Key Technology Research and Development Program of the Ministry of Science and Technology of China under grant number 2018YFC0407502, China Postdoctoral Science Foundation under grant number 2019M660780, Science and Technology Project of Sichuan Province under grant number 2020YFS0441, and a Chengdu University of Information Technology funded project under grant number KYT201738.

Data Availability Statement: All satellite data used in the study are available for free download through https://earthexplorer.usgs.gov/ (accessed on 10 May 2018).

Acknowledgments: The authors would like to thank the editors and the anonymous reviewers for their crucial comments, which improved the quality of this paper.

Conflicts of Interest: The authors declare no conflict of interest.

References

1. Costanza, R.; Anderson, S.J.; Sutton, P.; Mulder, K.; Mulder, O.; Kubiszewski, I.; Wang, X.; Liu, X.; Pérez-Maqueo, O.; Martinez, M.L. The global value of coastal wetlands for storm protection. *Glob. Environ. Change* **2021**, *70*, 102328. [CrossRef]
2. Narayan, S.; Beck, M.W.; Wilson, P.; Thomas, C.J.; Guerrero, A.; Shepard, C.C.; Reguero, B.G.; Franco, G.; Ingram, J.C.; Trespalacios, D. The Value of Coastal Wetlands for Flood Damage Reduction in the Northeastern USA. *Sci. Rep.* **2017**, *7*, 9463. [CrossRef] [PubMed]

3. Newton, A.; Icely, J.; Cristina, S.; Perillo, G.M.E.; Turner, R.E.; Ashan, D.; Cragg, S.; Luo, Y.M.; Tu, C.; Li, Y.; et al. Anthropogenic, Direct Pressures on Coastal Wetlands. *Front. Ecol. Evol.* **2020**, *8*, 144. [CrossRef]
4. Murray, N.J.; Phinn, S.R.; DeWitt, M.; Ferrari, R.; Johnston, R.; Lyons, M.B.; Clinton, N.; Thau, D.; Fuller, R.A. The global distribution and trajectory of tidal flats. *Nature* **2019**, *565*, 222–225. [CrossRef] [PubMed]
5. Cui, L.L.; Li, G.S.; Chen, Y.H.; Li, L.J. Response of Landscape Evolution to Human Disturbances in the Coastal Wetlands in Northern Jiangsu Province, China. *Remote Sens.* **2021**, *13*, 2030. [CrossRef]
6. Gaglio, M.; Aschonitis, V.G.; Gissi, E.; Castaldelli, G.; Fano, E.A. Land use change effects on ecosystem services of river deltas and coastal wetlands: Case study in Volano-Mesola-Goro in Po river delta (Italy). *Wetl. Ecol. Manag.* **2017**, *25*, 67–86. [CrossRef]
7. Wang, L.; Coles, N.; Wu, C.F.; Wu, J.P. Effect of Long-Term Reclamation on Soil Properties on a Coastal Plain, Southeast China. *J. Coast. Res.* **2014**, *30*, 661–669. [CrossRef]
8. Wang, X.G.; Yan, F.Q.; Su, F.Z. Changes in coastline and coastal reclamation in the three most developed areas of China, 1980–2018. *Ocean Coast Manage.* **2021**, *204*, 105542. [CrossRef]
9. Wang, X.X.; Xiao, X.M.; Zou, Z.H.; Chen, B.Q.; Ma, J.; Dong, J.W.; Doughty, R.B.; Zhong, Q.Y.; Qin, Y.W.; Dai, S.Q.; et al. Tracking annual changes of coastal tidal flats in China during 1986–2016 through analyses of Landsat images with Google Earth Engine. *Remote Sens. Environ.* **2020**, *238*, 110987. [CrossRef] [PubMed]
10. Chen, L.; Ren, C.Y.; Zhang, B.; Li, L.; Wang, Z.M.; Song, K.S. Spatiotemporal Dynamics of Coastal Wetlands and Reclamation in the Yangtze Estuary During Past 50 Years (1960s–2015). *Chin. Geogr. Sci.* **2018**, *28*, 386–399. [CrossRef]
11. Wang, X.X.; Xiao, X.M.; Xu, X.; Zou, Z.H.; Chen, B.Q.; Qin, Y.W.; Zhang, X.; Dong, J.W.; Liu, D.Y.; Pan, L.H.; et al. Rebound in China's coastal wetlands following conservation and restoration. *Nat. Sustain.* **2021**, *4*, 1076–1083. [CrossRef]
12. Tian, B.; Wu, W.T.; Yang, Z.Q.; Zhou, Y.X. Drivers, trends, and potential impacts of long-term coastal reclamation in China from 1985 to 2010. *Estuar. Coast. Shelf Sci.* **2016**, *170*, 83–90. [CrossRef]
13. Ma, Z.; Melville, D.S.; Liu, J.; Chen, Y.; Yang, H.; Ren, W.; Zhang, Z.; Piersma, T.; Li, B. Rethinking China's new great wall. *Science* **2014**, *346*, 912–914. [CrossRef] [PubMed]
14. Meng, W.Q.; He, M.X.; Hu, B.B.; Mo, X.Q.; Li, H.Y.; Liu, B.Q.; Wang, Z.L. Status of wetlands in China: A review of extent, degradation, issues and recommendations for improvement. *Ocean Coast. Manage.* **2017**, *146*, 50–59. [CrossRef]
15. Chen, Y.; Dong, J.; Xiao, X.; Zhang, M.; Tian, B.; Zhou, Y.; Li, B.; Ma, Z. Land claim and loss of tidal flats in the Yangtze Estuary. *Sci. Rep.* **2016**, *6*, 24018. [CrossRef] [PubMed]
16. Yang, H.; Ma, M.; Thompson, J.R.; Flower, R.J. Protect coastal wetlands in China to save endangered migratory birds. *Proc. Natl. Acad. Sci. USA.* **2017**, *114*, E5491–E5492. [CrossRef] [PubMed]
17. Sun, X.; Li, Y.F.; Zhu, X.D.; Cao, K.; Feng, L. Integrative assessment and management implications on ecosystem services loss of coastal wetlands due to reclamation. *J. Clean Prod.* **2017**, *163*, S101–S112. [CrossRef]
18. Liu, Q.; Mou, X. Interactions between Surface Water and Groundwater: Key Processes in Ecological Restoration of Degraded Coastal Wetlands Caused by Reclamation. *Wetlands* **2016**, *36*, S95–S102. [CrossRef]
19. Zhang, H.; Wu, P.B.; Fan, M.M.; Zheng, S.Y.; Wu, J.T.; Yang, X.H.; Zhang, M.; Yin, A.J.; Gao, C. Dynamics and driving factors of the organic carbon fractions in agricultural land reclaimed from coastal wetlands in eastern China. *Ecol. Indic.* **2018**, *89*, 639–647. [CrossRef]
20. Wang, W.; Liu, H.; Li, Y.Q.; Su, J.L. Development and management of land reclamation in China. *Ocean Coast. Manage.* **2014**, *102*, 415–425. [CrossRef]
21. Choi, C.Y.; Jackson, M.V.; Gallo-Cajiao, E.; Murray, N.J.; Clemens, R.S.; Gan, X.J.; Fuller, R.A. Biodiversity and China's new Great Wall. *Divers. Distrib.* **2018**, *24*, 137–143. [CrossRef]
22. Lin, W.J.; Wu, J.; Lin, H.J. Contribution of unvegetated tidal flats to coastal carbon flux. *Glob. Change Biol.* **2020**, *26*, 3443–3454. [CrossRef] [PubMed]
23. Yan, X.L.; Liu, M.; Zhong, J.Q.; Guo, J.T.; Wu, W. How Human Activities Affect Heavy Metal Contamination of Soil and Sediment in a Long-Term Reclaimed Area of the Liaohe River Delta, North China. *Sustainability* **2018**, *10*, 338. [CrossRef]
24. Hua, J.; Feng, Y.; Jiang, Q.; Bao, X.; Yin, Y. Shift of bacterial community structure along different coastal reclamation histories in Jiangsu, Eastern China. *Sci. Rep.* **2017**, *7*, 10096. [CrossRef]
25. Jin, Y.W.; Yang, W.; Sun, T.; Yang, Z.F.; Li, M. Effects of seashore reclamation activities on the health of wetland ecosystems: A case study in the Yellow River Delta, China. *Ocean Coast. Manage.* **2016**, *123*, 44–52. [CrossRef]
26. Donohue, I.; Hillebrand, H.; Montoya, J.M.; Petchey, O.L.; Pimm, S.L.; Fowler, M.S.; Healy, K.; Jackson, A.L.; Lurgi, M.; McClean, D.; et al. Navigating the complexity of ecological stability. *Ecol. Lett.* **2016**, *19*, 1172–1185. [CrossRef]
27. Kéfi, S.; Domínguez-García, V.; Donohue, I.; Fontaine, C.; Thébault, E.; Dakos, V. Advancing our understanding of ecological stability. *Ecol. Lett.* **2019**, *22*, 1349–1356. [CrossRef] [PubMed]
28. Grimm, V.; Wissel, C. Babel, or the ecological stability discussions: An inventory and analysis of terminology and a guide for avoiding confusion. *Oecologia* **1997**, *109*, 323–334. [CrossRef] [PubMed]
29. Li, W.; Tan, R.; Wang, J.; Fan, D.; Yang, Y. Effects of anthropogenic disturbance on richness-dependent stability in Napahai plateau wetland. *Sci. Bull.* **2013**, *55*, 4120–4125. [CrossRef]
30. Li, H.F.; Li, L.F.; Su, F.L.; Wang, T.L.; Gao, P. Ecological stability evaluation of tidal flat in coastal estuary: A case study of Liaohe estuary wetland, China. *Ecol. Indic.* **2021**, *130*, 108032. [CrossRef]

31. Li, X.H.; Lei, S.G.; Liu, Y.; Chen, H.; Zhao, Y.B.; Gong, C.A.; Bian, Z.F.; Lu, X.G. Evaluation of Ecological Stability in Semi-Arid Open-Pit Coal Mining Area Based on Structure and Function Coupling during 2002–2017. *Remote Sens.* **2021**, *13*, 5040. [CrossRef]
32. Parparov, A.; Gal, G. Quantifying Ecological Stability: From Community to the Lake Ecosystem. *Ecosystems* **2017**, *20*, 1015–1028. [CrossRef]
33. Zhang, Z.S.; Xue, Z.S.; Lyu, X.G.; Tong, S.Z.; Jiang, M. Scaling of soil carbon, nitrogen, phosphorus and C:N:P ratio patterns in peatlands of China. *Chin. Geogr. Sci.* **2017**, *27*, 507–515. [CrossRef]
34. Mukherjee, K. Wetland habitat stability assessment in hydro-geomorphological (HGM) and surface water availability (SWA) conditions in a lower Gangetic floodplain region of Eastern India. *Ecol. Indic.* **2020**, *119*, 106842. [CrossRef]
35. Ren, P.; Hong, B.T.; Cheng, W.X.; Zhou, J.M. Stability evaluation of forest ecosystem and study of spatial differential features in the upper Yangtze River. *Geogr. Res.* **2013**, *32*, 1017–1024.
36. Zhou, Z.B.; Xu, X.W.; Lei, J.Q.; Li, S.G. Study on the ecological stability of tarim desert highway shelterbelt. *Sci. Bull.* **2006**, *51*, 126–312. [CrossRef]
37. Wu, J.G. Landscape Ecology—Concepts and Theories. *Chinese J. Ecol.* **2000**, *19*, 42–52.
38. Ma, T.T.; Li, X.W.; Bai, J.H.; Cui, B.S. Impacts of Coastal Reclamation on Natural Wetlands in Large River Deltas in China. *Chin. Geogr. Sci.* **2019**, *29*, 640–651. [CrossRef]
39. Yu, H.; Liu, X.M.; Kong, B.; Li, R.P.; Wang, G.X. Landscape ecology development supported by geospatial technologies: A review. *Ecol. Inform.* **2019**, *51*, 185–192. [CrossRef]
40. Xu, X.; Yang, T. Mathematical model study on overall impact of Sanmenwan Bay reclamation project. *J. Mar. Sci.* **2006**, *4*, 49–59.
41. Zhang, C.; Zheng, J.; Dong, X.; Cao, K.; Zhang, J. Morphodynamic response of Xiaomiaohong tidal channel to a coastal reclamation project in Jiangsu Coast, China. *J. Coast. Res.* **2013**, *65*, 630–635. [CrossRef]
42. Yu, J.; Bao, X.W.; Ding, Y.; Zhang, W.; Zhou, L.L. The impact of large-scale reclamation on hydro-dynamic environment—A case study of Xinghua Bay. *J. Ocean Univ. China* **2016**, *15*, 583–592. [CrossRef]
43. Zhou, Y.K.; Ning, L.X.; Bai, X.L. Spatial and temporal changes of human disturbances and their effects on landscape patterns in the Jiangsu coastal zone, China. *Ecol. Indic.* **2018**, *93*, 111–122. [CrossRef]
44. Bao, J.L.; Gao, S. Traditional coastal management practices and land use changes during the 16–20th centuries, Jiangsu Province, China. *Ocean Coast. Manage.* **2016**, *124*, 10–21. [CrossRef]
45. Li, L.J.; Li, G.S.; Cui, L.L.; He, L.; Chen, Y.H. Method for modelling ecological competition based on Pareto optimality: A case study of coastal wetlands in Jiangsu Province, China. *Ecol. Indic.* **2021**, *129*, 107946. [CrossRef]
46. Ma, T.T.; Li, X.W.; Bai, J.H.; Cui, B.S. Tracking three decades of land use and land cover transformation trajectories in China's large river deltas. *Land Degrad. Dev.* **2019**, *30*, 799–810. [CrossRef]
47. Liu, S.L.; Zhao, Q.H.; Wen, M.X.; Deng, L.; Dong, S.K.; Wang, C. Assessing the impact of hydroelectric project construction on the ecological integrity of the Nuozhadu Nature Reserve, southwest China. *Stoch. Environ. Res. Risk Assess.* **2013**, *27*, 1709–1718. [CrossRef]
48. Li, H.L.; Peng, J.; Liu, Y.X.; Hu, Y.N. Urbanization impact on landscape patterns in Beijing City, China: A spatial heterogeneity perspective. *Ecol. Indic.* **2017**, *82*, 50–60. [CrossRef]
49. Zhu, C.M.; Zhang, X.L.; Zhou, M.M.; He, S.; Gan, M.Y.; Yang, L.X.; Wang, K. Impacts of urbanization and landscape pattern on habitat quality using OLS and GWR models in Hangzhou, China. *Ecol. Indic.* **2020**, *117*, 106654. [CrossRef]
50. Kirwan, M.L.; Megonigal, J.P. Tidal wetland stability in the face of human impacts and sea-level rise. *Nature* **2013**, *504*, 53–60. [CrossRef]
51. Tobler, W.R. A Computer Movie Simulating Urban Growth in the Detroit Region. *Econ. Geogr.* **1970**, *46*, 234–240. [CrossRef]
52. Fan, Q.D.; Ding, S.Y. Landscape pattern changes at a county scale: A case study in Fengqiu, Henan Province, China from 1990 to 2013. *Catena* **2016**, *137*, 152–160. [CrossRef]
53. Di, X.H.; Hou, X.Y.; Wang, Y.D.; Wu, L. Spatial-temporal Characteristics of Land Use Intensity of Coastal Zone in China During 2000–2010. Chin. *Geogr. Sci.* **2015**, *25*, 51–61. [CrossRef]
54. Cavallero, L.; Lopez, D.R.; Raffaele, E.; Aizen, M.A. Structural-functional approach to identify post-disturbance recovery indicators in forests from northwestern Patagonia: A tool to prevent state transitions. *Ecol. Indic.* **2015**, *52*, 85–95. [CrossRef]
55. Wang, Y.; Cadotte, M.W.; Chen, Y.; Fraser, L.H.; Zhang, Y.; Huang, F.; Luo, S.; Shi, N.; Loreau, M. Global evidence of positive biodiversity effects on spatial ecosystem stability in natural grasslands. *Nat. Commun.* **2019**, *10*, 3207. [CrossRef]
56. Pennekamp, F.; Pontarp, M.; Tabi, A.; Altermatt, F.; Alther, R.; Choffat, Y.; Fronhofer, E.A.; Ganesanandamoorthy, P.; Garnier, A.; Griffiths, J.I.; et al. Biodiversity increases and decreases ecosystem stability. *Nature* **2018**, *563*, 109–112. [CrossRef]
57. Hu, T.; Fan, J.; Hou, H.; Li, Y.; Li, Y.; Huang, K. Long-term monitoring and evaluation of land development in a reclamation area under rapid urbanization: A case-study in Qiantang New District, China. *Land Degrad. Dev.* **2021**, *32*, 3259–3271. [CrossRef]
58. Lin, Q.Y.; Yu, S. Losses of natural coastal wetlands by land conversion and ecological degradation in the urbanizing Chinese coast. *Sci. Rep.* **2018**, *8*, 16046. [CrossRef]
59. Jim, C.Y. Sustainable urban greening strategies for compact cities in developing and developed economies. *Urban Ecosyst.* **2012**, *16*, 741–761. [CrossRef]

land

Article

Using Time-Series Remote Sensing Images in Monitoring the Spatial–Temporal Dynamics of LULC in the Msimbazi Basin, Tanzania

Herrieth Machiwa [1,2], Joseph Mango [1,3], Dhritiraj Sengupta [1] and Yunxuan Zhou [1,*]

1 State Key Laboratory of Estuarine and Coastal Research, East China Normal University, 500 Dongchuan Road, Shanghai 200241, China; machiwa.herrieth@udsm.ac.tz (H.M.); jsmmango@stu.ecnu.edu.cn (J.M.); dhritiraj@sklec.ecnu.edu.cn (D.S.)
2 Department of Computer Science and Engineering, College of Information and Communication Technologies, University of Dar es Salaam, P.O. Box 33335, Dar es Salaam 14112, Tanzania
3 Department of Transportation and Geotechnical Engineering, University of Dar es Salaam, P.O. Box 35131, Dar es Salaam 14113, Tanzania
* Correspondence: zhouyx@sklec.ecnu.edu.cn; Tel.: +86-138-1882-8655

Abstract: The basins containing rivers and wetlands are very significant to the surrounding dwellers in various ways, altogether aiming at boosting the economy for most developing countries. Unfortunately, the benefits are frequently overlooked and lead to basin mismanagement and degradation posed by increasing population. This study used population and satellite data to quantify the extent of land-use and land-cover changes along the Msimbazi valley between 1990 and 2019. Geographic information system and remote sensing techniques were used in the analysis and processing of remotely sensed images acquired in 1990, 2000, 2010 and 2019. The results reveal that the dominant area is built-up land that occupied 39.3% of the total in 1990 and gradually increased to 42.6% in 2000, 54.1% in 2010 and 65.5% in 2019. Moreover, forest and agriculture that in 1990 had been the second and third largest in size, respectively, had been decreasing throughout the entire period. The population increase had been threatening wetland vegetation during the initial 10 years (1990 to 2000); however, the wetland vegetation showed subsequent improvement after the implementation of some government initiatives. Other land cover, such as bush land and grassland, showed minority status with inconsistent changes in either increase or decrease. These findings imply that the Msimbazi Basin suffers much from uncoordinated human activities that consequently degrade its fertility. This degradation can be observed as well from the population distribution maps that show that a huge stress is being exerted along the riverine due to population growth and urbanization. The study also highlights that a lack of intensive management plans that are supported by clear legal commitments for optimal and sustainable resource utilization contributes to wetland deterioration.

Keywords: remote sensing; spatial–temporal changes; human activities; wetland

Citation: Machiwa, H.; Mango, J.; Sengupta, D.; Zhou, Y. Using Time-Series Remote Sensing Images in Monitoring the Spatial–Temporal Dynamics of LULC in the Msimbazi Basin, Tanzania. *Land* **2021**, *10*, 1139. https://doi.org/10.3390/land10111139

Academic Editor: Richard C. Smardon

Received: 23 September 2021
Accepted: 22 October 2021
Published: 26 October 2021

1. Introduction

Globally, river and wetland ecosystems provide significant benefits for humans. Accordingly, different conventions have been formulated to define their conservation for sustainable use. The Ramsar Convention of 1971, for example, defines wetlands as areas of marsh, fen, peat land or water, whether natural or artificial, permanent or temporary, with water that is static, flowing, fresh, brackish or salt, including areas of marine water, the depth of which at low tide does not exceed six meters [1,2]. These areas are recognized as the most important environmental components due to their substantial benefits biologically, ecologically and economically [3,4]. Some biological and ecological benefits derived from wetlands include providing habitats for specific flora and fauna, mitigating flooding, minimizing erosion, controlling pollution and regulating climate [5]. Furthermore, the wetlands support human livelihoods in numerous ways, such as by furnishing water for domestic use and land suitable for socio-economic functions such as agriculture, grazing, industries and settlements [6]. Generally, such benefits are also obtained from the ecosystems with rivers that in some cases connect with wetlands [7].

Despite the substantial ecological, economic and biological contributions of rivers and wetlands, their values have nonetheless been seriously stressed by anthropogenic disturbances [8,9]. Human activities, when carried out arbitrarily, exert pressure on the resource—hence, resulting in adverse impact on rivers and wetlands, such as land degradation, soil erosion, biodiversity loss, pollution and frequent floods [10]. Such impact has been reported in the Yellow River delta of Shandong Province in China [11], the Nakivubo and Lubigi wetlands in Uganda [5,6], and East Kolkata wetland in India [12]. Similar problems have also been happening in the Msimbazi Basin, which contains wetlands with mangroves and marshlands and the river flowing through many areas in Dar es Salaam, Tanzania [13–15]. Furthermore, due to industrial activities, the studies by [16–19] have detected the presence of a high level of heavy metal pollution in the soil, water and vegetables grown in this area. These adverse effects not only affect the humans but also contribute to the degradation of the basin's potential. Such lands need protection, and for this purpose, Tanzanian National Land Policy (1995) states, for example, that measures will be taken to prevent building and encourage development that is environmentally friendly and beneficial for the local community. Nonetheless, despite the presence of such directives, the studies that have been carried out to date show that many areas in this basin are already inhabited, and hence, the extent of development and trends regarding any additional land-use activity need to be monitored from time to time in order to determine the quality of the constructions and their ecological ramifications [9,20].

This research was designed to join the efforts of restoring the destroyed wetlands and to thus benefit from what their ecosystem provides [21–23]. Contrary to the existing studies that focused mainly on the consequences arising from intensive land utilization, this study aimed to determine spatiotemporal dynamics in land use and land cover along the Msimbazi valley between 1990 and 2019. Further, it aimed to realize whether the causes and effects of changes comply with other policies and legal frameworks governing river and wetland resource utilization. These objectives were achieved through the processing of the Landsat remote sensing imagery and population data acquired for four periods of the studied time from GloVis and the National Bureau of Statistics (NBS), respectively. Ultimately, this study is important for providing significant information to land-use planners, policy-makers, decision-makers and other environmental stakeholders for sustainable land management in the Msimbazi valley and all other areas containing river and wetlands ecosystems.

2. Materials and Methods

2.1. Description of the Study Area

Msimbazi Basin is part of the Wami-Ruvu Basin that contains rivers and wetlands towards the coast of the Indian Ocean in Tanzania (Figure 1). It is located in Dar es Salaam city, and its boundary coverage is estimated at 162 square kilometers. The main river in this basin is also called Msimbazi (approximately 35 km long), and it traverses through many areas in Dar es Salaam from the Kisarawe district in the Coast region to its discharge into the Indian ocean. Topographically, the highest altitude of this basin is about 308 m (from MSL) recorded at the Pugu Hills in Kisarawe, and geologically, it is characterized by quaternary and Neogene deposits. The wetlands area within the basin is very important as a biodiverse habitat for endemic species. According to the latest Population and Housing Census (PHC) carried out in 2012, the population at this area of study was 2.5 million, and most live in unplanned settlements with inadequate infrastructure services such as sanitation and solid-waste management. Their economic activities include commercial, industrial (e.g., textile, breweries and meat plants) and agricultural pursuits that supply Dar es Salaam city with most of its vegetables and fruits [24]. The climatic condition of this area is humid tropical and characterized by two seasons: dry and wet. The average annual rainfall ranges from 800 to 1400 mm, and the area has bimodal seasons, i.e., long rains (March–May) and short rains (October–December). The mean daily temperature varies between 18 °C and 33 °C. The mean annual evaporation rate is 2104 mm, and humidity lies between 67% and 96%.

2.2. Data Source

2.2.1. Population Data

Due to the nature of this study, the population in the study area was considered to be the main influence on land-cover changes. This association is given in the sense that an increase in population varies proportionally with human activities that result in variations in land use that ultimately make changes to the land cover. Therefore, based on the same focus for determining spatiotemporal changes of LULC, this study obtained population information from the national censuses conducted in 1988, 2002 and 2012 at NBS Tanzania. Available online: https://www.nbs.go.tz/index.php/en/census-surveys/population-and-housing-census (accessed on 18 January 2021). These population data were obtained along with the shape-files of the valley system and their administrative units. Due to the lack of the recent population census, the 2019 population data were obtained through projection of the 2012 population data by an arithmetical method [13]. In order to fulfill requirements of the formula, the existing population (Po) in 2012, growth rate (r = 5.6%) and the time interval for estimation (n = 7) were identified and used to compute the 2019 population (Pn) using the formula shown below:

$$P_n = P_o\left(1 + \frac{r}{100}\right)^n$$

2.2.2. LULC Data

In order to associate trends of the LULC, four remotely satellite datasets overing the study area were identified and downloaded from the Global Visualization Viewer (GloVis) at Glovis. Available online: https://glovis.usgs.gov/ (accessed on 15 March 2021). Landsat-5TM (Thematic Mapper) data for 1990, 2000 and 2010 and Landsat-8OLI (Operational Land Imager) data for 2019 were acquired to determine the extent of LULC spanning 29 years. All of these images have a spatial resolution of 30 m, with multispectral coverage from the visible to the middle infrared radiation fields of the electromagnetic spectrum. In ensuring that the used data were of not poor quality, images were acquired in the dry season (June to October). During this season, the chances of obtaining cloud-free data are higher due to the low cloud coverage and scant ground surface reflectance changes. Moreover, the use of Sentinel-1 data could be effective in this study of coastal wetland by providing cloud-free independent images. This is because they carry C-band synthetic aperture radar (SAR) that can provide images under all weather conditions, day or night. Unfortunately, since it was launched in April 2014, the data were available only for the 2019 time period.

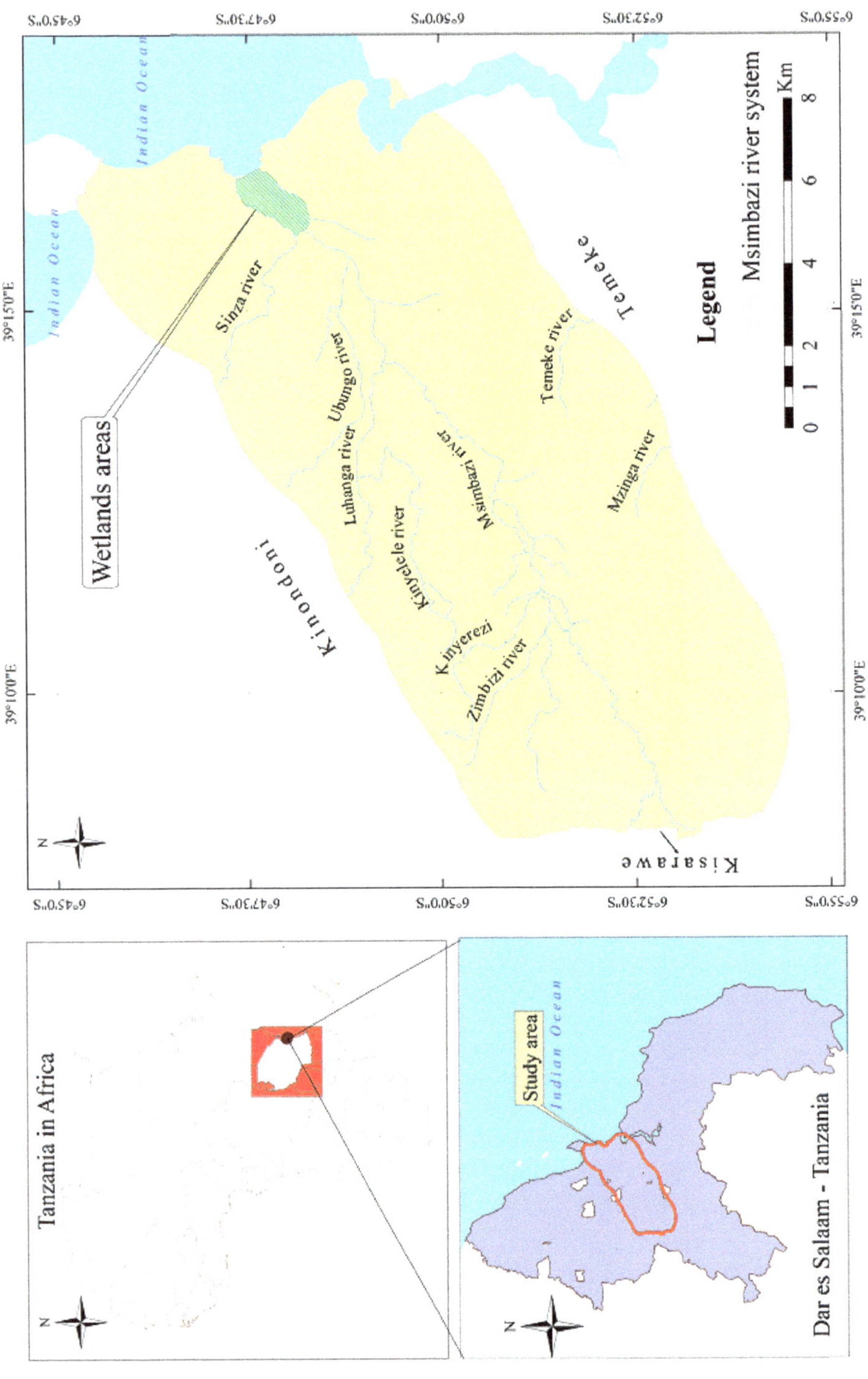

Figure 1. Map of Dar es Salaam showing location of the study area.

2.3. Image Processing and Classification

The obtained population and satellite data were processed using different approaches based on their status and the direction of the study. The population data from the National Bureau of Statistics (NBS) was not integrated in the shape-file of their administrative units; thus, its format hindered spatial association for easy interpretation via maps. To overcome the formatting challenge, ArcGIS 10.5 software (Environmental Systems Research Institute, West Redlands, CA, USA) was used to explore the data, unify names and update population values for subsequent mapping.

The satellite data for identifying land-use patterns was acquired by separate bands in the WGS84 coordinate system. Thus, in order to make them suitable for deriving intended results, preprocessing of images was done. In this phase, layer stacking, correction of the atmospheric effects and realignments of images were carried out. Layer stacking was performed to combine several single-band images to generate a single multi-layer image. The histogram equalization technique in ERDAS Imagine version 9.1 software was used to enhance the quality of each of the combined images for simplifying visual interpretability. Thereafter, images were corrected using the ENVI 5.1 FLAASH module to eliminated any atmospheric effects. All images were projected using the Universal Transverse Mercator (UTM) coordinate system of the WGS 1984, zone 37S datum so as to align them with other data of the study area.

The pre-processed image was then classified to generate the land-use land-cover status of the area. The image classification process was undertaken by maximum likelihood algorithm using a supervised classification technique. Generally, the classification involved selection and digitization of known pixels "training sites" defined by the user, which guided the software in categorizing all pixels into respective land-cover classes based on the spectral signatures. The classification process was performed using ArcGIS 10.5 software. The area was classified into seven land classes: agriculture, built-up land, forest, bushland, grassland, water and wetland vegetation. A description of these land-cover classes is presented in Table 1.

Table 1. Land-use land-cover classes used in the classification.

Land-Use Class	Description
Agriculture	Land used for agriculture, including paddy fields, irrigated and dry farmland, vegetation and fruit gardens, etc.
Grassland	Land covered by grasses mainly used for grazing.
Forest	Natural and secondary forest covered with trees, including woodlands and dense and open forests.
Bushland	Land that is dominated by bushes.
Wetland vegetation	Land consisting of shallow water bodies and wetland plants, such as mangroves, and salt marshes.
Water	Land covered by water bodies such as rivers, lakes and ponds.
Built-up land	Land that was modified by human activity, including residential, industrial, transportation and other infrastructures.

During land-cover classification, the annual change rates were calculated to show the rate of change per year within each specific 10-year (decadal) interval by using the formula below:

$$\frac{\left(\frac{LC_f - LC_i}{LC_i}\right) \times 100}{t_f - t_i}$$

where LC_f is the land coverage area for the final, LC_i is the land coverage area for the initial year, t_f is the final time period and t_i is the initial time period.

2.4. Accuracy Assessment for Land-Cover Classification

The next step after the classification process was to determine how the classified land-use land-cover classification aligned with reality. This objective was realized using a more accurate image of 2016 with 10 m spatial resolution from Sentinel-2 that was downloaded and used to extract and compare its pixel-based information with the classified imagery. In the first step of this post-classification process, 1000 accuracy assessment points were specified and randomly distributed to extract pixel values of the classified imagery. Next, the obtained results of the point shapefiles were updated with the ground-truth values from the Sentinel image. Last, the final files with complete pixel values were used to compute confusion matrixes, providing overall accuracy results, as presented in Section 3.1, below.

3. Results and Discussions

3.1. Accuracy Assessment Results

The accuracy results show that all imagery—i.e., for the years 1990, 2000, 2010 and 2019—were correctly classified, with kappa coefficients of 79, 91, 90 and 85%, respectively. According to the classification schemes by Anderson [25], these results, presented in detail in Tables 2–5, provide evidence of sufficient reliability for assessing the detected land-cover changes in the studied area. Furthermore, the other overall accuracy results—for example, for the year 2019—show similarity with the image of 2018 that was classified and assessed using the same approach as used in a study by Jamila et al. [7]. This matching implies that, in addition to the compliance with standards, the results of accuracy obtained in this study are also common in other research that involved the Msimbazi Basin. When observing all classes individually, the areas of agriculture and water, especially for the images of 2010 and 2019, showed lower accuracy results compared with others. In general, many reasons for the imprecision in the accuracy for some classes in LULC studies can be argued, including the occurrence of natural disasters such as floods due to seasonal variations [26]. However, in this case, the researchers view this cause as being insignificant because all images were taken during the dry season, i.e., June to October. Rather, the invasion and instability of land-use activities, particularly in such prime areas of water and agriculture, could be the main reason for frequent land cover changes, which when compared with static references provide significant mismatches of their reflectance.

Table 2. Accuracy assessment results of LULC classification for image of 1990.

LULC	Ground Truth Pixels									
	Wetland Vegetation	**Bushland**	**Forest**	**Grassland**	**Agriculture**	**Built-Up**	**Water**	**Total**	**U-Accuracy**	**Kappa**
Wetland vegetation	4	0	0	0	0	0	0	4	1	-
Bushland	7	19	0	0	0	11	0	35	0.513514	-
Forest	21	1	208	0	0	30	0	260	0.8	-
Grassland	0	0	0	30	0	6	0	36	0.833333	-
Agriculture	4	1	2	3	8	11	0	29	0.275862	-
Built-up	5	1	0	0	2	611	3	622	0.982315	-
Water	2	0	0	0	0	0	12	14	0	-
Total	43	22	210	33	10	669	15	1002	0	-
P-Accuracy	0.903023	0.863636	0.99	0.909091	0.8	0.9133	0.8	0	0.89022	-
Kappa	-	-	-	-	-	-	-	-	-	0.792278

Table 3. Accuracy assessment results of LULC classification for image of 2000.

LULC	Ground Truth Pixels									
	Wetland Vegetation	**Bushland**	**Forest**	**Grassland**	**Agriculture**	**Built-Up**	**Water**	**Total**	**U-Accuracy**	**Kappa**
Wetland vegetation	44	1	0	0	0	1	0	46	0.956522	-
Bushland	1	14	2	1	0	3	0	21	0.666667	-
Forest	1	1	218	1	0	8	0	229	0.951965	-
Grassland	1	0	2	41	0	9	1	54	0.759259	-
Agriculture	0	1	0	3	3	4	0	11	0.272727	-
Built-up	1	1	2	0	1	627	1	633	0.990521	-
Water	0	0	1	0	0	0	9	10	0.9	-
Total	48	18	225	46	4	652	11	1004	0	-
P-Accuracy	0.916667	0.777778	0.968889	0.891304	0.75	0.961656	0.818	0	0.952191	-
Kappa	-	-	-	-	-	-	-	-	-	0.910515

Table 4. Accuracy assessment results of LULC classification for image of 2010.

LULC	Ground Truth Pixels									
	Wetland Vegetation	**Bushland**	**Forest**	**Grassland**	**Agriculture**	**Built-Up**	**Water**	**Total**	**U-Accuracy**	**Kappa**
Wetland vegetation	39	0	0	0	0	1	1	41	0.95122	-
Bushland	0	21	1	3	0	3	0	28	0.75	-
Forest	2	0	216	2	0	8	1	229	0.943231	-
Grassland	1	1	0	24	0	3	0	29	0.827586	-
Agriculture	0	0	2	0	7	8	2	19	0.368421	-
Built-up	1	1	3	o	3	639	1	648	0.986111	-
Water	0	0	0	0	0	0	8	8	1	-
Total	43	23	222	29	10	663	12	1002	0	-
P-Accuracy	0.906977	0.913043	0.972973	0.827586	0.7	0.963801	0.583333	0	0.951098	-
Kappa	-	-	-	-	-	-	-	-	-	0.904584

Table 5. Accuracy assessment results of LULC classification for image of 2019.

LULC	Ground Truth Pixels									
	Wetland Vegetation	**Bushland**	**Forest**	**Grassland**	**Agriculture**	**Built-Up**	**Water**	**Total**	**U-Accuracy**	**Kappa**
Wetland vegetation	29	0	0	0	0	1	0	30	0.966667	-
Bushland	0	19	2	0	0	1	0	22	0.863636	-
Forest	3	0	190	0	0	4	1	198	0.959596	-
Grassland	0	2	6	32	0	20	0	60	0.533333	-
Agriculture	1	0	4	1	6	1	1	14	0.428571	-
Built-up	0	3	13	3	3	645	2	669	0.964126	-
Water	0	0	0	0	0	0	8	8	1	-
Total	33	24	215	36	9	672	12	1001	0	-
P-Accuracy	0.87978	0.791667	0.883721	0.888889	0.666667	0.959821	0.666667	0	0.928072	-
Kappa	-	-	-	-	-	-	-	-	-	0.857554

3.2. *Trend and Extent of Land-Use Land-Cover Change*

The LULC maps for 1990, 2000, 2010 and 2019 generated from the Landsat images are presented in Figure 2. The distribution and extent of changes are shown in Table 6 and Figure 3. From the table, it can be clearly observed that built-up land has dominated and increased tremendously from 6348 ha (39.3%) in 1990 to 10,612ha (65.6%) in 2019. Further analysis revealed that in 1990, the land covered by built-up land and forest comprised 66.2% of all land use, whereas in the most recent period (2019), 66.5% of land cover came from built-up land alone. This implies that the increase in built-up land has generally been to satisfy the growing population's demands at the expense of the forests [27]. The majority of people depend on the forest's resources—i.e., firewood, charcoal, timber and medicinal herbs—for sustaining their livelihood. These findings also comport with land-cover change detection along the Tanzanian coast researched by Wang [28].

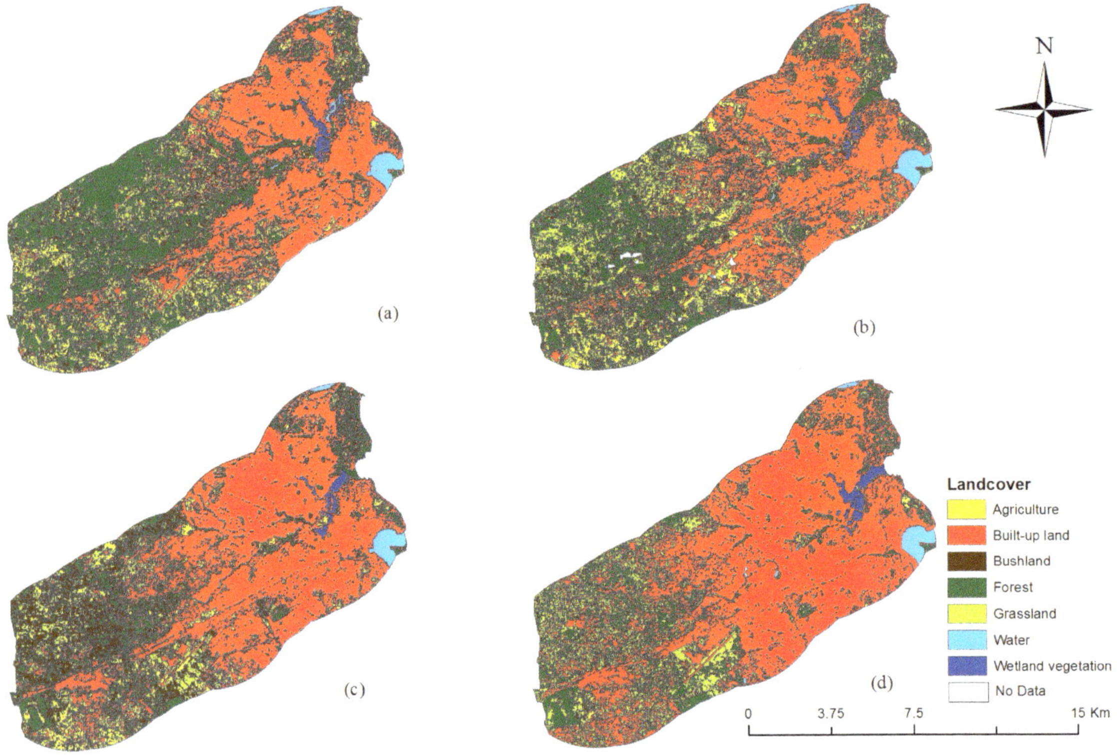

Figure 2. Land-use land-cover maps for (**a**) 1990, (**b**) 2000, (**c**) 2010 and (**d**) 2019.

Table 6. Land-use land-cover status and change between 1990 and 2019.

Land-Use Types	Land-Cover								Annual Change Rate		
	Year: 1990		Year: 2000		Year: 2010		Year: 2019		1990–2000	2000–2010	2010–2019
	Ha	%	Ha	%	Ha	%	Ha	%	%	%	%
Agriculture	2450	15.2	2396	14.8	1430	8.8	509	3.1	−0.2	−4.0	−7.2
Built-up	6348	39.3	6887	42.6	8742	54.1	10,612	65.6	0.8	2.7	2.4
Bushland	1793	11.1	1222	7.6	2287	14.1	363	2.2	−3.2	8.7	−9.3
Forest	4344	26.9	3327	20.6	2483	15.4	1917	11.9	−2.3	−2.5	−2.5
Grassland	882	5.5	2019	12.5	920	5.7	2388	14.8	12.9	−5.4	17.7
Water	226	1.4	211	1.3	183	1.1	192	1.2	−0.7	−1.3	0.5
Wetland vegetation	123	0.8	104	0.6	123	0.8	184	1.1	−1.5	1.8	5.5
TOTAL	16,166	100.0	16,166	100.0	16,168	100.0	16,165	100.0			

Despite the decrease in forest land cover, the wetland vegetation still increased gradually from 104 ha in 2000 to 184 ha in 2019. This situation has shown similarity with the mangrove dynamics study conducted along the mainland coast by Wang [29]. It is possible that one of the reasons is due to its location being at the river mouth, which is prone to the effects of the sea. In this respect, the area is found waterlogged almost in all seasons of the year; thus, the area cannot provide a chance for the construction of dwellings. Apart from its geographic position, the government's initiatives under its national Integrated Coastal Management (ICM) strategies of 2002 have contributed by protecting mangroves (wetland vegetation) that were earlier stressed when used for domestic energy (e.g., firewood and charcoals) by inhabitants [29].

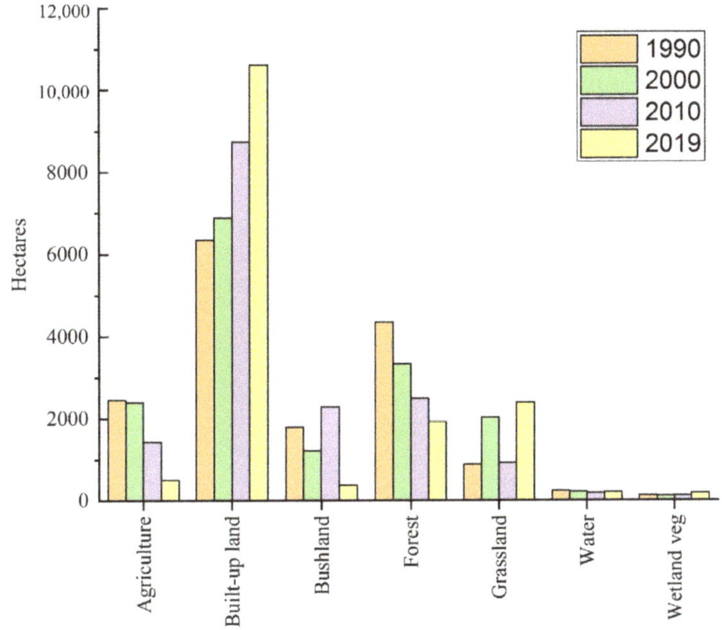

Figure 3. Chart showing the extent of land-use land-cover changes between 1990 and 2019.

Conversely, the areas occupied by agriculture were shown to decrease significantly, with the greatest change over the latter period from 2010 to 2019, in which a sharp decrease of 7% was realized (Figure 3). Agriculture activity by urban dwellers is getting scarce elsewhere in developing countries due to land scarcity arising from rapid population growth [30]. Likewise, the decrease in agricultural land use in our study is greatly associated with urbanization (Figures 2a–d and 3) where people access surrounding land for establishing settlements. In addition, agricultural land in the area was found to be polluted with heavy metals from the industrial discharges that ultimately degraded the soil fertility, thus making the land unfit for cultivation [15,31–33]. With time, the abandoned unfertile land transformed into bushland and grassland [34]. The area covered with grassland increased at a rate of 12.9% per year and 17.7% per year between the periods 1990 and 2000 and between 2010 and 2019, respectively. Possibly, this increase is linked to the slash and burn practices under shifting cultivation, increasingly cutting down trees and expanding urban areas [35,36].

3.3. *Urbanization and Its Implication on the Wetland*

Considering the analysis results of LULC, built-up land has drawn much attention in this study. In most peri-urban areas, the expansion of built-up land is merely attributed to increases in population. As such, the increasing population in conjunction with rapid urban expansion and industrial activities mostly contributes to threats to the natural environment and the entire ecosystem [37,38]. Environmental unsustainability occurs when the demand rate for resources becomes higher than the rate of resources provision required for eco-friendly economic growth. In Dar es Salaam, the population growth rate is one of the highest among other cities in sub-Saharan Africa [39]; it is a city in which most people are known to concentrate in the Kinondoni district, which the Msimbazi River traverses. This is reflected in Figure 4, which shows how the population of the Msimbazi area is increasing in parallel with the Dar es Salaam population. Census data from National Bureau of Statistics has shown that Msimbazi had a total population of 1.1 million in 1988, which was projected to be an estimated 2019 population of about 3.7 million (Table 7).

Generally, this tremendous increase in the population is a consequence of three factors: high birth and fertility rates, reclassification of rural land into urban areas and rural-to-urban migration [40–42]. The majority of people migrate into urban areas in search of a better livelihood, social services and economic opportunities [43,44].

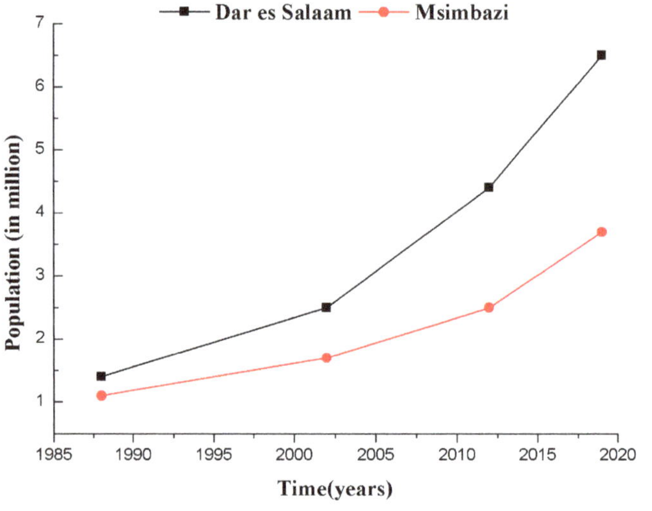

Figure 4. Population growth trend of Dar es Salaam and Msimbazi (1988–2019).

Table 7. Population census and growth rate of Msimbazi (1988–2019).

Population (in Millions)				Average Growth Rate		
Year: 1988	**Year: 2002**	**Year: 2012**	**Year: 2019**	**1988–2002**	**2002–2012**	**2012–2019**
1.1	1.7	2.5	3.7	3.0	3.8	5.7

Looking at this vast demographic growth on the one hand has positive effects on socio-economic development. On the other hand, the population growth seemed to account for the stresses on the riverine system, as observed in zones A, B and C in Figure 5. The reasons behind this trend could be that the Msimbazi area is characterized by informal settlements [45], and most of its dwellers use the benefits of water availability and soil fertility to engage in irrigation farming along the riverbanks [46,47]. Spatially, it is observed that the eastward area extends towards the city center, thus providing many business opportunities that attract people to migrate and live in the nearby surrounding area [48,49]. Certainly, the business opportunities have increased the demand for agricultural products (vegetables and fruits) in town markets, and as a result, over time, many people along the river have continued to cultivate and stress the valley system for economic gain [15,46]. In fact, stresses arising from continuing intense pressure exerted on the land are diverse depending on the nature of disturbances. For instance, unplanned spatial expansion and over-cultivation along the valley clogs the drainage systems, increases surface run-off and leads to frequent floods [50,51]. Furthermore, because of the potential escalation of environmental pollution (air, water and land), the agricultural and marine products of these resources will therefore be harmful for human consumption [14,31,52].

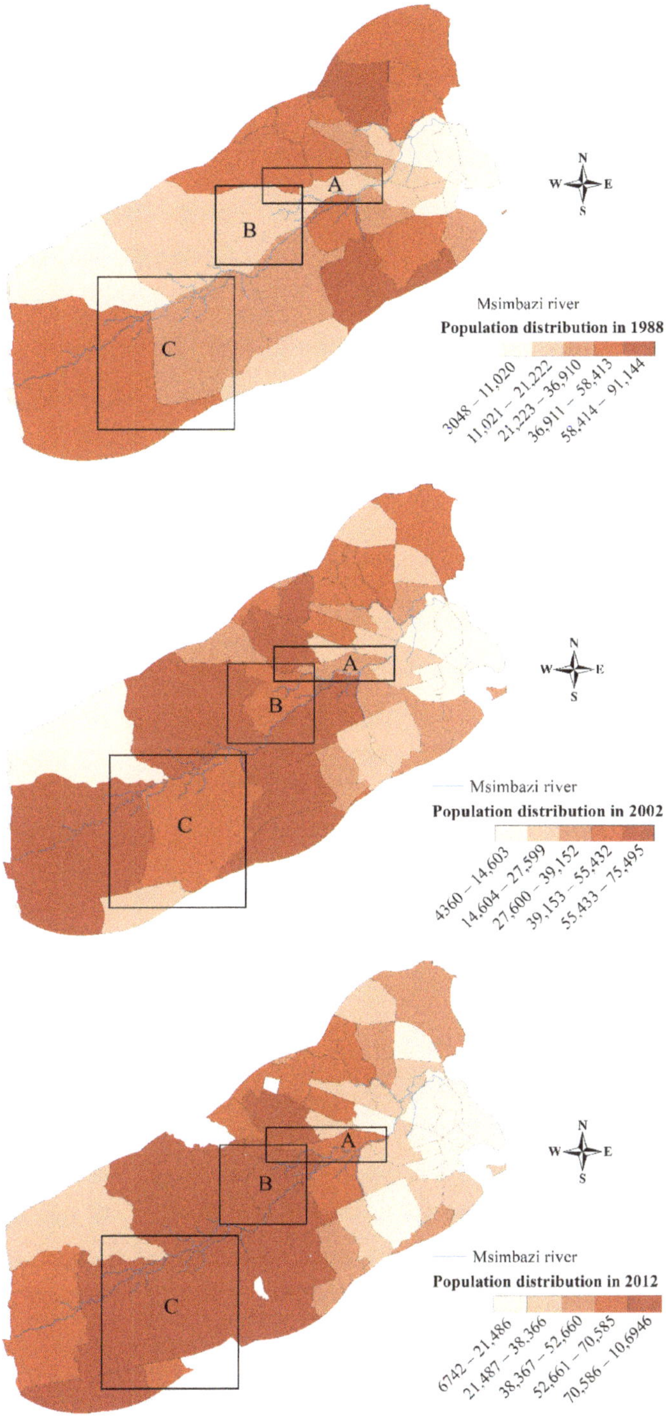

Figure 5. Population distribution maps by wards in 1988, 2002 and 2012.

3.4. Policy and Legal Insights into Wetlands Management

Despite the stresses caused by socio-economic gains, the study area lacks a comprehensive legal framework for governing sustainable resource utilization. This could be justified based on the fact that the management of wetlands in Tanzania depends on directives from multi-sectoral organs with stakes in the resources contained therein [47,53,54]. In this respect, other parts of the wetlands that are of little or no concern to those organs (sectors) are easily vulnerable to anthropogenic activities that degrade the wetlands through pollution, erosion and the loss of its biodiversity. Currently, the Wildlife Division (WD) under the Ministry of Natural Resources and Tourism is in charge of overseeing all wetlands in the country. However, its scope of management is limited only to the areas found within wildlife reserves [47,53], i.e., other wetlands in rural or urban areas (Msimbazi, in this case) receive almost no attention from the WD.

Notwithstanding the lack of specific policy on wetland management, the adopted policy since the year 2000 for the conservation of Ramsar sites seems to recognize few wetlands (about 5.5% of all wetlands) falling within the criteria stipulated by the convention [53]. This implies that even if the WD could extend its scope to include all Ramsar sites, a large percentage of the wetlands would still continue to rely on multi-sectoral directives that essentially focus on providing sustainable use of specific resources owned by the wetlands. The water sector for example uses its Water Policy of 2007 and Resource Management Act No.11 of 2009 to define wetlands as sources of water that need to be protected together with their aquatic biodiversity. Similar objectives are realized from the environmental sector that uses its Environment Policy of 1997 and Management Act No. 20 of 2004 to recognize wetlands as fragile ecosystems that play an important role in water systems. Generally, these two sectors focus on the management and conservation of wetlands as sources of water, and therefore they have limited scope when it comes to protecting the buffer zones of 60 or 120 m [45].

It is for this reason that the land sector within either urban or district councils appears to be the main concern for managing wetlands at-large and beyond the jurisdiction of the WD and other dedicated sectors such as Forestry and Fishery. The National Land Policy of 1995 and its subsequent Land and Village Acts of 1999 have set guidelines for sustainable land use in urban and rural areas. In line with this objective, the guideline clearly defined wetlands as the "wastelands" that are not useful for social and economic development. However, in other sections, the guideline seemed to present a contradiction by encouraging developments that benefit the public and the local community [54]. Hence, this study argues that to some extent, such loopholes weaken implementation of the law, and consequently offer opportunities for people to encroach upon and stress the wetlands, as explained earlier in the study.

4. Conclusions

In the findings of this research, various changes on the land-cover pattern in this area of study are demonstrated through the decrease and increase in certain classes of land cover. Built-up land has been dominant among others and has shown a steady increase at the expense of forest and agriculture. Conversion of agricultural land is likely the result of soil and water pollution due to rapid urbanization with unplanned settlements. Population growth and urban expansion are the main drivers of wetland degradation and other weather-related disasters, such that continuing exposure to drought, floods and pollution will most probably have an adverse impact on the health and livelihood of vulnerable communities. Moreover, findings also indicate that wetland restoration and sustainability have been a challenge for environmental officials due to the deficiencies in policy and law that specifically govern wetlands utilization. This calls for assorted approaches among decision makers and various stakeholders for strengthening sustainable utilization through establishing effective policy and legislation, as well as educating the local communities on how to wisely use wetlands resources. In general, these findings provide the key for sustaining the river and wetland systems in urban areas, even though more promising results could be achieved if all the datasets were available for the exact times of the studied period. There was a slight deviation in the time period for the population data, and the resulting lack of data led to a projection of values for the year 2019. Despite the gap, all data revealed realistic characterizations of the study area. Overall, this study recommends the preservation of wetland and the restriction of certain activities, such as cultivation and building along wetland peripheries, for the benefit of both the current and the future population.

Author Contributions: Conceptualization, H.M., Y.Z., J.M. and D.S.; investigation, H.M. and Y.Z.; software, H.M., D.S. and J.M.; validation, J.M., D.S. and H.M.; methodology, J.M. and H.M.; data curation, H.M., D.S. and J.M.; formal analysis, H.M., J.M. and Y.Z.; writing—original draft preparation, H.M. and J.M.; writing—review & editing, H.M. and Y.Z.; visualization, H.M., D.S. and J.M.; supervision, Y.Z.; funding acquisition, Y.Z.; All authors have read and agreed to the version of the manuscript.

Funding: This work was funded by the NSFC Project Number: 41476151 and "China-Africa Universities 20+20 Cooperation Plan" by the Ministry of Education of China.

Institutional Review Board Statement: Not applicable/study not involving humans or animals.

Informed Consent Statement: Not applicable/study not involving humans.

Data Availability Statement: All data used in this study are contained within the article.

Acknowledgments: We appreciate the support given by our colleagues during data collection, as well as their valuable suggestions that helped in improving the manuscript. The first author would like to thank the Chinese Scholarship Council for sponsoring her studies in China.

Conflicts of Interest: The authors declare no conflict of interest.

References

1. Hu, S.; Niu, Z.; Chen, Y.; Li, L.; Zhang, H. Global wetlands: Potential distribution, wetland loss, and status. *Sci. Total Environ.* **2017**, *586*, 319–327. [CrossRef]
2. World Resources Institute. *Millennium Ecosystem Assessment: Ecosystems and Human Well-Being: Wetlands and Water*; World Resources Institute: Washington, DC, USA, 2005.
3. Mwakaje, A. Wetlands, livelihoods and sustainability in Tanzania. *Afr. J. Ecol.* **2009**, *47*, 179–184. [CrossRef]
4. Uluocha, N.O.; Okeke, I.C. Implications of wetlands degradation for water resources management: Lessons from Nigeria. *GeoJournal* **2004**, *61*, 151–154. [CrossRef]
5. Isunju, J.B.; Kemp, J. Spatiotemporal analysis of encroachment on wetlands: A case of Nakivubo wetland in Kampala, Uganda. *Environ. Monit. Assess.* **2016**, *188*, 203. [CrossRef] [PubMed]
6. Omagor, J.G. Effect of human wetland encroachment on the degradation of Lubigi Wetland System, Kampala City Uganda. *Environ. Ecol. Res.* **2018**, *6*, 562–570.
7. Ngondo, J.; Mango, J.; Liu, R.; Nobert, J.; Dubi, A.; Cheng, H. Land-Use and Land-Cover (LULC) Change Detection and the Implications for Coastal Water Resource Management in the Wami–Ruvu Basin, Tanzania. *Sustainability* **2021**, *13*, 4092. [CrossRef]
8. Schuyt, K.D. Economic consequences of wetland degradation for local populations in Africa. *Ecol. Econ.* **2005**, *53*, 177–190. [CrossRef]
9. Liu, G.; Zhang, L.; Zhang, Q.; Musyimi, Z.; Jiang, Q. Spatio–temporal dynamics of wetland landscape patterns based on remote sensing in Yellow River Delta, China. *Wetlands* **2014**, *34*, 787–801. [CrossRef]
10. Kalisa, D.; Majule, A.; Lyimo, J.G. Role of wetlands resource utilization on community livelihoods: The case of Songwe River Basin, Tanzania. *Afr. J. Agric. Res.* **2013**, *8*, 6457–6467.
11. Wang, M.; Qi, S.; Zhang, X. Wetland loss and degradation in the Yellow River Delta, Shandong Province of China. *Environ. Earth Sci.* **2012**, *67*, 185–188. [CrossRef]
12. Mondal, B.; Dolui, G.; Pramanik, M.; Maity, S.; Biswas, S.S.; Pal, R. Urban expansion and wetland shrinkage estimation using a GIS-based model in the East Kolkata Wetland, India. *Ecol. Indic.* **2017**, *83*, 62–73. [CrossRef]
13. Bushesha, M.S.; Mbura, J.A. Identification of Reasons for and Socio-Economic Impacts of Persistent Floods in Dar Es Salaam. *World J. Soc. Sci. Res.* **2015**, *2*, 180. [CrossRef]
14. De Risi, R.; De Paola, F.; Turpie, J.; Kroeger, T. Life Cycle Cost and Return on Investment as complementary decision variables for urban flood risk management in developing countries. *Int. J. Disaster Risk Reduct.* **2018**, *28*, 88–106. [CrossRef]
15. John, R.; Magina, F.B.; Kemwita, E.F. From Msimbazi River Valley to Mabwepande Settlement: The Resettlement Process and Its Challenges. *Curr. Urban Stud.* **2019**, *7*, 399–426. [CrossRef]

16. Mwegoha, W.J.S.; Kihampa, C. Heavy metal contamination in agricultural soils and water in Dar es Salaam city, Tanzania. *Afr. J. Environ. Sci. Technol.* **2010**, *4*, 763–769.
17. Leonard, L.S.; Mwegoha, W.J.S.; Kihampa, C. Heavy metal pollution and urban agriculture in Msimbazi river valley: Health risk and public awareness. *Int. J. Plant Anim. Environ. Sci.* **2012**, *2*, 107–118.
18. Shimba, M.J.; Mkude, I.T.; Jonah, F.E. Impacts of waste on macroinvertebrate assemblages of Msimbazi River, Tanzania. *Int. J. Biodivers. Conserv.* **2018**, *10*, 106–116.
19. Sawe, S.F.; Shilla, D.A.; Machiwa, J.F. Assessment of enrichment, geo-accumulation and ecological risk of heavy metals in surface sediments of the Msimbazi mangrove ecosystem, coast of Dar es Salaam, Tanzania. *Chem. Ecol.* **2019**, *35*, 834–844. [CrossRef]
20. Wang, X.; Ning, L.; Yu, J.; Xiao, R.; Li, T. Changes of urban wetland landscape pattern and impacts of urbanization on wetland in Wuhan City. *Chin. Geogr. Sci.* **2008**, *18*, 47–53. [CrossRef]
21. Ehrenfeld, J.G. Evaluating wetlands within an urban context. *Urban Ecosyst.* **2000**, *4*, 69–85. [CrossRef]
22. Morin, T.; Bohrer, G.; Naor-Azrieli, L.; Mesi, S.; Kenny, W.; Mitsch, W.; Schäfer, K. The seasonal and diurnal dynamics of methane flux at a created urban wetland. *Ecol. Eng.* **2014**, *72*, 74–83. [CrossRef]
23. Kim, K.-G.; Lee, H.; Lee, N.-H. Wetland restoration to enhance biodiversity in urban areas: A comparative analysis. *Landsc. Ecol. Eng.* **2011**, *7*, 27–32. [CrossRef]
24. Sawe, S.F.; Shilla, D.A.; Machiwa, J.F. Lead (Pb) contamination trends in Msimbazi estuary reconstructed from 210Pb-dated sediment cores (Msimbazi River, Tanzania). *J. Environ. Forensics* **2021**, *22*, 99–107. [CrossRef]
25. Anderson, J.R.; Hardy, E.E.; Roach, J.T.; Witmer, R.E. *A Land Use and Land Cover Classification System for Use with Remote Sensor Data*; US Government Printing Office: Washington, DC, USA, 1976; 964p.
26. Najibi, N.; Devineni, N. Recent trends in the frequency and duration of global floods. *Earth Syst. Dyn.* **2018**, *9*, 757–783. [CrossRef]
27. Nzunda, N.; Munishi, P.; Soka, G.E.; Monjare, J.F. Influence of socio-economic factors on land use and vegetation cover changes in and around Kagoma forest reserve in Tanzania. *Ethiop. J. Environ. Stud. Manag.* **2013**, *6*, 480–488. [CrossRef]
28. Wang, Y.; Tobey, J.; Bonynge, G.; Nugranad, J.; Makota, V.; Ngusaru, A.; Traber, M. Involving geospatial information in the analysis of land-cover change along the Tanzania coast. *Coast. Manag.* **2005**, *33*, 87–99. [CrossRef]
29. Wang, Y.; Bonynge, G.; Nugranad, J.; Traber, M.; Ngusaru, A.; Tobey, J.; Hale, L.; Bowen, R.; Makota, V. Remote Sensing of Mangrove Change Along the Tanzania Coast. *Mar. Geod.* **2003**, *26*, 35–48. [CrossRef]
30. Forkuor, G.; Cofie, O. Dynamics of land-use and land-cover change in Freetown, Sierra Leone and its effects on urban and peri-urban agriculture—A remote sensing approach. *Int. J. Remote Sens.* **2011**, *32*, 1017–1037. [CrossRef]
31. Bahemuka, T.; Mubofu, E.B. Heavy metals in edible green vegetables grown along the sites of the Sinza and Msimbazi rivers in Dar es Salaam, Tanzania. *Food Chem.* **1999**, *66*, 63–66. [CrossRef]
32. Kihampa, C.; Mwegoha, W.J. Heavy metals accumulation in vegetables grown along the Msimbazi River in Dar es Salaam, Tanzania. *Int. J. Biol. Chem. Sci.* **2010**, *4*, 6. [CrossRef]
33. Chanzi, G. Heavy Metal Pollution Assessment along Msimbazi River, Tanzania. *J. Sci. Res. Rep.* **2017**, *17*, 1–8. [CrossRef]
34. Rautiainen, A.; Virtanen, T.; Kauppi, P.E. Land cover change on the Isthmus of Karelia 1939–2005: Agricultural abandonment and natural succession. *Environ. Sci. Policy* **2016**, *55*, 127–134. [CrossRef]
35. Kashaigili, J.; Majaliwa, A. Integrated assessment of land use and cover changes in the Malagarasi river catchment in Tanzania. *Phys. Chem. Earth Parts A/B/C* **2010**, *35*, 730–741. [CrossRef]
36. Hyandye, C.; Mandara, C.G.; Safari, J. GIS and logit regression model applications in land use/land cover change and distribution in Usangu catchment. *Am. J. Remote Sens.* **2015**, *3*, 6–16. [CrossRef]
37. Kebede, A.S.; Nicholls, R.J. Exposure and vulnerability to climate extremes: Population and asset exposure to coastal flooding in Dar es Salaam, Tanzania. *Reg. Environ. Chang.* **2012**, *12*, 81–94. [CrossRef]
38. Athukorala, D.; Estoque, R.C.; Murayama, Y.; Matsushita, B. Impacts of Urbanization on the Muthurajawela Marsh and Negombo Lagoon, Sri Lanka: Implications for Landscape Planning towards a Sustainable Urban Wetland Ecosystem. *Remote Sens.* **2021**, *13*, 316. [CrossRef]
39. Parsa, A.; Nakendo, F.; McCluskey, W.J.; Page, M.W. Impact of formalization of property rights in informal settlements: Evidence from Dar es Salaam city. *Land Use Policy* **2011**, *28*, 695–705. [CrossRef]
40. Kangalawe, R.Y.; Lyimo, J.G. Population dynamics, rural livelihoods and environmental degradation: Some experiences from Tanzania. *Environ. Dev. Sustain.* **2010**, *12*, 985–997. [CrossRef]
41. Hambati, H. Weathering the storm: Disaster risk and vulnerability assessment of informal settlements in Mwanza city, Tanzania. *Int. J. Environ. Stud.* **2013**, *70*, 919–939. [CrossRef]
42. Buhaug, H.; Urdal, H. An urbanization bomb? Population growth and social disorder in cities. *Glob. Environ. Chang.* **2013**, *23*, 1–10. [CrossRef]
43. Kombe, W.J. Land use dynamics in peri-urban areas and their implications on the urban growth and form: The case of Dar es Salaam, Tanzania. *Habitat Int.* **2005**, *29*, 113–135. [CrossRef]
44. Kayombo, M.C.; Mayo, A.W. Assessment of Microbial Quality of Vegetables Irrigated with Polluted Waters in Dar es Salaam City, Tanzania. *Environ. Ecol. Res.* **2018**, *6*, 229–239. [CrossRef]
45. Kironde, J.M.L. Governance deficits in dealing with the plight of dwellers of hazardous land: The case of the Msimbazi River Valley in Dar es Salaam, Tanzania. *Curr. Urban Stud.* **2016**, *4*, 303–328. [CrossRef]

46. Palela, E. The Impacts of Anthropogenic Factors on Urban Wetlands: The Case of Msimbazi Valley. Master's Thesis, University of Dar es Salaam, Dar es Salaam, Tanzania, 2000.
47. Collier, P.; Jones, P. Transforming dar es salaam into a city that work. *Tanzan. Path Prosper.* **2017**, *86*, 86–104.
48. Sauka, S. *Climate Resilience in Developing Cities: Msimbazi Basin, Dar es Salaam*; Report on Policy Insights; South African Institute of International Affairs: Johannesburg, South Africa, 2019.
49. Turpie, J.; Kroeger, T.; De Risi, R.; de Paola, F.; Letley, G.; Forsythe, K.; Day, L. *Return on Investment in Green Urban Development: Amelioration of Flood Risk in the Msimbazi River Catchment*; World Bank: Dar Es Salaam, Tanazania, 2017.
50. Kikwasi, G.; Mbuya, E. Vulnerability analysis of building structures to floods: The case of flooding informal settlements in Dar es salaam, Tanzania. *Int. J. Build. Pathol. Adapt.* **2019**, *37*, 2398–4708. [CrossRef]
51. Aselina, M.A. Levels of Industrial Pollutants and Their Effects on Water Resources and Livelihoods Along Msimbazi Sub catchment. Ph.D. Thesis, Kenyatta University, Dar es Salaam, Tanzania, September 2014.
52. Materu, S.F.; Urban, B.; Heise, S. A critical review of policies and legislation protecting Tanzanian wetlands. *Ecosyst. Health Sustain.* **2018**, *4*, 310–320. [CrossRef]
53. Mombo, F.; Speelman, S.; Hella, J.; Van Huylenbroeck, G. How characteristics of wetlands resource users and associated institutions influence the sustainable management of wetlands in Tanzania. *Land Use Policy* **2013**, *35*, 8–15. [CrossRef]
54. MLHUD. *National Land Policy*; The Ministry of Lands and Human Settlements: Dar es Salaam, Tanzania, 1995.

land

MDPI

Article

Millennial-Scale Carbon Storage in Natural Pine Forests of the North Carolina Lower Coastal Plain: Effects of Artificial Drainage in a Time of Rapid Sea Level Rise

Maricar Aguilos [1,*], Charlton Brown [1], Kevan Minick [1], Milan Fischer [2], Omoyemeh J. Ile [1], Deanna Hardesty [1], Maccoy Kerrigan [1], Asko Noormets [3] and John King [1]

1 Department of Forestry and Environmental Resources, North Carolina State University, Raleigh, NC 27695, USA; cfbrown@ncsu.edu (C.B.); kjminick@ncsu.edu (K.M.); ojile@ncsu.edu (O.J.I.); dmhardes@ncsu.edu (D.H.); mdkerrig@ncsu.edu (M.K.); john_king@ncsu.edu (J.K.)
2 Global Change Research Institute of the Czech Academy of Sciences, 603 00 Brno, Czech Republic; fischer.m@czechglobe.cz
3 Department of Ecology and Conservation Biology, Texas A&M University, College Station, TX 77843, USA; noormets@tamu.edu
* Correspondence: mmaguilo@ncsu.edu

Citation: Aguilos, M.; Brown, C.; Minick, K.; Fischer, M.; Ile, O.J.; Hardesty, D.; Kerrigan, M.; Noormets, A.; King, J. Millennial-Scale Carbon Storage in Natural Pine Forests of the North Carolina Lower Coastal Plain: Effects of Artificial Drainage in a Time of Rapid Sea Level Rise. *Land* **2021**, *10*, 1294. https://doi.org/10.3390/land10121294

Academic Editor: Richard C. Smardon

Received: 8 November 2021
Accepted: 24 November 2021
Published: 25 November 2021

Publisher's Note: MDPI stays neutral with regard to jurisdictional claims in published maps and institutional affiliations.

Abstract: Coastal forested wetlands provide important ecosystem services along the southeastern region of the United States, but are threatened by anthropogenic and natural disturbances. Here, we examined the species composition, mortality, aboveground biomass, and carbon content of vegetation and soils in natural pine forests of the lower coastal plain in eastern North Carolina, USA. We compared a forest clearly in decline (termed "ghost forest") adjacent to a roadside canal that had been installed as drainage for a road next to an adjacent forest subject to "natural" hydrology, unaltered by human modification (termed "healthy forest"). We also assessed how soil organic carbon (SOC) accumulation changed over time using ^{14}C radiocarbon dating of wood sampled at different depths within the peat profile. Our results showed that the ghost forest had a higher tree density at 687 trees ha^{-1}, and was dominated by swamp bays (*Persea palustric*), compared to the healthy forest, which had 265 trees ha^{-1} dominated by pond pine (*Pinus serotina* Michx). Overstory tree mortality of the ghost forest was nearly ten times greater than the healthy forest ($p < 0.05$), which actually contributed to higher total aboveground biomass (55.9 ± 12.6 Mg C ha^{-1} vs. 27.9 ± 8.7 Mg ha^{-1} in healthy forest), as the dead standing tree biomass (snags) added to that of an encroaching woody shrub layer during ecosystem transition. Therefore, the total aboveground C content of the ghost forest, 33.98 ± 14.8 Mg C ha^{-1}, was higher than the healthy forest, 24.7 ± 5.2 Mg C ha^{-1} ($p < 0.05$). The total SOC stock down to a 2.3 m depth in the ghost forest was 824.1 ± 46.2 Mg C ha^{-1}, while that of the healthy forest was 749.0 ± 170.5 Mg C ha^{-1} ($p > 0.05$). Carbon dating of organic sediments indicated that, as the sample age approaches modern times (surface layer year 2015), the organic soil accumulation rate (1.11 to 1.13 mm year^{-1}) is unable to keep pace with the estimated rate of recent sea level rise (2.1 to 2.4 mm year^{-1}), suggesting a causative relationship with the ecosystem transition occurring at the site. Increasing hydrologic stress over recent decades appears to have been a major driver of ecosystem transition, that is, ghost forest formation and woody shrub encroachment, as indicated by the far higher overstory tree mortality adjacent to the drainage ditch, which allows the inland propagation of hydrologic/salinity forcing due to SLR and extreme storms. Our study documents C accumulation in a coastal wetland over the past two millennia, which is now threatened due to the recent increase in the rate of SLR exceeding the natural peat accumulation rate, causing an ecosystem transition with unknown consequences for the stored C; however, much of it will eventually be returned to the atmosphere. More studies are needed to determine the causes and consequences of coastal ecosystem transition to inform the modeling of future coastal wetland responses to environmental change and the estimation of regional terrestrial C stocks and flux.

Keywords: ghost forest; forested wetland; aboveground biomass; soil carbon; carbon dating

1. Introduction

Wetlands occupy 4–6% of Earth's land area and store C in the range of 202–535 Gt [1,2]. The importance of wetlands in providing habitats for diverse plants and animals, and a range of other ecosystem services, is well-recognized, but a sufficient understanding of the aboveground and belowground carbon cycling of wetlands is not well-understood [3,4]. Wetlands are a significant type of ecosystem of the southeastern USA coastal plain. However, natural and anthropogenic disturbances threaten the ecological integrity and C storage function of coastal wetland ecosystems.

Climate change threatens wetland ecosystems by exacerbating environmental stressors, which could have effects on the biodiversity, wetland productivity, and resilience to other stressors [5–8]. How climate change affects such wetland ecosystem functions has been given relatively little attention [9,10]. Climate change is expected to accelerate sea level rise (SLR) to between 0.4 and 1.2 m by 2100 [6], and alter the frequency and intensity of coastal storms [11]. Recently, it was recognized that the formation of "ghost forests", created by the submergence of low-lying land, is one of the most obvious indicators of climate change [12]. SLR also increases saltwater intrusion into freshwater ecosystems [13–15] and alters the distribution and quantity of carbon within coastal wetlands, causing species shifts and landward migration, contributing to the direct loss of wetland area [10]. Shifts in species composition in response to SLR may cause lower marsh species to replace upper marsh species as wetlands move landward and as the continental margin adjusts [16].

In addition to these climatic impacts, the draining of coastal wetlands across the southeastern USA persisted well into the 1970s and has only slowed in the past 30 years [12]. During that time, the US government decided to drain the wetlands to decrease the groundwater level for flood control and more viable crop and timber production, construction of access roads, and other land uses [17]. Eastern North Carolina was extensively ditched over many years, and one area formerly managed for timber is now the Alligator River National Wildlife Refuge (ARNWR) in Dare County, administered by the US Fish and Wildlife Service [18]. Typical lower coastal plain road construction consisted of digging a ditch connected to a major tributary, sound or other waterway to provide drainage, and using the resulting "borrow" material as the bed for the adjacent road. Such roads run throughout ARNWR and across the southeastern lower coastal plain [18].

The effect of ditching on groundwater hydrology and drainage for a given site varies depending on the location relative to the ditch. Visual observations, whether from the ground or remotely-sensed (Figure 1), indicate that in eastern NC, vegetation on the side of the road with a ditch is often dead or dying, while on the other side the vegetation is much healthier/abundant/productive. Ditches were implicated in allowing saltwater to move inland into ecosystems that are not salt-tolerant [12], yet disentangling the hydrologic effects from the salinity effects of coastal forest decline is still a developing science.

Wetland productivity is mediated by complex bio-geomorphic feedbacks, whereby plants accumulate organic matter and trap inorganic sediments to maintain their elevation in relation to local groundwater or adjacent water bodies [19,20]. The soil accretion capacity of coastal wetlands is a key factor in the overall resilience to environmental change [20–22]. Quantifying primary biomass production and soil organic carbon (SOC) storage in wetland forests is therefore critical for predicting the fate of coastal wetlands and the associated ecosystem services as the continental margin adjusts to SLR.

Wetlands contain 20–30% of the estimated 1500 Pg of global soil carbon [4] making them a significant global storehouse for C. The anoxic condition of wetland soils slows down the decomposition process, resulting in the accumulation of organic matter. Thus, wetlands can store a large amount of carbon, making them an important sink of atmospheric carbon. Wetland C stocks vary in response to factors, including salinity, regional climate, water chemistry, soil type, lateral and vertical variability in soil properties, and vegetation, which complicates quantification of process rates [23,24]. These complexities limit our understanding of the quantity and distribution of stored carbon in wetlands [2,4]. Therefore, a more accurate approach to wetland soil carbon accounting must be used to develop models relating the C dynamics to temporal trends in wetlands. Techniques such as radiocarbon dating markers to determine the geochronology of a series of plant macrofossils are useful tools to estimate carbon accumulation rates, and were used to integrate the ages of fossils from 2000–3000 years ago [25]. The radiocarbon dating of plant material found at depth in peat profiles near ARNWR was also used to recreate historical patterns of past vegetation dynamics/marsh accretion as a means of sea level reconstruction over the recent geologic past [19,25].

Figure 1. Time series of Google Earth-Landsat images and a ground-based picture in 2015 of the field site of the current study in the northern part of the Alligator River National Wildlife Refuge, Dare County, North Carolina, USA. In 1998 and 2008, virtually all of the satellite image area hosted healthy pond pine forest, whereas, by 2015, significant areas of ghost forest were present (grey color) and coastal marsh (beige color) was expanding. The bank of East Lake is visible in the lower left (dark color). Satellite images show a perspective of approximately 1163 m altitude. Field site elevation is approximately 30 cm AMSL. Photo credit: John King.

In this study, we quantified the vegetation structure and C storage of two adjacent natural wetland pine forests as affected by millennial-scale SLR and century-scale artificial drainage at ARNWR, in eastern North Carolina, USA. We estimated the biomass of vegetation at different levels of the overstory canopy and evaluated tree mortality and species composition. We also estimated the total carbon stock in vegetation and soils and how these changed as a function of distance from local water bodies/open water. Further, we assessed the rate of SOC accumulation using ^{14}C radiocarbon dating on wood samples collected at different depths within the peat profile down to the underlying Pleistocene mineral sedimentary deposits, spanning approximately 1800 years BP, and related findings to local sea level reconstructions. We hypothesized that the ghost forest, which has a visibly high mortality (e.g., Figure 1), would have lower total C stocks (in vegetation and soils) compared to adjacent healthy forests. We also hypothesized that ecosystem C storage would be positively correlated with the distance from open water, due to protection from salinity and/or hydrologic stress. Finally, we hypothesized that the recent decades-long visible increase in ghost forests would be associated with recent decadal-scale change in the rate of local SLR as revealed by the ^{14}C dating estimation of marsh accretion over the past two millennia.

2. Methods

2.1. Study Site

This study was conducted in the northern reaches of the Alligator River National Wildlife Refuge (ARNWR) in Dare County, North Carolina (Figure 2), which spans over 61,600 hectares [18]. The wetlands physiography of ARNWR is a product of post-Wisconsinian glaciation, as coastal rivers cut channels through the landscape at a time of lower sea level that consequently filled with marine sediments overlain by accumulating organic peat [26,27]. The soil at this site is classified as a Belhaven series histosol (loamy, mixed, dystic, thermic Terric Haplosaprists) with a highly decomposed organic matter layer underlain by loamy marine sediments [28]. This part of the refuge, adjacent to the water bodies of Albemarle Sound and East Lake, was historically dominated by extensive pond pine forest due to this species' tolerance of hydric soil conditions and dependence on fire for regeneration [18,29]. An abandoned fire road with adjacent drainage ditch, built in 1965 (Ed Sawyer Road) (Figures 1 and 2), provided an ideal location to observe the effects of altered hydrology and the distance from surrounding water bodies on pond pine forest carbon storage and ecosystem transition. Common overstory species at the study site include swamp bay (*Persea palustris*), pond pine (*Pinus serotina*), red maple (*Acer rubrum*) and sweetgum (*Liquidambar styraciflua*), in addition to numerous understory shrubs and herbaceous plants. The elevation is approximately 30 cm above mean sea level, average annual precipitation is 1163 ± 49 mm (1981–2018) and average annual temperature is 15.7 °C ± 0.4 °C (2009–2018).

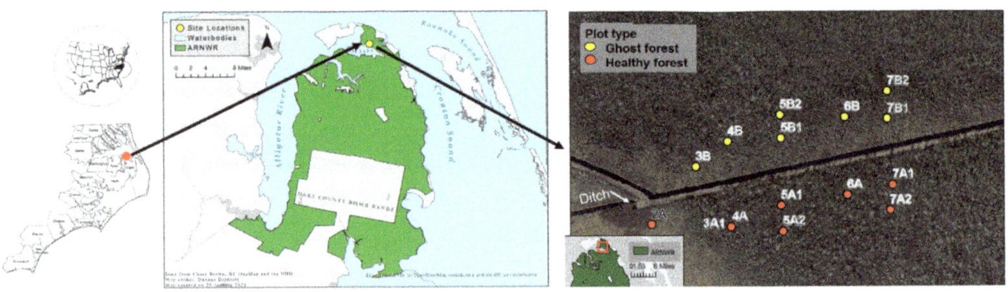

Figure 2. Location of the study site in the northern part of Alligator River National Wildlife Refuge in eastern North Carolina, USA. On the right panel, red circles denoted by their corresponding plot codes represent sampling plots of the healthy pond pine forest and yellow circles are ghost forest sampling locations. Care was taken to sample vegetation at least 100 m away from the ditch and roadside to minimize edge effects. East Lake/Alligator River is to the west of the study site, while the Albemarle/Croatan Sound is to the east.

2.2. Overstory Tree Biomass and C Content

Vegetation sampling occurred throughout 2014 and 2015. Paired, circular (15 m radius) vegetation sampling plots were installed in healthy forest and ghost forest along both sides of the fire road, starting near the bank of East Lake and progressing inland (Figure 2). Overstory tree species were identified and live/dead status was determined. All tree heights (living and dead) were measured with a Nikon laser hypsometer, and diameters were measured at breast height (1.3 m above ground level) with a D-tape. Dead trees were assigned to a decay class using USDA Forest Service Forest Inventory and Analysis (FIA) protocols [30] based on visible characteristics (Table 1). Overstory standing tree biomass was estimated using species-specific allometric relationships [31] for all overstory species (Table 1). Allometric regressions for estimating biomass of pond pine do not exist in the literature, representing a pressing research need, so equations were used for the taxonomically closely-related pitch pine (*Pinus resinosus* L.) [32]. Live tree C content was calculated as 50% of biomass [32], and that of dead trees as the calculated live C content adjusted by the % remaining by decay class.

Table 1. Allometric equations and decay classes used to estimate live biomass and C content of overstory tree species at Alligator River National Wildlife Refuge in eastern, North Carolina, USA [30,32].

Live Biomass Regressions		
Species	**Allometric Equation**	**Notes**
Pond pine (*P. serotina*)	y = exp(2.0171 + 2.3373 × ln(dbh))	Equation for *P. rigida*
Red maple (*A. rubrum*)	y = 2.52363 × $(dbh^2)^{1.9648}$	
Sweetgum (*L.sty raciflua*)	y = 1.822108 × $(dbh^2)^{1.2635}$	
Bay (*Persea* spp.)	y = −13.388 + 6.82 × (dbh^2)	
FIA decay class	Qualification	C remaining (%)
0	Alive	100
1	Dead, small branches attached, knife does not penetrate	97
2	Dead, no small branches, has large branches, little penetration	97
3	Dead, no large branches, height intact, more penetration	86
4	Dead, top broken, easy penetration, highly decayed	53

2.3. Mid-Story and Understory Biomass and C Content

Mid-story biomass was sampled using 5 m radius circular plots nested within each overstory plot. Diameter of woody plants up to 10 cm diameter was measured at 10 cm above the ground. On-site, species-specific, allometric equations were developed to estimate mid-story, woody plant biomass. Fifty mid-story plants ranging in size and species were destructively harvested, and heights, diameters and total fresh weight were measured directly. Wood samples were cut from each plant to determine fresh-to-dry weight ratios, applied to the green biomass estimates. For small mid-story plants, biomass was estimated using the exponential equation $y = 0.0944\exp^{(0.7102x)}$ ($R^2 = 0.89$), where x is the diameter-at-breast-height (dbh). For large mid-story plants, the linear equation $y = 1.3229x - 2.3518$ ($R^2 = 0.78$) was used. Understory, primarily herbaceous, plant biomass was estimated by destructively harvesting 1 × 1 m clip plots in the center of the mid-story plots, dried to constant mass at 65 °C, and weighed. Carbon content of all mid-story and understory plants was assumed to be 50% of live biomass.

2.4. Soil Sampling

Soil samples were taken with a McCauley auger at 10 cm increments to a depth of 1 m at one location in each overstory plot. Samples were sieved (<2 mm) and dried to constant mass at 65 °C, and weighed. Bulk density was determined by dividing mass by volume for each depth increment. Each soil increment was analyzed separately for carbon concentration, content and stable C isotopes using a Picarro G2201-isotopic CO_2/CH_4 analyzer with combustion module at the Tree Physiology and Ecosystem Science Laboratory, North Carolina State University, USA. Percent soil carbon and bulk density were then used to estimate the total soil carbon content. The soil bulk density and carbon content were determined using the following formula:

$$\textit{Soil bulk density}\left(\mathrm{g\ cm^{-3}}\right) = \frac{\textit{Ovendry sample mass}\ (\mathrm{g})}{\textit{Sample volume}\ (\mathrm{cm^3})}$$

$$\textit{Soil carbon content}\left(\mathrm{g\ C\ cm^{-2}}\right) = \textit{bulk density}\left(\mathrm{g\ cm^{-3}}\right) \times \textit{soil depth}\ (\mathrm{cm}) \times \%\ \textit{organic C}$$

The total soil carbon pool was determined by summing the carbon mass of each of the sampled soil depths extrapolated into per hectare basis (Mg C ha^{-1}).

2.5. Belowground Wood Sampling

Wood fragments were recovered from soil samples at various depths within the 1 m peat profile sampled with the McCauley auger, and down to 2.3 m depth using a bucket auger. This was the maximum peat depth encountered at the site, furthest inland from the shoreline, below which lay Pleistocene mineral sedimentary deposits. The deepest wood sample collected was embedded in the mineral sediments, in what appeared to be a buried soil A-horizon, suggesting presence of an upland forest at some point in the past. We used the radiocarbon dating method to determine the age of the buried wood samples [33]. Sixteen of these samples were sent off for ^{14}C dating at the National Ocean Sciences Accelerator Mass Spectrometry Facility at the Woods Hole Oceanographic Institution.

2.6. Data Analysis

We used ANOVA and Tukey's HSD Test for comparison between paired plots, testing the level of significant differences at a 95% confidence level. Smoothed-curve fittings were carried out with locally weighted logarithmic regression in *ggplot2* package [34]. Contour plots were used to present a 2-dimensional surface by plotting the aboveground C and SOC as contours in x-axis (latitude) and y-axis (longitude). Contour plots were generated using *plotly* [35], *tidyverse* [36], and *reshape2* [37] packages. All analyses were processed in R version 4.1.1 [38].

3. Results

3.1. Overstory Species Composition, Tree Density, Mortality and Diameter

Overstory species composition, tree density, and mortality differed significantly between the ghost forest and the healthy forest (Table 2). The ghost forest had a tree density of 687 trees ha^{-1} (Table 2), characterized by 56% swamp bay, 38% pond pine, and the remaining 5% was red maple. Sixty percent of the overstory (or 412 trees ha^{-1}) in the ghost forest was standing dead or dying, half of which were swamp bay, 43% were pond pine, and the remainder were red maple. The most prominent decay class of swamp bay and pond pine in the ghost forest was class 2, characterized as dead with only large branches left. In contrast, the healthy forest had a tree density of 265 trees ha^{-1} (Table 2), with 72% pond pine, 24% red maple and 4% sweetgum. The amount of standing dead trees in the healthy forest was only 17% (42 trees ha^{-1}). Most dead trees were pond pine of decay class 1 (dead, small branches attached). Interestingly, there were no swamp bay trees in the healthy forest, a species that dominated the overstory of the ghost forest (e.g., species replacement during ecosystem transition). While the stand density of the ghost forest was over 2.5 times higher than in the healthy forest, mortality was nearly ten times greater (Table 2). The ghost forest was also characterized by having a significantly smaller diameter and height of trees than the healthy forest ($p < 0.05$; Figure 3).

Table 2. Overstory species composition, tree density, and decay class of dead trees in healthy forest and ghost forest in the northern part of the Alligator River National Wildlife Refuge in eastern North Carolina. Different letters after a total value denote significant differences at $p < 0.05$ between ghost and healthy forest.

Location	General Species Composition	No. of Trees Per ha^{-1}			* Decay Class of Standing Dead Trees
		Healthy/No Decay	Standing Dead/Dying	Total (Healthy + Standing Dead)	
Ghost forest	Swamp bay	176	210	388	2
	Maple	12	24	36	1
	Pitch Pine	85	178	263	2
	Total	273 ^{a}	412 ^{a}	687 ^{a}	

Table 2. *Cont.*

Location	General Species Composition	No. of Trees Per ha^{-1}			* Decay Class of Standing Dead Trees
		Healthy/No Decay	Standing Dead/Dying	Total (Healthy + Standing Dead)	
Healthy forest	Maple	63	2	65	1
	Pitch Pine	148	42	190	1
	Sweetgum	10	0	10	
	Total	220 ^{b}	44 ^{b}	265 ^{b}	

* Decay class 0—alive; 1—dead but small branches attached; 2—dead with only large branches left, 3—large branches gone; 4—top broken off, very decayed.

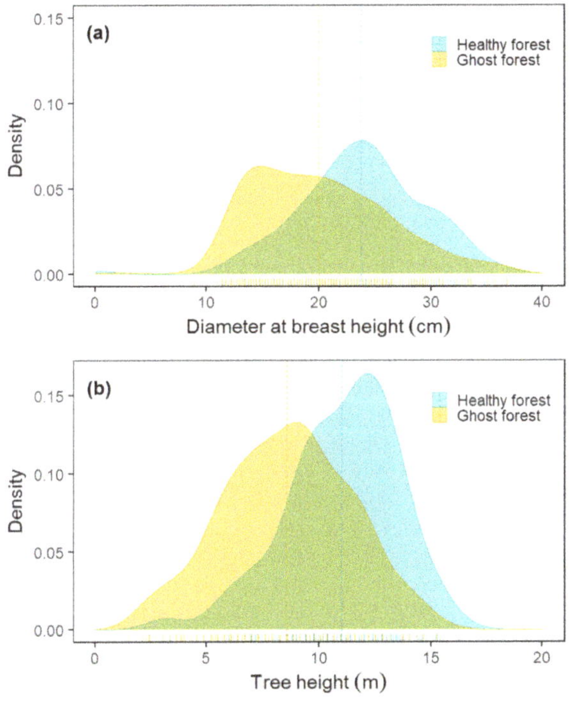

Figure 3. Diameter (**a**) and height (**b**) distributions of all species in ghost forest and healthy forest in the northern part of the Alligator River National Wildlife Refuge in eastern North Carolina. Dashed lines represent the means of each population sample.

3.2. Overstory, Mid-Story and Understory Biomass

Ghost forest and healthy forest differed significantly in total overstory and midstory biomass, but not in understory biomass (Table 3). On a per tree basis, the healthy forest had a higher average overstory tree biomass (0.11 Mg ha^{-1}), which is 18% higher than in the ghost forest (0.09 Mg ha^{-1}) ($p < 0.05$). However, given the higher tree density in the ghost forest (Table 1), the overall total overstory biomass in the ghost forest was 55.97 $\pm$ 12.61 Mg ha^{-1} compared to 27.99 $\pm$ 8.75 Mg ha^{-1} in the healthy forest (Table 3). Almost 44% of the total overstory biomass in the ghost forest was in standing dead trees, however, with a total dead biomass of 25.06 $\pm$ 6.26 Mg ha^{-1}. In contrast, of the 27.99 Mg ha^{-1} total overstory biomass in the healthy forest, a total of 88% was in live trees (24.81 $\pm$ 7.25 Mg ha^{-1}). The remaining 11% was in standing dead trees, primarily pine (Table 3). Interestingly, the mid-story biomass (3.65 $\pm$ 2.70 Mg ha^{-1}) of the ghost forest was only 25% of the amount in the healthy forest (14.06 ± 7.30 Mg ha^{-1}), possibly due to the dense shrub canopy hindering the development of a mid-story in the ghost forest. Finally, understory biomass in the ghost forest averaged 8.3 Mg ha^{-1} compared to 7.4 Mg ha^{-1} in the healthy forest, but the difference was not significant due to the relatively large variance.

Table 3. Summary of biomass in each vegetation strata by species in healthy forest and ghost forest in the northern part of the Alligator National Wildlife Refuge in eastern North Carolina. Different letters after a biomass value denote significant differences at $p < 0.05$ between healthy forest and ghost forest.

	Species	Overstory Biomass (Mg ha^{-1})			Mid-Story Biomass (Mg ha^{-1})	Under-Story Biomass (Mg ha^{-1})
		Alive	Standing Dead	Total BIOMASS (Alive + Dead)		
Ghost forest	Swamp bay	16.35	10.08	26.43		
	Maple	2.69	1.42	4.11		
	Pine	11.87	13.57	25.43		
	Total	*30.91 ^{a} ± 6.96*	*25.06 ^{a} ± 6.26*	*55.97 ^{a} ± 12.61*	*3.65 ^{a} ± 2.70*	*8.32 ^{a} ± 4.1*
Healthy forest	Maple	8.45	0.23	8.68		
	Pine	15.42	2.96	18.38		
	Sweetgum	0.93	0.00	0.93		
	Total	*24.81 ^{b} ± 7.25*	*3.19 ^{b} ± 1.65*	*27.99 ^{b} ± 8.75*	*14.06 ^{b} ± 7.30*	*7.41 ^{a} ± 1.28*

The ecosystem transition from the healthy forest to the ghost forest resulted in a major shift in overstory species composition and forest structure, from dominance/co-dominance by pond pine/red maple to that of woody bay shrubs/(declining) pond pine, and resulting shifts in biomass distribution (Table 3). In the healthy forest, the total live overstory biomass averaged 24.8 Mg ha^{-1}, with pond pine and red maple accounting for 62% and 34%, respectively. In ghost forest, the total live overstory biomass of 30.9 Mg ha^{-1} was comprised of 53% bay shrubs and 38% pond pine, although the pond pine was in rapid decline (Figure 1). With this transition, the structure of the forest changed from median diameter/height of 24 cm/11.5 m in the healthy forest to 20 cm/8.5 m in the ghost forest (Figure 3), in addition to the increased density. Finally, the proportion of dead overstory biomass to total overstory biomass shifted from 0.16 in the healthy forest to 0.60 in the ghost forest, signifying impending large losses of ecosystem C as the dead material decomposed in the coming years.

3.3. Ecosystem Carbon Stocks in Vegetation and Soils

As with biomass, the total ecosystem C stocks differed significantly between the healthy forest and ghost forest, as did the distribution between different compartments (Figure 4). The total aboveground C content of ghost forest, 33.9 Mg C ha^{-1}, was 9.2 Mg C ha^{-1} greater than that of the healthy forest at 24.7 Mg C ha^{-1}. The C content of dead trees in the ghost forest contributed to almost 44% of the total aboveground C, while that contribution in the healthy forest was only 11%. The percentage contribution of overstory live tree C to the total aboveground C was similar in both forests, at 45–50%, but the amount was double in the ghost forest at 27.9 Mg C ha^{-1}, compared to healthy forest at 13.9 Mg C ha^{-1}. The percentage contribution of the mid-story to total aboveground C differed greatly between the healthy forest (28%) and ghost forest (5%), as did the amounts, at 7.0 and 1.8 Mg C ha^{-1}, respectively (Figure 4). Finally, the understory C content of both forests was similar at 3.7 Mg C ha^{-1} in the healthy forest and 4.2 Mg C ha^{-1} in the ghost forest, contributing 15% and 12% to total aboveground C, respectively.

Figure 4. Estimates of aboveground (overstory, mid-story, and understory) vegetation and soil carbon stocks (0–100 cm depth) of healthy forest and ghost forest in the northern part of Alligator River National Wildlife Refuge in eastern North Carolina. Vegetation C stocks were calculated from biomass estimates, and thus have the same levels of statistical significance (Table 3).

3.4. Change in Overstory C Stocks with Distance from Open Water

The visible ecosystem transition occurring at our study site (Figures 1 and 2), was accompanied by a progression of increasing vegetation biomass C and soil organic C with a distance from open water. Averaged over all sampling plots, the total C in overstory trees averaged 13.6 Mg C ha^{-1} in the ghost forest, compared to 25.0 Mg C ha^{-1} in healthy forests (Figure 5a,c). There was a correlation (R^2 = 0.52) between the increasing live overstory C and increasing distance from open water in the healthy forest but not with the ghost forest (R^2 < 0.01; Figure 6). There was a strong relationship between the increasing soil organic C (SOC) with increasing distance from open water in the ghost forest (R^2 = 0.60) compared to the healthy forest (R^2 = 0.30; Figure 5c,d and Figure 6).

3.5. Soil Carbon Stocks

The mean soil organic carbon stocks across the plots in the ghost forest was 412 Mg C ha^{-1}, which was 37.5 Mg C ha^{-1} greater than the healthy forest: 374.5 Mg C ha^{-1} (Figure 7). In both forests, the soil C content in the top 50 cm was less than the lower 50 cm. Across the entire vertical 1 m soil profile, the ghost forest had a higher soil C than the healthy forest. The mean soil C in the top 50 cm in the ghost forest was 367.2 Mg C ha^{-1} and, in the bottom 50 cm, 456.9 Mg C ha^{-1}, whereas the healthy forest had 324.7 Mg C ha^{-1} and 424.3 Mg C ha^{-1}, for top 50 cm and below 50 cm depths, respectively.

3.6. Wood Sample Carbon Dating and Soil Accumulation

The radio-isotope of C (^{14}C) in wood samples recovered from the peat during soil sampling was used to age the soil profile and calculate peat accumulation rates (Table 4). Wood samples ranged in age from ca. AD 235 to ca. AD 1870, with a fairly strong relationship between sample age and peat soil depth in soil (Figure 8a). It is notable that the deepest (−234 cm) and oldest (AD 235) wood sample was recovered from the A-horizon of a buried mineral soil profile, while all the rest were recovered from the overlying peat.

As the sample age approached modern times, we found a decreasing soil accumulation rate (y-axis near 0, Table 4; Figure 8b). The rate of soil accumulation over time ranged from 2.59 to 7.92 mm year^{-1}. To obtain the annual soil carbon accumulation rate, we divided the average carbon stock between the ghost forest and healthy forest (393 Mg C ha^{-1}) by the age of the wood samples following the methods of [25]. The resulting average annual carbon accumulation rate was roughly 0.21–1.67 Mg C ha^{-1} year^{-1}.

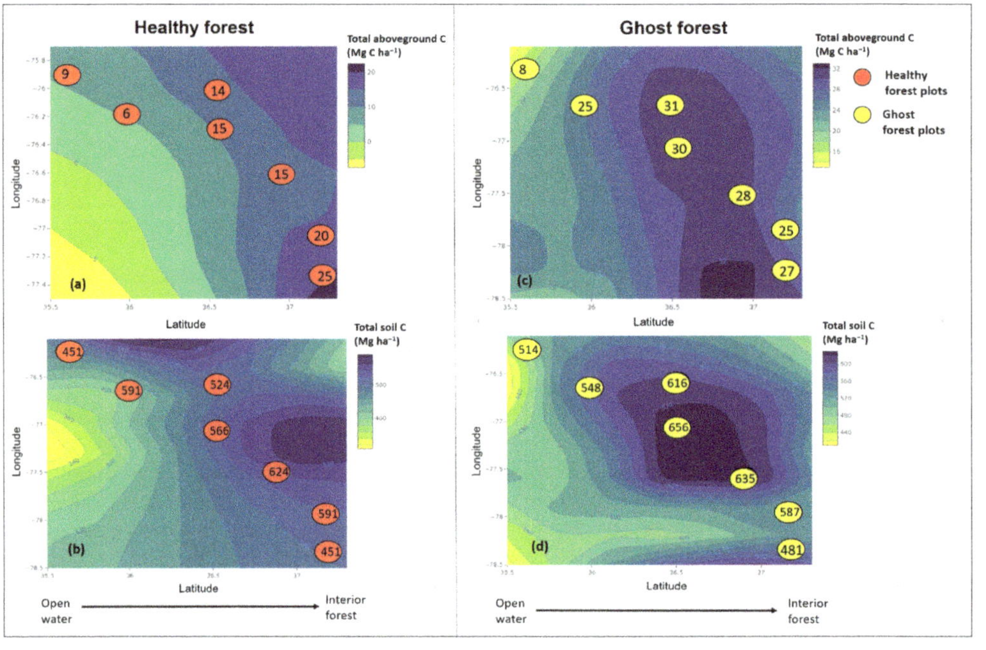

Figure 5. Map of variation in total aboveground C content (**a**,**b**) and total soil C content (**c**,**d**) as a function of distance from the bank of East Lake (open water) in healthy forest (left panels) and ghost forest (right panels) in the northern part of Alligator River National Wildlife Refuge in eastern North Carolina. Red circles denoted by their corresponding plot codes represent sampling plots of the healthy pond pine forest, and yellow circles are ghost forest sampling plot locations. Values inside each circle are the total aboveground C (**a**,**b**) and total soil C (**c**,**d**).

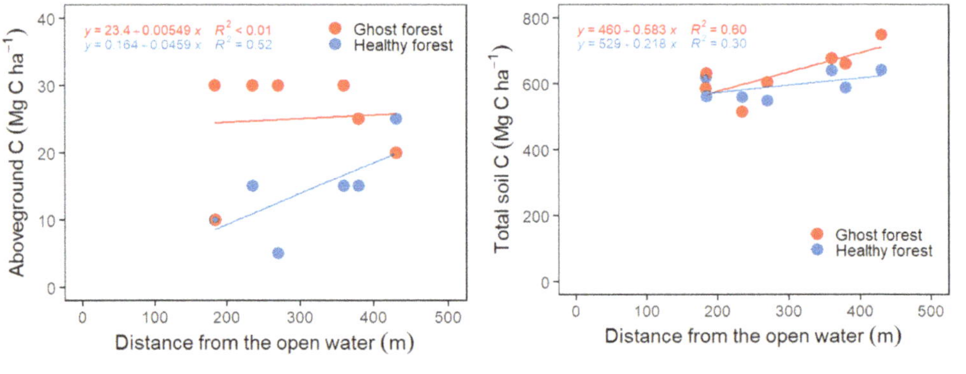

Figure 6. Relationships between aboveground C (**left**) and total soil C (**right**) with the distance from the open water in the healthy and ghost forests.

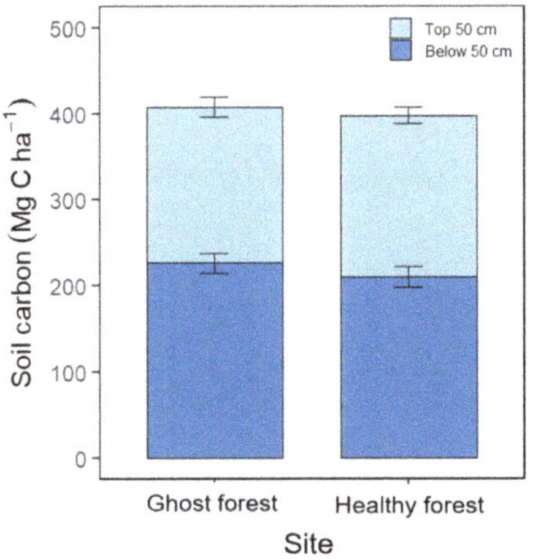

Figure 7. Soil carbon content at the ghost and healthy forest at the top 50 cm and below 50 cm.

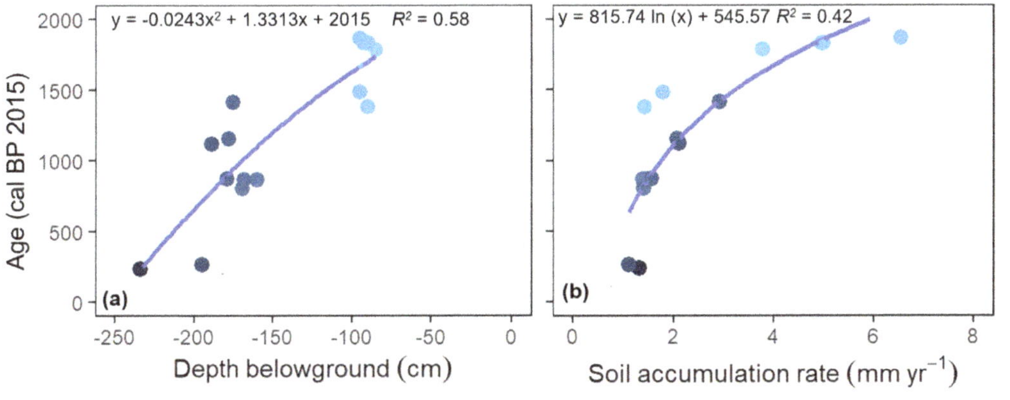

Figure 8. The relationship between the (**a**) depth of wood samples collected at the study sites and age (calendar year BP 2015), and (**b**) soil accumulation rate and age of the study site determined using the radiocarbon (^{14}C) analysis for calendar year BP 2015. A total of 16 underground wood samples were used in the analysis.

Table 4. Depth of soil where 16 wood samples were collected, the estimated age (approximate calendar year AD) and calculated soil accumulation rate using the method of McTigue et al. (2019) [26].

Soil Depth (cm)	Age (Apprx. Calendar Year BP 2015)	Soil Accumulation Rate (mm/Year)
−85	1790	3.78
−90	1380	1.42
−90	1835	5.00
−92	1830	4.97
−95	1485	1.79
−95	1870	6.55
−160	865	1.39
−168	865	1.46
−169	805	1.40
−175	1415	2.92

Table 4. *Cont.*

Soil Depth (cm)	Age (Apprx. Calendar Year BP 2015)	Soil Accumulation Rate (mm/Year)
−178	1155	2.07
−179	875	1.57
−189	1120	2.11
−190	1775	7.92
−195	265	1.11
−234	235	1.31

4. Discussion

4.1. Species Composition and Forest Mortality

Historical ditching and drainage along lower coastal plain roads and other infrastructure alter site hydrology, and potentially saltwater exposure regimes, in ways that affect contemporary ecosystem function, with implications for responses to future environmental change. The ditch, which provides drainage for the road, may also allow the inland intrusion of fresh- or saltwater during storm surges further than would normally occur. In contrast, the roadbed can act as a dam, blocking floodwater from inundating ecosystems on the other side. In the current study, this altered hydrology due to a roadbed–ditch system resulted in the conversion of what we termed healthy pond pine forest into actively transitioning ghost forest. The strong differences in species composition between the ghost forest and healthy forest was the result of this disturbance to the hydrologic and, potentially, the salinity regimes. The composition of wetland vegetation is strongly controlled by the wetland water level and hydroperiod [39,40]. Physical and environmental stressors may exclude several plant species, and the duration and intensity of flooding serves as environmental determinants of plant species composition [9]. Swamp bay, pond pine, red maple, and sweetgum are typical overstory tree species found within the Alligator River National Wildlife Refuge [41]. The healthy forest studied here was dominated by a pond pine overstory with the other species as associated sub-canopy co-dominants typical of much of the refuge. However, the adjacent ghost forest, just on the other side of the roadbed–ditch, was characterized by the almost complete mortality of the pine overstory, and the replacement of the understory by dense thickets of swamp bays and other woody shrubs. This type of ecosystem transition (replacement of forest by shrub-dominated systems) was also typical over much of the refuge, and was due to the shrubs having a higher flood tolerance, even when compared to the moderately flooded tolerant pond pine, as was reported for other similar systems [3].

According to personnel from the refuge and the ARNWR Comprehensive Management Plan [18], the roadbed–ditch system was constructed at the study site in 1965. However, it is interesting to note that the remotely-sensed signs of forest decline only began to appear after the mid-2000s (Figure 1). This suggests the environmental stress factors leading to large-scale ecosystem transition became severe only relatively recently. Although pond pine is considered an invader species in wetland coastal plain settings, it is unusual for this species to mature in waterlogged areas [42]. This supports our hypothesis that the 67% overstory mortality in the ghost forest could, at least in part, be due to the increasing inundation at the site in recent years. The decreasing canopy cover of the dying pond pine results in high understory light levels, allowing for very flood-tolerant swamp bays to flourish, changing the species composition. The continuing dominance of a pond pine overstory in the healthy forest implicates the roadbed–ditch system as a causative factor driving this ecosystem transition by slowing site drainage, thus lengthening the inundation period (hydroperiod), leading to tree mortality. However, the 54% mortality of swamp bays and other woody shrubs in the ghost forest, indicates that the hydrological stress is exceeding the tolerance limits of even these very flood-tolerant plants as part of the ecosystem transition to ghost forest. This suggests that the current shrub-dominance is only a transitory state, and that further ecosystem transition occurs with continuing increases in the severity of the driving environmental stress factors.

In addition to hydrologic stress, saltwater intrusion was implicated as a factor driving the formation of ghost forests along the southeast Atlantic and Gulf coasts [12]. In the early stages of saltwater intrusion, live trees may suffer reduced annual growth, and then forest distress becomes more visible. Young trees die conspicuously and new tree recruitment ceases [43,44]. This cessation of recruitment skews tree age distributions towards older age classes until these older remaining trees eventually die, forming relict trees that will become ghost forests. Although we did not measure salinity exposure regimes in the current study, accordingly, trees are most vulnerable to salinity stress during germination and as seedlings [45]. Additionally, even the most flood-tolerant species require periodic soil aeration for successful seed germination [3]. Therefore, the mortality occurring in our study may have been driven by the synergistic impacts of salinity and inundation, a similar condition reported by other studies [45–47]. High mortality rates and the subsequent large amount of forest carbon losses in coastal forests was attributed to sea level rise, drought [48], hurricane/storms [10,13], and hydrological connectivity from dams, ditching, and canals [49]. The increasingly apparent forest mortality and ecosystem transition, of which the current ghost forests appear to be at a seral stage, suggest that the environmental drivers recently surpassed the threshold of tolerance of coastal pond pine in eastern North Carolina, accelerated by the effects of human infrastructure on hydrology and, potentially, salinity regimes.

4.2. *Aboveground Biomass and Carbon Content of Vegetation*

The proliferation of swamp bays and the decline of overstory pines with ecosystem transition to ghost forests has implications for aboveground biomass and the carbon content of these coastal ecosystems. The capacity of forest ecosystems to store C in biomass varies depending on the species composition, age, and population density of each layer of the plant community, N:P ratio, total nutrient supply, morphological and physiological traits of plants, the amplitude of water level, water pH and electrical conductivity [39,50]. Despite the fact that the healthy forest had a higher average aboveground biomass on a per tree basis, the ghost forest, with a higher number trees of smaller diameter, had a higher total aboveground biomass than that of the road side. The intraspecific competition of trees in higher population densities usually led to a higher total aboveground carbon content [51].

We did not measure the water table depth in this study; however, a carbon flux monitoring station within ARNWR recorded a prolonged hydroperiod (2009–present) during non-growing seasons [52–54]. This monitoring station is located inland of the ARNWR [44]. Since our study plots were located in the outer part of ARNWR and were closer to a body of water (East Lake), we supposed our study sites would have experienced more severe, prolonged flooding than the carbon monitoring site. Hydrological conditions significantly influence wetland functioning [55]. Water level affects oxygen availability, decomposition rates, and nutrient mobilization from the sediment [56,57] and thus affects biomass production [58,59]. The water level also influences the ability of seeds to germinate and establish [60].

A healthy forest has a multi-layered structure that includes overstory, mid-story, and understory vegetation. Forest structure is a very critical component affecting natural tree regeneration and forest stability [61,62]. However, our results showed an imbalance in forest structure in both the ghost and healthy forest. The overstory trees have a far greater density than the mid-story and understory stratum. In the ghost forest, only a small percentage of regeneration species live past the mid-story stage, as shown by low biomass in the mid-story level (Table 3). An increase in overstory structure induces a reduction in mid-story or understory vegetation due to competition for resources such as light, nutrients, and space. Maintaining enough understory vegetation biomass should be considered in the management goals of any forest ecosystem.

A higher overstory canopy density and biomass in both the ghost and healthy forests in our study ultimately has a negative effect on the understory layer. This negative impact significantly limits the restoration of proper stand structure in this wetland forest. However, there are also positive effects with rising overstory tree density as this increases the tree number, allocation fraction of photosynthetic products in stems, and the aboveground net primary productivity [63]. As a result of the interaction between the positive and negative effects of tree density, the aboveground biomass of trees increases with a higher tree density but at a declining rate after a period of time [51].

4.3. Soil Carbon

Our results confirmed the relevance of freshwater wetlands within ARNWR as a C sink. Total carbon (C) stocks in the study sites ranged from 749.0 $\pm$ 170.5 Mg ha^{-1} in the healthy forest to 824.1 $\pm$ 46.2 Mg ha^{-1} in the ghost forest. These values are higher than those published for other eastern Mountains and the Upper Midwest, averaging 478 $\pm$ 58 Mg ha^{-1} to 539 $\pm$ 47 Mg ha^{-1} in the top 100 cm soil depth [4]; 533Mg C ha^{-1} in organic wetlands of southeastern Atlantic coastal plains [64]; and 586 Mg C ha^{-1}, 100 cm soil depth [65] in forested wetlands in New England, USA. Our total soil C was also higher than (195 $\pm$ 25 Mg ha^{-1}) that reported in the Interior Plains of the Dakota Praire [66,67] and 198 $\pm$ 21 and 216 $\pm$ 30 Mg ha^{-1} in the Coastal Plains and West [4]; 340 $\pm$ 94 Mg ha^{-1} in the Gulf of Mexico measured to soil depths of 100 and 200 cm [16]; 109 Mg C ha^{-1} in flats and levees; and 193 Mg C ha$-$1 in mineral wetlands of the southeastern Atlantic coastal plains [64].

The depth of peat horizons is mainly determined by the activity of soil organisms, the repeated interruption of humus accumulation by water dynamics, soil micro-relief, the distance to bodies of water, production and decomposition dynamics, organic and mineral depositional history [2,16], which might cause the low soil C stocks found in top 0–50 cm soil horizons in both the ghost and healthy forests. By contrast, the higher soil carbon below (50–100 cm depth) in our study indicates the episodic burial and preservation of soil C due to anaerobic soil conditions that slow the mineralization of the buried C [64]. However, a study reported that soil layers below 30 cm deep contain substantial cumulative reservoirs of carbon, with 65% of the total wetland soil carbon stored between 30 and 120 cm [4]. It must be noted that we only determined C stocks of up to a 100 cm depth and that total stocks may be even higher at deeper layers. Soil sampling in deeper layers is likely needed for the more effective evaluation and modeling of wetland soil carbon dynamics.

4.4. Soil Accumulation over Time

The organic sediment of the soils we analyzed was deposited over 1870 years, which indicated the soil's capacity to sequester C in the long-term (soil-carbon-memory effect) [2]. Fixed C may be released as CO_2 by microorganisms during decomposition or stored belowground, where anaerobic decomposition occurs slowly, and recalcitrant C may remain for millennia [16]. The ongoing delivery of sediment due to flooding leads to sediment deposition, the burial of organic matter, and the vertical accretion of marsh surfaces, thus allowing carbon to accumulate over long periods [4]. Increasing rates of sea level rise may also contribute to soil accretion by increasing the duration of inundation and increasing sediment deposition on marsh surfaces [68]. Anthropogenic impacts such as the presence of ditches and dikes also have a strong association with affecting soil carbon deposition [4]. A slightly lower soil carbon stored in the healthy forest might be due to the presence of ditches increasing stream and base flows, thereby increasing the annual discharge, which leads to lower (that is, drier) groundwater levels. This lower groundwater level over time could increase soil carbon oxidation and affect soil carbon stores [4]. More stored carbon in the ghost forest might be due to the occurrence of constantly saturated conditions, allowing for more efficient C mineralization at the soil surface and limiting microbial respiration at depth [64], and some contribution of dying overstory trees to soil C inputs.

Our mean soil accumulation rate was 2.59 mm year^{-1}, yet we found a decreasing trend as our site aged closer to modern times some 235 to 265 years ago (1.11 mm year^{-1} to 1.31 mm year^{-1}) (Table 4). This decline in trend may occur because the vertical accretion rates overlap with the range of sea level rise rates. Numerical models predict that the maximum vertical accretions rates 2100 years ago were generally 5–30 mm year^{-1} [69]. Once these maximum limits are achieved, wetland drowning can be observed [12,70]. When these threshold rates of sea level rise are exceeded, wetlands must migrate laterally into submerging uplands to survive [12].

Vegetation at our site persisted for at least 1870 years, resisting drowning to sea level rise for many centuries because natural ecogeomorphic feedbacks allowed coastal habitats to vertically accrete and keep pace with the sea level [71]. However, a study reported that sea levels have risen at an average rate of 2.1 mm year^{-1}, representing the steepest century-scale increase of the past two millennia [72], and this rate was initiated between AD 1865 and 1892. Another study also reported a rise in sea level from 2.2 mm year^{-1} to 2.4 mm year^{-1} around the 18th century [73]. If we follow the result of these studies with a sea level rise rate of 2.1 mm year^{-1} to 2.4 mm year^{-1}, this means that the sediment accumulation in our study sites of less than 1.5 mm year^{-1}, during the turn of 18th century, was less than the estimated rate of sea level rise (2.1–2.4 mm year^{-1}). These estimates suggest that our site may have already been slightly submerged long before and has transitioned from a depositional to an erosive environment and experienced a drowning effect [12], slowly migrating laterally towards uplands to survive as it approaches modern times. Recent global modeling suggests wetland migration into submerging uplands is the single most significant factor influencing wetland area over time, and that global wetland area could increase by up to 60% by 2100 for a 1.1 m sea level rise [74]. Our study site also appeared to be undergoing a transition from organic to mineral accumulation based on the decreased organic matter and sediment accumulation heading towards the open water. This stratigraphic sequence was indicative of submerging marsh [16].

Our estimated carbon accumulation rate (CAR) of 0.21–1.67 Mg ha^{-1} year^{-1} aged 235–1870 years old at both the ghost and healthy forests (Table 4 aligns with other studies conducted elsewhere in the United States. Johnson et al., 2007 [75] collected and aged a New England salt marsh core to 3700 cal BP and reported a CAR of 0.4 Mg ha^{-1} year^{-1}. Brevik and Homburg, 2004 [76] collected a core in Southern California that formed over the course of 5000 years and measured a CAR of 0.3 Mg ha^{-1} year^{-1}. Drexler, 2011 [77] measured CAR in cores >6000 years old and found CARs from 0.38 to 0.79 Mg ha^{-1} year^{-1}. A study also found a CAR of 0.39 Mg ha^{-1} year^{-1} when integrated over ~2400 years at 2.2 m depth of salt marsh [25]. These results suggest that carbon storage in marshes can be enhanced by SLR-driven OM burial. We caution that our sampling sites only covered a few plots with a small sampling size, and our estimates came with some degree of uncertainty. Telford et al., 2004 [78] compared different age–depth modeling techniques for radiocarbon-dated sediments and found that uncertainty is often underestimated, and the errors were surprisingly large, especially when there were few dated levels.

5. Conclusions

Our study quantifies the impact of roadside ditching and drainage on ecosystem transition from pond pine to shrub pocosin along the lower coastal plain of North Carolina. Ditches and the roadbeds protect forest on the roadside but accelerates the ecosystem transition to a ghost forest on the ditch side, which has implications for the aboveground C storage and cycling. Our study helped us to better understand how coastal wetland soil carbon accumulation changed over time and the response mechanism of vegetation to these changing hydrological perturbations. Our investigation contributes to providing information on the current status and past trends in coastal wetland soil dynamics that may provide valuable insights into the future responses of the wetland ecosystems to sea level rise and climate change. The mismanagement of the wetland ecosystem may convert the present soil C pools into net C sources in the future. Therefore, conservation strategies for the wetland ecosystem should, to a greater extent, incorporate the functions of disturbed and undisturbed wetland forests as C sinks. Flood control structures prevented the migration of outer bank marshes into inland wetlands. The management of wetland ecosystems must therefore consider the cost of constructing more conventional flood control structures to the cost of damages associated with flooding and the danger of inland forested wetlands becoming ghost forests. Choices between defending the coast from sea level rise and facilitating ecosystem transition play a critical role in determining the fate and function of coastal wetland ecosystems. Accurate carbon accounting in wetlands is vital to reduce the risk of climate change contributions by identifying and protecting wetland landscapes that hold disproportionately large soil C stocks.

Author Contributions: Conceptualization, C.B. and J.K.; methodology, C.B. and J.K.; software, C.B. and M.A.; validation, K.M. and M.F.; formal analysis, M.A., O.J.I., D.H. and M.K.; Investigation, C.B., K.M. and M.F.; resources, J.K.; data curation, M.A., C.B., O.J.I., D.H. and M.K.; writing—original draft preparation, M.A. and C.B.; writing—review and editing, J.K., O.J.I., K.M., D.H. and M.K.; visualization, M.A.; supervision, A.N. and J.K.; project administration, A.N. and J.K.; funding acquisition, J.K. All authors have read and agreed to the published version of the manuscript.

Funding: Primary funding was provided by the USDA NIFA (Multi-agency A.5 Carbon Cycle Science Program) award 2014-67003-22068. Additional funding was provided by the DOE NICCR award 08-SC-NICCR-1072, the USDA Forest Service award 13-JV-11330110-081, and the DOE LBNL award DE-AC02-05CH11231. Work by MF was supported by project SustES—Adaptation strategies for sustainable ecosystem services and food security under adverse environmental conditions (CZ.02.1.01/0.0/0.0/16_019/0000797).

Institutional Review Board Statement: Not applicable.

Informed Consent Statement: Not applicable.

Data Availability Statement: Not applicable.

Conflicts of Interest: The authors declare no conflict of interest.

References

1. Mitra, S.; Wassmann, R.; Vlek, P.L.G. An Appraisal of Global Wetland Area and Its Organic Carbon Stock. *Curr. Sci.* **2005**, *88*, 25–35.
2. Cierjacks, A.; Kleinschmit, B.; Babinsky, M.; Kleinschroth, F.; Markert, A.; Menzel, M.; Ziechmann, U.; Schiller, T.; Graf, M.; Lang, F. Carbon Stocks of Soil and Vegetation on Danubian Floodplains. *J. Plant Nutr. Soil Sci.* **2010**, *173*, 644–653. [CrossRef]
3. Casey, W.P.; Ewel, K.C. Patterns of Succession in Forested Depressional Wetlands in North Florida, USA. *Wetlands* **2006**, *26*, 147–160. [CrossRef]
4. Nahlik, A.M.; Fennessy, M.S. Carbon Storage in US Wetlands. *Nat. Commun.* **2016**, *7*, 13835. [CrossRef]
5. Day, J.W.; Christian, R.R.; Boesch, D.M.; Yáñez-Arancibia, A.; Morris, J.; Twilley, R.R.; Naylor, L.; Schaffner, L.; Stevenson, C. Consequences of Climate Change on the Ecogeomorphology of Coastal Wetlands. *Estuaries Coasts* **2008**, *31*, 477–491. [CrossRef]
6. Horton, B.P.; Rahmstorf, S.; Engelhart, S.E.; Kemp, A.C. Expert Assessment of Sea-Level Rise by AD 2100 and AD 2300. *Quat. Sci. Rev.* **2014**, *84*, 1–6. [CrossRef]
7. Cormier, N.; Krauss, K.W.; Conner, W.H. Periodicity in Stem Growth and Litterfall in Tidal Freshwater Forested Wetlands: Influence of Salinity and Drought on Nitrogen Recycling. *Estuaries Coasts* **2013**, *36*, 533–546. [CrossRef]
8. Ensign, S.H.; Hupp, C.R.; Noe, G.B.; Krauss, K.W.; Stagg, C.L. Sediment Accretion in Tidal Freshwater Forests and Oligohaline Marshes of the Waccamaw and Savannah Rivers, USA. *Estuaries Coasts* **2014**, *37*, 1107–1119. [CrossRef]
9. Truus, L. Estimation of Above-Ground Biomass of Wetlands. In *Biomass and Remote Sensing of Biomass*; IntechOpen: London, UK, 2011. [CrossRef]
10. Stagg, C.L.; Schoolmaster, D.R.; Piazza, S.C.; Snedden, G.; Steyer, G.D.; Fischenich, C.J.; McComas, R.W. A Landscape-Scale Assessment of Above- and Belowground Primary Production in Coastal Wetlands: Implications for Climate Change-Induced Community Shifts. *Estuaries Coasts* **2017**, *40*, 856–879. [CrossRef]
11. Rasmussen, C.; Southard, R.J.; Horwath, W.R. Litter Type and Soil Minerals Control Temperate Forest Soil Carbon Response to Climate Change. *Glob. Chang. Biol.* **2008**, *14*, 2064–2080. [CrossRef]
12. Kirwan, M.L.; Gedan, K.B. Sea-Level Driven Land Conversion and the Formation of Ghost Forests. *Nat. Clim. Chang.* **2019**, *9*, 450–457. [CrossRef]
13. Hopkinson, C.S.; Lugo, A.E.; Alber, M.; Covich, A.P.; Van Bloem, S.J. Forecasting Effects of Sea-Level Rise and Windstorms on Coastal and Inland Ecosystems. *Front. Ecol. Environ.* **2008**, *6*, 255–263. [CrossRef]

14. Neumann, B.; Vafeidis, A.T.; Zimmermann, J.; Nicholls, R.J. Future Coastal Population Growth and Exposure to Sea-Level Rise and Coastal Flooding—A Global Assessment. *PLoS ONE* **2015**, *10*, e0118571. [CrossRef]
15. White, E.; Kaplan, D. Restore or Retreat? Saltwater Intrusion and Water Management in Coastal Wetlands. *Ecosyst. Health Sustain.* **2017**, *3*, e01258. [CrossRef]
16. Elsey-Quirk, T.; Seliskar, D.M.; Sommerfield, C.K.; Gallagher, J.L. Salt Marsh Carbon Pool Distribution in a Mid-Atlantic Lagoon, USA: Sea Level Rise Implications. *Wetlands* **2011**, *31*, 87–99. [CrossRef]
17. Dreyer, G.D.; Niering, W.A. *Bulletin No. 34: Tidal Marshes of Long Island Sound: Ecology, History and Restoration. Human Impacts on Tidal Wetlands: History and Regulations*; Connecticut College Arboretum: New London, CT, USA, 1995; ISBN 1878899058.
18. Bryant, M.R.; Jerome, P.; Andrew, J.; Hamilton, S. *Alligator River National Wildlife Refuge Comprehensive Conservation Plan*; Fish and Wildlife Service, US Department of the Interior: Atlanta, GA, USA, 2008.
19. McKee, K.L. Biophysical Controls on Accretion and Elevation Change in Caribbean Mangrove Ecosystems. *Estuar. Coast. Shelf Sci.* **2011**, *91*, 475–483. [CrossRef]
20. Morris, J.T.; Sundareshwar, P.V.; Nietch, C.T.; Kjerfve, B.; Cahoon, D.R. Responses of Coastal Wetlands to Rising Sea Level. *Ecology* **2002**, *83*, 2869–2877. [CrossRef]
21. Kirwan, M.L.; Guntenspergen, G.R. Response of Plant Productivity to Experimental Flooding in a Stable and a Submerging Marsh. *Ecosystems* **2015**, *18*, 903–913. [CrossRef]
22. Pennings, S.C.; Grant, M.B.; Bertness, M.D. Plant Zonation in Low-Latitude Salt Marshes: Disentangling the Roles of Flooding, Salinity and Competition. *J. Ecol.* **2005**, *93*, 159–167. [CrossRef]
23. Holmquist, J.R.; Windham-Myers, L.; Bliss, N.; Crooks, S.; Morris, J.T.; Megonigal, J.P.; Troxler, T.; Weller, D.; Callaway, J.; Drexler, J.; et al. Accuracy and Precision of Tidal Wetland Soil Carbon Mapping in the Conterminous United States. *Sci. Rep.* **2018**, *8*, 9478. [CrossRef] [PubMed]
24. Bridgham, S.D.; Megonigal, J.P.; Keller, J.K.; Bliss, N.B.; Trettin, C. The Carbon Balance of North American Wetlands. *Wetlands* **2006**, *26*, 889–916. [CrossRef]
25. McTigue, N.; Davis, J.; Rodriguez, A.B.; McKee, B.; Atencio, A.; Currin, C. Sea Level Rise Explains Changing Carbon Accumulation Rates in a Salt Marsh Over the Past Two Millennia. *J. Geophys. Res. Biogeosci.* **2019**, *124*, 2945–2957. [CrossRef]
26. Riggs, S.R.; Ames, D. *Drowning the North Carolina Coast: Sea-Level Rise and Estuarine Dynamics*; NC Sea Grant: Raleigh, NC, USA, 2003.
27. Riggs, S.R.; Ames, D.V.; Culver, S.J.; Mallinson, D.J. Review Reviewed Work(s): The Battle for North Carolina's Coast: Evolutionary History, Present Crisis, and Vision for the Future. *Southeast. Geogr.* **2012**, *52*, 242–244. [CrossRef]
28. Minick, K.J.; Kelley, A.M.; Miao, G.; Li, X.; Noormets, A.; Mitra, B.; King, J.S. Microtopography Alters Hydrology, Phenol Oxidase Activity and Nutrient Availability in Organic Soils of a Coastal Freshwater Forested Wetland. *Wetlands* **2019**, *39*, 263–273. [CrossRef]
29. Burns, R.M.; Honkala, B.H. (Eds.) Bramlett DL Pond pine (Pinus serotinaMichx). In *The Silvics of North America—Agricultural Handbook 654*; USDA Forest Service: Washington, DC, USA, 1990.
30. FIA. *Phase 3Field Guide—Down Woody Materials, Section 25*; Down Woody Materials; USDA: Kansas City, MO, USA, 2007.
31. Jenkins, J.C.; Chojnacky, D.C.; Heath, L.S.; Birdsey, R.A. *Comprehensive Database of Diameter-Based Biomass Resgressions for North American Tree Species*; United States Department of Agriculture, Forest Service, Northeastern Research Station: Kansas City, MO, USA, 2003.
32. Preston, R.J.; Braham, R.R. *North American Trees*; Iowa State Press: Ames, IA, USA, 2002.
33. Hajdas, I.; Hendriks, L.; Fontana, A.; Monegato, G. Evaluation of Preparation Methods in Radiocarbon Dating of Old Wood. *Radiocarbon* **2017**, *59*, 727–737. [CrossRef]
34. Wickham, H. *Ggplot2: Elegant Graphics for Data Analysis*; Springer: New York, NY, USA, 2016; ISBN 978-3-319-24277-4.
35. Sievert, C. *Interactive Web-Based Data Visualization with R, Plotly, and Shiny*; Chapman and Hall/CRC: Boca Raton, FL, USA, 2020; ISBN 9781138331457.
36. Wickham, H.; Averick, M.; Bryan, J.; Chang, W.; McGowan, L.; François, R.; Grolemund, G.; Hayes, A.; Henry, L.; Hester, J.; et al. Welcome to the Tidyverse. *J. Open Source Softw.* **2019**, *4*, 1686. [CrossRef]
37. Wickham, H. Reshaping Data with the Reshape Package. *J. Stat. Softw.* **2007**, *12*, 1–20.
38. R Core Team. *R: A Language and Environment for Statistical Computing*; R Foundation for Statistical Computing: Vienna, Austria, 2021.
39. Ilomets, M.; Truus, L.; Pajula, R.; Sepp, K. Species Composition and Structure of Vascular Plants and Bryophytes on the Water Level Gradient within a Calcareous Fen North Estonia. *Est. J. Ecol.* **2010**, *59*, 19–38. [CrossRef]
40. Wilcox, D.A.; Nichols, S.J. The Effects of Water-Level Fluctuations on Vegetation in a Lake Huron Wetland. *Wetlands* **2008**, *28*, 487–501. [CrossRef]
41. Aguilos, M.; Mitra, B.; Noormets, A.; Minick, K.; Prajapati, P.; Gavazzi, M.; Sun, G.; McNulty, S.; Li, X.; Domec, J.C.; et al. Long-Term Carbon Flux and Balance in Managed and Natural Coastal Forested Wetlands of the Southeastern USA. *Agric. For. Meteorol.* **2020**, *288–289*, 108022. [CrossRef]
42. Craine, S.I.; Orians, C.M. Pitch Pine (Pinus Rigida Mill.) Invasion of Cape Cod Pond Shores Alters Abiotic Environment and Inhibits Indigenous Herbaceous Species. *Biol. Conserv.* **2004**, *116*, 181–189. [CrossRef]

43. Williams, K.; Ewel, K.C.; Stumpf, R.P.; Putz, F.E.; Workman, T.W. Sea-Level Rise and Coastal Forest Retreat on the West Coast of Florida, USA. *Ecology* **1999**, *80*, 2045–2063. [CrossRef]
44. Begin, Y. The Effects of Shoreline Transgression on Woody Plants, Upper St Lawrence Estuary, Quebec. *J. Coast. Res.* **1990**, *6*, 815–827.
45. Pezeshki, S.R.; Delaune, R.D.; Patrick, W.H. Flooding and Saltwater Intrusion: Potential Effects on Survival and Productivity of Wetland Forests along the U.S. Gulf Coast. *For. Ecol. Manag.* **1990**, *33–34*, 287–301. [CrossRef]
46. Lennard-Barrett, E.G. The Interaction between Waterlogging and Salinity in Higher Plants: Causes, Consequences and Implications. *Plant Soil* **2003**, *253*, 35–54. [CrossRef]
47. Conner, W.H. *The Effect of Salinity and Waterlogging on Growth and Survival of Baldcypress and Chinese Tallow Seedlings*; Coastal Education & Research Foundation, Inc.: Lawrence, KS, USA, 1994; Volume 10, pp. 1045–1049. Available online: https://www.Jstor.Org/Stable/4298295 (accessed on 15 August 2021).
48. Desantis, L.R.G.; Bhotika, S.; Williams, K.; Putz, F.E. Sea-Level Rise and Drought Interactions Accelerate Forest Decline on the Gulf Coast of Florida, USA. *Glob. Chang. Biol.* **2007**, *13*, 2349–2360. [CrossRef]
49. Poulter, B.; Goodall, J.L.; Halpin, P.N. Applications of Network Analysis for Adaptive Management of Artificial Drainage Systems in Landscapes Vulnerable to Sea Level Rise. *J. Hydrol.* **2008**, *357*, 207–217. [CrossRef]
50. Mendoza-Ponce, A.; Galicia, L. Aboveground and Belowground Biomass and Carbon Pools in Highland Temperate Forest Landscape in Central Mexico. *Forestry* **2010**, *83*, 497–506. [CrossRef]
51. Ile, O.J.; Aguilos, M.; Morkoc, S.; Minick, K.; Domec, J.-C.; King, J.S. Productivity of Low-Input Short-Rotation Coppice American Sycamore (*Platanus Occidentalis* L.) Grown at Different Planting Densities as a Bioenergy Feedstock over Two Rotation Cycles. *Biomass Bioenergy* **2021**, *146*, 105983. [CrossRef]
52. Aguilos, M.; Sun, G.; Noormets, A.; Domec, J.C.; McNulty, S.; Gavazzi, M.; Prajapati, P.; Minick, K.J.; Mitra, B.; King, J. Ecosystem Productivity and Evapotranspiration Are Tightly Coupled in Loblolly Pine (*Pinus Taeda* L.) Plantations along the Coastal Plain of the Southeastern U.S. *Forests* **2021**, *12*, 1123. [CrossRef]
53. Aguilos, M.; Sun, G.; Noormets, A.; Domec, J.C.; McNulty, S.; Gavazzi, M.; Minick, K.; Mitra, B.; Prajapati, P.; Yang, Y.; et al. Effects of Land-Use Change and Drought on Decadal Evapotranspiration and Water Balance of Natural and Managed Forested Wetlands along the Southeastern US Lower Coastal Plain. *Agric. For. Meteorol.* **2021**, *303*, 108381. [CrossRef]
54. Minick, K.J.; Mitra, B.; Li, X.; Fischer, M.; Aguilos, M.; Prajapati, P.; Noormets, A.; King, J.S. Wetland Microtopography Alters Response of Potential Net CO_2 and CH_4 Production to Temperature and Moisture: Evidence from a Laboratory Experiment. *Geoderma* **2021**, *402*, 115367. [CrossRef]
55. Ketcheson, S.J.; Price, J.S.; Carey, S.K.; Petrone, R.M.; Mendoza, C.A.; Devito, K.J. Constructing Fen Peatlands in Post-Mining Oil Sands Landscapes: Challenges and Opportunities from a Hydrological Perspective. *Earth-Sci. Rev.* **2016**, *161*, 130–139. [CrossRef]
56. Lamers, L.P.M.; van Diggelen, J.M.H.; Op Den Camp, H.J.M.; Visser, E.J.W.; Lucassen, E.C.H.E.T.; Vile, M.A.; Jetten, M.S.M.; Smolders, A.J.P.; Roelofs, J.G.M. Microbial Transformations of Nitrogen, Sulfur, and Iron Dictate Vegetation Composition in Wetlands: A Review. *Front. Microbiol.* **2012**, *3*, 156. [CrossRef]
57. Harpenslager, S.F.; van den Elzen, E.; Kox, M.A.R.; Smolders, A.J.P.; Ettwig, K.F.; Lamers, L.P.M. Rewetting Former Agricultural Peatlands: Topsoil Removal as a Prerequisite to Avoid Strong Nutrient and Greenhouse Gas Emissions. *Ecol. Eng.* **2015**, *84*, 159–168. [CrossRef]
58. Sarneel, J.M.; Geurts, J.J.M.; Beltman, B.; Lamers, L.P.M.; Nijzink, M.M.; Soons, M.B.; Verhoeven, J.T.A. The Effect of Nutrient Enrichment of Either the Bank or the Surface Water on Shoreline Vegetation and Decomposition. *Ecosystems* **2010**, *13*, 1275–1286. [CrossRef]
59. Dee, S.M.; Ahn, C. Plant Tissue Nutrients as a Descriptor of Plant Productivity of Created Mitigation Wetlands. *Ecol. Indic.* **2014**, *45*, 68–74. [CrossRef]
60. Overbeek, C.C.; Harpenslager, S.F.; van Zuidam, J.P.; van Loon, E.E.; Lamers, L.P.M.; Soons, M.B.; Admiraal, W.; Verhoeven, J.T.A.; Smolders, A.J.P.; Roelofs, J.G.M.; et al. Drivers of Vegetation Development, Biomass Production and the Initiation of Peat Formation in a Newly Constructed Wetland. *Ecosystems* **2020**, *23*, 1019–1036. [CrossRef]
61. Suchar, V.A.; Crookston, N.L. Understory Cover and Biomass Indices Predictions for Forest Ecosystems of the Northwestern United States. *Ecol. Indic.* **2010**, *10*, 602–609. [CrossRef]
62. Kumar, P.; Chen, H.Y.H.; Thomas, S.C.; Shahi, C. Linking Resource Availability and Heterogeneity to Understorey Species Diversity through Succession in Boreal Forest of Canada. *J. Ecol.* **2018**, *106*, 1266–1276. [CrossRef]
63. Litton, C.M.; Ryan, M.G.; Knight, D.H. Effects of Tree Density and Stand Age on Carbon Allocation Patterns in Postfire Lodgepole Pine. *Ecol. Appl.* **2004**, *14*, 460–475. [CrossRef]
64. Ricker, M.C.; Lockaby, B.G. Soil Organic Carbon Stocks in a Large Eutrophic Floodplain Forest of the Southeastern Atlantic Coastal Plain, USA. *Wetlands* **2015**, *35*, 291–301. [CrossRef]
65. Davis, A.A.; Stolt, M.H.; Compton, J.E. Spatial Distribution of Soil Carbon in Southern New England Hardwood Forest Landscapes. *Soil Sci. Soc. Am. J.* **2004**, *68*, 895–903. [CrossRef]
66. Gleason, R.A.; Laubhan, M.K.; Euliss, N.H. *Ecosystems Services Derived from Wetland Conservation Practices in the United States Prairie Pothole Region with an emphasis on the U.S. Department Of Agriculture Conservation Reserve and Wetlands Reserve Programs*; U.S. Geological Survey Professional Paper 1745; U.S. Geological Survey: Liston, VA, USA, 2008; 58p, ISBN 9781411320178.

67. Johnston, C.A. Wetland Losses Due to Row Crop Expansion in the Dakota Prairie Pothole Region. *Wetlands* **2013**, *33*, 175–182. [CrossRef]
68. Mudd, S.M.; Howell, S.M.; Morris, J.T. Impact of Dynamic Feedbacks between Sedimentation, Sea-Level Rise, and Biomass Production on near-Surface Marsh Stratigraphy and Carbon Accumulation. *Estuar. Coast. Shelf Sci.* **2009**, *82*, 377–389. [CrossRef]
69. Kirwan, M.L.; Temmerman, S.; Skeehan, E.E.; Guntenspergen, G.R.; Fagherazzi, S. Overestimation of Marsh Vulnerability to Sea Level Rise. *Nat. Clim. Chang.* **2016**, *6*, 253–260. [CrossRef]
70. Crosby, S.C.; Sax, D.F.; Palmer, M.E.; Booth, H.S.; Deegan, L.A.; Bertness, M.D.; Leslie, H.M. Salt Marsh Persistence Is Threatened by Predicted Sea-Level Rise. *Estuar. Coast. Shelf Sci.* **2016**, *181*, 93–99. [CrossRef]
71. Kirwan, M.L.; Megonigal, J.P. Tidal Wetland Stability in the Face of Human Impacts and Sea-Level Rise. *Nature* **2013**, *504*, 53–60. [CrossRef]
72. Kemp, A.C.; Kegel, J.J.; Culver, S.J.; Barber, D.C.; Mallinson, D.J.; Leorri, E.; Bernhardt, C.E.; Cahill, N.; Riggs, S.R.; Woodson, A.L.; et al. Extended Late Holocene Relative Sea-Level Histories for North Carolina, USA. *Quat. Sci. Rev.* **2017**, *160*, 13–30. [CrossRef]
73. Kopp, R.E. Does the Mid-Atlantic United States Sea Level Acceleration Hot Spot Reflect Ocean Dynamic Variability? *Geophys. Res. Lett.* **2013**, *40*, 3981–3985. [CrossRef]
74. Schuerch, M.; Spencer, T.; Temmerman, S.; Kirwan, M.L.; Wolff, C.; Lincke, D.; McOwen, C.J.; Pickering, M.D.; Reef, R.; Vafeidis, A.T.; et al. Future Response of Global Coastal Wetlands to Sea-Level Rise. *Nature* **2018**, *561*, 231–234. [CrossRef]
75. Johnson, B.J.; Moore, K.A.; Lehmann, C.; Bohlen, C.; Brown, T.A. Middle to Late Holocene Fluctuations of C3 and C4 Vegetation in a Northern New England Salt Marsh, Sprague Marsh, Phippsburg Maine. *Org. Geochem.* **2007**, *38*, 394–403. [CrossRef]
76. Brevik, E.C.; Homburg, J.A. A 5000 Year Record of Carbon Sequestration from a Coastal Lagoon and Wetland Complex, Southern California, USA. *Catena* **2004**, *57*, 221–232. [CrossRef]
77. Drexler, J.Z. Peat Formation Processes through the Millennia in Tidal Marshes of the Sacramento-San Joaquin Delta, California, USA. *Estuaries Coasts* **2011**, *34*, 900–911. [CrossRef]
78. Telford, R.J.; Heegaard, E.; Birks, H.J.B. All Age-Depth Models Are Wrong: But How Badly? *Quat. Sci. Rev.* **2004**, *23*, 1–5. [CrossRef]

land

Article

Effects of Grazer Exclusion on Carbon Cycling in Created Freshwater Wetlands

Delanie M. Spangler, Anna Christina Tyler and Carmody K. McCalley *

Thomas H. Gosnell School of Life Sciences, Rochester Institute of Technology, Rochester, NY 14623-5603, USA; DelanieMS@gmail.com (D.M.S.); actsbi@rit.edu (A.C.T.)
* Correspondence: ckmsbi@rit.edu

Abstract: Wetland ecosystems play a significant role in the global carbon cycle, and yet are increasingly threatened by human development and climate change. The continued loss of intact freshwater wetlands heightens the need for effective wetland creation and restoration. However, wetland structure and function are controlled by interacting abiotic and biotic factors, complicating efforts to replace ecosystem services associated with natural wetlands and making ecologically-driven management imperative. Increasing waterfowl populations pose a threat to the development and persistence of created wetlands, largely through intensive grazing that can shift vegetation community structure or limit desired plant establishment. This study capitalized on a long-term herbivore exclusion experiment to evaluate how herbivore management impacts carbon cycling and storage in a created wetland in Western New York, USA. Vegetation, above- and belowground biomass, soil carbon, carbon gas fluxes and decomposition rates were evaluated in control plots with free access by large grazers and in plots where grazers had been excluded for four years. Waterfowl were the dominant herbivore at the site. Grazing reduced peak growing season aboveground biomass by over 55%, and during the summer, gross primary productivity doubled in grazer exclusion plots. The shift in plant productivity led to a 34% increase in soil carbon after exclusion of grazers for five growing seasons, but no change in belowground biomass. Our results suggest that grazers may inhibit the development of soil carbon pools during the first decade following wetland creation, reducing the carbon sequestration potential and precluding functional equivalence with natural wetlands.

Keywords: grazing; created wetlands; freshwater marshes; carbon cycling

Citation: Spangler, D.M.; Tyler, A.C.; McCalley, C.K. Effects of Grazer Exclusion on Carbon Cycling in Created Freshwater Wetlands. *Land* **2021**, *10*, 805. https://doi.org/10.3390/land10080805

Academic Editor: Richard C. Smardon

Received: 1 July 2021
Accepted: 30 July 2021
Published: 31 July 2021

1. Introduction

Freshwater wetlands are among the most ecologically and economically valuable ecosystems in the world, providing ecosystem services such as habitat for migratory waterfowl [1], nutrient cycling [2,3], and carbon storage [4–6]. Urban and agricultural development is threatening wetlands, resulting in the need for restoration and creation to prevent loss of key ecosystem functions. Wetland ecosystems, however, are driven by complex interactions between biotic and abiotic factors, including hydrology, nutrient cycling, competition, and grazing, which influence ecosystem structure and function and pose challenges to successful restoration efforts (e.g., [7–9]). A greater understanding of the interplay among biotic and abiotic drivers of function, the trajectory of wetland development over time, and which management tools can be leveraged to maximize desired outcomes is required for more successful wetland restoration [10].

Emergent vegetation, such as *Typha* spp., is a key driver of carbon (C) cycling in freshwater wetlands [11,12]. Plants fix inorganic C from the atmosphere through photosynthesis, store organic C in above- and belowground biomass, and transfer carbon to sediments through decomposition and root exudation. Soil carbon is often stored for long periods of time due to anaerobic soil conditions [13,14]. High photosynthetic activity, coupled with anaerobic conditions, means that wetlands can be substantial carbon sinks [6,15,16]. However, the magnitude of soil carbon content is a key functional difference between natural and created wetlands, with an average shortcoming in soil C of 51.7% relative to natural wetlands within two decades of creation in the United States [17]. Globally, created wetlands require more than 20 years to reach C levels of natural wetlands [9], and estimates of time to reach comparable C levels in the United States range from 30 to 300 years [18], with variation among wetland types [17]. This highlights the need to identify wetland creation and management approaches that accelerate soil development and carbon uptake and storage.

Herbivory has the potential to exacerbate plant and carbon cycling differences between created and natural wetlands. Hydrology plays a key role in this top-down dynamic in freshwater wetlands, with stable hydrologic regimes attracting waterfowl, in particular the Canada goose (*Branta canadensis*) and ducks (*Anas* spp.), to wetlands for nesting and feeding [19–21]. Created wetlands often feature deep standing water and young palatable vegetation, which offers desirable habitat for migratory waterfowl [22]. This, coupled with rising waterfowl populations [23], can cause created wetlands to be particularly vulnerable to intensive grazing [21]. Herbivory in wetlands can cause a top-down cascade that shifts ecosystem structure and function [24,25]. Although primary functional traits of wetlands are determined by physico-chemical factors, especially hydrology and nutrient availability, interaction between bottom-up and top-down factors may ultimately control plant community composition and biomass [26–29].

Excessive grazing by waterfowl in wetlands can cause a decrease in total plant cover [21,30], and also in plant biomass [29,31]. Preferential grazing, especially when heavy, may also shift plant community composition [21,32–34]. Often waterfowl target younger, more palatable plants, leading to a shift in plant species composition [35–37] and changes in the quality and quantity of plant litter. When intensive grazing reduces plant cover or changes plant species composition, net primary productivity (NPP) and organic carbon brought into and stored in soil may decrease [38–41]. In grasslands, grazing may also increase decomposition, by changing both litter quality and the soil environment [42]. In wetlands, plants also impact soil processes through oxygen transport into sediments [43,44], and therefore intensive grazing of emergent plants could further alter belowground carbon cycling by reducing aerobic microsites. While the impact of waterfowl on plant cover, biomass and species composition has been studied in freshwater wetlands, the impact on CO_2 fluxes, net ecosystem exchange (NEE), and overall carbon storage is less well understood.

One of the key limitations to many studies on herbivore impacts in wetlands is that experiments often last only one or two seasons, limiting the ability to observe long-term changes in carbon cycling and storage. The current study utilizes a long-term herbivore exclosure experiment and builds upon Lodge and Tyler [21], who found that during the first 2 years of grazer exclusion, plant cover increased by 55% and peak growing season plant diversity by 30% in a permanently flooded wetland, suggesting that, in some cases, grazer management could be used to establish desired plant communities during wetland creation. This result raised the question of whether shifts in plant communities following grazer exclusion also promote the development of C stocks in created wetlands. The current study expands on the long-term grazer exclusion experiment by quantifying C pools and fluxes and addressing whether changes in plant cover following herbivore management result in long-term changes in wetland carbon cycling and storage. The overarching objective of this study was to better understand the impacts of grazers on carbon cycling, with the intent to help managers develop management practices that promote carbon sequestration in created wetlands. We hypothesized that a reduction in total plant biomass by grazers would lead to decreased photosynthetic carbon uptake (GPP), lower net ecosystem exchange (NEE), and ultimately decreased soil carbon storage.

2. Methods

2.1. Site Description

This experiment took place between June 2017 and the end of April 2019, using experimental plots established in 2014 by Lodge et al. [21,45]. Plots were located at High Acres Nature Area (HANA) in a series of natural and created wetlands in Western New York, USA (43°5′ N, 77°23′ W) owned and managed by Waste Management of New York, LLC. This study was conducted in a 1.87 ha shallow emergent marsh, the North Pool of the Western Wetland complex at HANA. The land was previously used as a gravel depository, but was abandoned in the 1960s, left to fallow, and converted to an emergent wetland in 2009. Prior to its use as a gravel depository, the site was used for agricultural purposes. The wetland is fed through the subsurface by the adjacent remnant quarry pond, and contains a culvert in the south end which controls water flow to the pond directly south of the area, allowing control of water levels and consistent standing water year-round [21]. Soils in the North Pool wetland have relatively low organic matter, soil nutrients (nitrate, ammonium and total phosphorus); dominant plant species include broadleaf arrowhead (*Sagittaria latifolia*), pickerelweed (*Pontedaria cordata*), and white pond lily (*Nymphaea odorata*) [21].

In June of 2014, sixteen pairs of plots were established at the site, with each pair consisting of a 1 m^2 hardware mesh caged plot and an uncaged plot marked with poles (Figure 1). As described by Lodge et al. [21], plots were arranged randomly in four blocks of four pairs. A three-sided cage-control plot was also included in each block to ensure that the response variables were unaffected by the cages themselves. Because there was no cage effect over the first two years of the experiment, the cage controls were not used in this study.

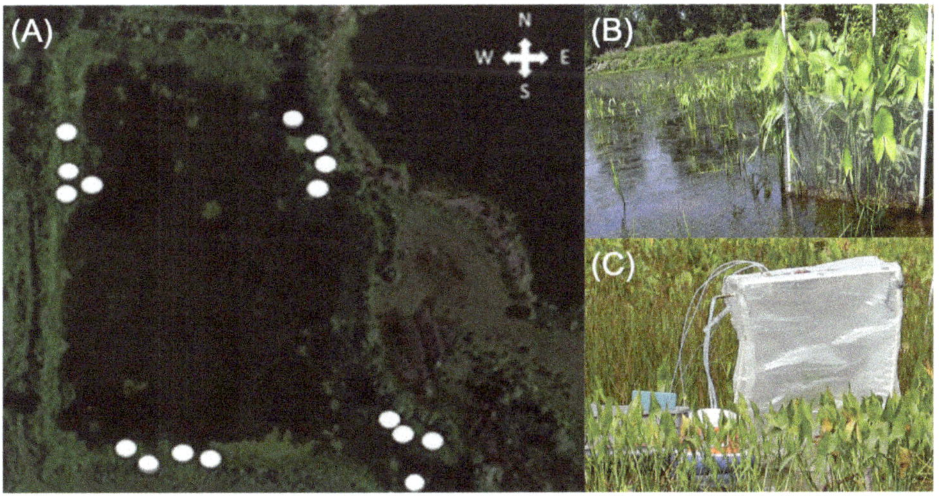

Figure 1. Experimental design at the North Pool of the Western Wetland complex at High Acres Nature Area in Western New York. (**A**) The site was divided into four blocks and the white markers indication locations of pairs of caged and uncaged plots, (**B**) cages were 1 m^2 and constructed of hardware cloth (photo credit: Kimberly Lodge), and (**C**) gas flux chambers were placed over the plots during measurements and were constructed from pvc pipe, clear polycarbonate sheeting and polyethylene greenhouse film that was rolled into the water and secured at the sediment surface (photo credit: Benjamin Hamilton.)

2.2. Grazer Abundance

We quantified waterfowl abundance based on observations by trained researchers and volunteers from June 2017 through November 2018. Species, abundance, date, and time of day were recorded on every visit with the frequency of observations varying among seasons (n = 2 to 19; higher observations in summer and fall). Grazer point counts were converted to density in units of individuals per ha and compiled by season (winter, spring, summer, and fall). Evidence of other herbivores, including deer, muskrats, and beavers, was present, but individuals were rarely observed.

2.3. Vegetation Cover, Biomass and Elemental Composition

Vegetation surveys were conducted every six weeks between early June and August 2017 and May and August 2018. Surveys included estimation of total plant cover within the plot as well as stem counts for each species and total grazer damage [46,47]. Percent cover was estimated by at least two observers per plot. We quantified damage by estimating the total leaf area removed by large grazers, relative to the extant abundance of each species [34,48]. The stem height, leaf height, and leaf width were measured for five individuals of each species for use in aboveground biomass estimation. Water depth was measured in conjunction with all vegetation surveys by averaging three measurements per plot. We estimated aboveground biomass for the eight plant species that comprised >95% of total cover roughly every six weeks during the growing seasons of 2017 and 2018 using species-specific allometric equations. We selected representative culms growing outside of experimental plots at the peak of the growing season (June–July), cut individuals from the bottom of the stem at the soil surface and immediately measured stem height, leaf height, and leaf width. We determined the per stem dry mass after drying at 60 °C and created regression curves based on the best-fit allometric relationship for each species (Supplemental Information, Table S1). Stem density and allometric characteristics measured during the vegetation surveys were used to calculate the total plot-level biomass for each species at each time point [48–50] and summed for each plot. We assessed C and nitrogen (N) composition of the five most abundant plant species from samples collected in August for use in the decomposition study (below). Plants were air-dried, ground to homogeneity using an electric coffee mill, and analyzed using a Perkin Elmer 2400 CHNS Elemental Analyzer. The total aboveground C in plant biomass was calculated for each species using the estimated biomass and measured C content and then summed for each plot.

A single soil core (6 cm diameter × 20 cm depth) was collected from each plot using an auger for the determination of belowground biomass and the elemental composition of roots and rhizomes. Cores were sieved (1 mm mesh) to remove soil particles, and roots and rhizomes were weighed after drying at 60 °C [36]. The C and N composition was measured as above and the total biomass (to 20 cm) and tissue C content used to calculate total C in belowground biomass.

2.4. Soil Characteristics and Elemental Composition

In October 2018 we used a syringe corer (2.5 cm diameter × 10 cm depth) to collect samples in triplicate from each plot for bulk density and elemental composition. Bulk density was calculated based on the mass of the soil core after drying at 60 °C and the initial core volume. Visible roots and rhizomes were removed and cores were then homogenized using a mortar and pestle and C and N content was measured as above. Bulk density and the C content were used to calculate areal soil C in the top 10 cm of soil.

2.5. Decomposition

We identified the dominant plant species based on the plant surveys and used these to measure decomposition using the litterbag method [48]. Four species were selected that in combination contributed at least 60% of the total cover: broadleaf cattail (*Typha latifolia*), broadleaf arrowhead (*S. latifolia*), pickerelweed *P. cordata*, and white pond lily (*N. odorata*). Specimens were collected from outside experimental plots at the end of August and air dried in the laboratory. We filled 20 × 20 cm square bags constructed from polyester screen (approximate mesh size 1 × 1 mm) with 10 g dry litter and placed four litterbags of each species into 12 plots (6 caged and 6 uncaged) in September 2018. Bags were collected after 30, 61, 181, and 211 days. The remaining material was rinsed thoroughly with tap water to remove soil, dried at 60 °C, and weighed. The decomposition rate (k-value) of each species and treatment was calculated as the linear slope of the natural log of the percent original mass remaining versus days in the field [51,52].

2.6. Gas Fluxes

We measured CO_2 fluxes using the static chamber method [53]. Measurements were made during the peak growing season (June–July) and at the beginning of plant senescence (late August–September) in both 2017 and 2018. We measured fluxes in two of the four experimental blocks (eight pairs of caged and uncaged plots). The chamber was constructed from a PVC frame that covered a 1 m^2 area and was adjustable in height (1 to 1.7 m) depending on the height of the plant canopy. The top was fitted with a clear polycarbonate panel that covered roughly one-third of the area, and the remainder of the top and the sides were covered with clear polyethylene greenhouse film. The chamber fit over the permanently installed PVC corner posts used to mark the plot and during measurements the plastic film was rolled down into the water and secured at the sediment surface with a chain to prevent lateral exchange of water or gases. The polycarbonate top and clear plastic film used to construct the chamber allowed approximately 67% of photosynthetically active radiation to pass through. For dark measurements, we covered the chamber with an opaque tarp.

The polycarbonate top was fitted with a small radiative panel that attached to two gas tight bulkheads so that chilled water could be circulated continuously through the chamber while it was sealed. We continuously monitored temperature throughout the sampling periods both inside and outside the chamber, and maintained internal temperatures within 5 °C of external temperature [53] by adjusting the chilled water circulation. Two additional bulkheads connected the inflow and outflow of an infrared gas analyzer (LI-820) and air was continuosly circulated from the chamber, through the analyzer and back to the chamber using a small air pump. We measured changes in headspace CO_2 in both the light and the dark over approximately a 10 min period.

Fluxes were calculated from the slope of the first 5 min period after chamber closure and calculated based on the headspace of the chamber above the water. We estimated gross primary productivity (GPP) by subtracting the dark measurement (ecosystem respiration, ER) from the light measurement (net ecosystem exchange, NEE) for each plot. We estimated summer C budget values for GPP and ER from chamber fluxes, assuming a 24 h period for respiration and a 12 h period for photosynthesis. Because submerged plants may directly take up dissolved inorganic carbon (DIC) released from heterotrophs in the sediments and water column, which was not reflected in our measured changes of CO_2 in the chamber headspace, we have likely underestimated overall ER and GPP.

2.7. Statistical Analyses

We performed all statistical analyses using the JMP Pro 14 statistical software. We evaluated each dataset for homogeneity of variance and normality prior to statistical analysis. Heterogeneity among blocks within each site was analyzed by including block as a random factor in analyses. For data that met the requirements of normality, we used a one-way ANOVA (belowground biomass and soil elemental composition) or for variables that had a seasonal component (GPP, ER, NEE, aboveground biomass, plant cover) or a species component (decomposition) a full-factorial two-way ANOVA with treatment and month or species as fixed factors. When significant interactions were identified, we used the Tukey post-hoc test to determine the differences among means. For data that could not be successfully transformed to meet the normality assumptions, including grazer density and grazer damage, we used a one-way Kruskal–Wallis test, followed by a Mann–Whitney U test to compare means. The relationship between peak growing season (July) aboveground biomass and carbon gas fluxes (GPP and ER) was assessed using simple linear regression of data from both treatments and years.

3. Results

3.1. Grazing Pressure

Waterfowl were consistently abundant in the North Pool site throughout the course of the study with Canada goose (*Branta canadensis*), mallard ducks (*Anas platyrhynchos*), and common gallinules (*Gallinula galeata*) as the most common species. Density ranged from 2 to 18 individuals ha^{-1} d^{-1}. Peak abundance occurred in summer 2017, which was higher than spring and summer 2018 (Supplemental Information, Figure S1).

3.2. Hydrologic Conditions

The North Pool wetland was consistently flooded, with water depths >8 cm throughout the time of this study, ranging to a maximum of 50 cm (Supplemental Information, Figure S2). The ability to control water flow out of the culvert situated at the southern end of the wetland meant that the regional drought experienced in 2017 had minimal impacts on water levels. Water depth was similar between the two years, with growing season averages ranging from 24 to 29 cm.

3.3. Vegetation

A total of 12 species were found in experimental plots, with distinct differences in composition and abundance across seasons and treatments (Figure 2). Aboveground biomass showed a predictable seasonal pattern, with biomass increasing from spring into summer and peaking in mid to late summer (Table 1, Figure 2). Aboveground biomass in 2017, which was unusually hot and dry, was lower across seasons and treatments than in 2018. The exclosure treatment resulted in a significant reduction in aboveground biomass in both 2017 and 2018, and in 2018 there was a significant interaction between treatment and month, with the effect of grazer exclusion on aboveground biomass increasing from spring into summer (Table 1). During the height of the growing season (July), aboveground biomass was 3-fold (2017) and 2-fold (2018) higher in caged plots than uncaged plots. In the 2017, both grazer treatment and season were highly significant ($p < 0.0001$), with no interaction. However, in 2018, there was a significant interaction between season and treatment ($p = 0.003$), with the difference between caged and uncaged treatments increasing over the course of the growing season. Similar seasonal and treatment patterns were seen for plant cover (Supplemental Information, Figure S3, Table S2). The species selected for biomass assessment in both years comprised greater than 90% of the overall cover.

Table 1. Results of analysis of variance examining the effect of date (May, June, July, August or summer, fall) and grazing treatment (caged, uncaged) on aboveground biomass, gross primary productivity (GPP), ecosystem respiration (ER) and net ecosystem exchange (NEE) measured in 2017 and 2018. Significant *p*-values are bolded.

Variable	Date		Treatment		Dt x Tr	
	F	*p*	F	*p*	F	*p*
Aboveground biomass 2017	$F_{2,87} = 20.5$	**$p < 0.0001$**	$F_{1,87} = 17.0$	**$p < 0.0001$**	$F_{2,87} = 2.6$	$p = 0.08$
Aboveground biomass 2018	$F_{3,117} = 34.2$	**$p < 0.0001$**	$F_{1,117} = 35.7$	**$p < 0.0001$**	$F_{2,117} = 4.9$	**$p = 0.003$**
GPP 2017	$F_{1,25} = 23.2$	**$p < 0.0001$**	$F_{1,25} = 11.6$	**$p = 0.002$**	$F_{1,25} = 6.5$	**$p = 0.02$**
GPP 2018	$F_{1,26} = 7.7$	**$p = 0.01$**	$F_{1,26} = 4.0$	$p = 0.06$	$F_{1,26} = 3.9$	$p = 0.06$
ER 2017	$F_{1,27} = 7.7$	**$p = 0.01$**	$F_{1,27} = 6.4$	**$p = 0.02$**	$F_{1,27} = 0.01$	$p = 0.9$
ER 2018	$F_{1,26} = 0.003$	$p = 0.95$	$F_{1,26} = 5.1$	**$p = 0.03$**	$F_{1,26} = 3.1$	$p = 0.09$
NEE 2017	$F_{1,26} = 17.4$	**$p = 0.0003$**	$F_{1,26} = 7.5$	**$p = 0.01$**	$F_{1,26} = 7.5$	**$p = 0.01$**
NEE 2018	$F_{1,27} = 10.4$	**$p = 0.003$**	$F_{1,27} = 2.7$	$p = 0.1$	$F_{1,27} = 3.6$	$p = 0.07$

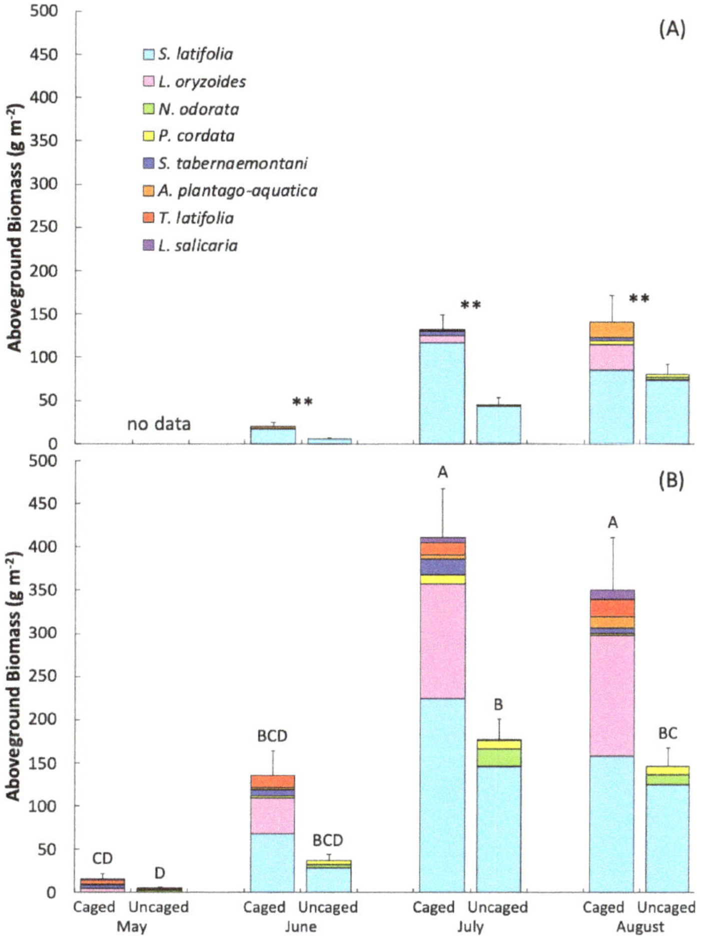

Figure 2. Aboveground biomass of dominant species during the (**A**) 2017 and (**B**) 2018 growing season for caged and uncaged treatments. Biomass of minor species, with no value > 2% of the total biomass in any season were excluded. These species are: *Potamogeton crispus*, *Echinochloa crus-galli*, *Polygonum persicaria*, and *Carex* sp. Error bars are the standard error for the total biomass in each season and treatment (n = 16). Stars indicate significant differences between grazer treatments (** $p < 0.001$) for 2017 where no interaction between season and treatment was found. Unique letters above bars in 2018 indicate statistically distinct values for the interaction between grazer treatment and season ($p < 0.01$).

The species composition of aboveground biomass shifted seasonally and between treatments (Figure 2). *S. latifolia* had the highest biomass across years and seasons, with the exception of spring 2018 when *Leersia oryzoides*, *T. latifolia*, *Alisma plantago-aquatica* were present but overall biomass was low. *S. latifolia* was particularly dominant in uncaged plots during the peak of the growing season, contributing >90% of the aboveground biomass in uncaged plots in 2017 and 75–85% of the biomass in uncaged plots in 2018. In 2018, *N. odorata* was also prominent in uncaged plots, contributing 11% of the peak growing season biomass, while being almost entirely absent from caged plots across both years. A larger number of species contributed to the aboveground biomass in caged plots, with *S. latifolia* contributing 50–90% of growing season biomass and *L. oryzoides* up to 40%, while being essentially absent (<0.05%) from uncaged areas. *T. latifolia* was present only in caged plots, and aside from May 2018, always low (<10%). *P. cordata* was present in both treatments across the growing season, but was a minor contributor to biomass. Additional species present in low abundance (<5% of total biomass) in caged plots included *Schoenoplectus tabernaemontani*, *Scirpus* sp., *Echinochloa crus-galli*, and *Lythrum salicaria*. Grazer damage was observed only for *P. cordata*, *S. latifolia*, and *N. odorata*. Of the species impacted by grazers, relative damage ranged from 17 to 25% (scaled to abundance) and there were no significant differences among species, although relative damage was highest for *S. latifolia* (Supplemental Information, Figure S4). We note, however, that these measurements do not take into account species completely excluded by grazers or those for which no measurement of grazer damage was assessed.

3.4. Soil and Belowground Biomass

Belowground plant biomass in the top 10 cm tended to be higher in caged plots (caged: 555 ± 127; uncaged: 338 ± 105 g m^{-2}). However, there was no significant effect of grazing ($F_{1,27}$ = 1.6 p = 0.2) and there was no difference in the C and N content between treatments ($F_{1,13}$ < 0.02 p = 0.9, Table 2). Soil C increased from 5.06 ± 0.25% to 6.42 ± 0.23% in the absence of grazers ($F_{1,30}$ = 12.3, p = 0.002, Table 2). Although not significant, this increase in %C was accompanied by increased %N ($F_{1,30}$ = 2.8, p = 0.1), resulting in no change in soil C:N (Table 2). Soil bulk density was not different between treatments (caged: 0.34 ± 0.03, uncaged: 0.32 ± 0.03) and the higher C content of soils within caged plots yielded 34% higher C storage in the top 10 cm of soil when grazers were excluded ($F_{1,30}$ = 6.4, p = 0.02).

Table 2. Elemental composition of dominant plant species at the study site and belowground biomass and soil from caged and uncaged plots, mean ± SE. Bold values indicate significant differences betwen caged an uncaged plots.

Variable	%C	%N	C:N
Vegetation			
S. latifolia	42.0 ± 0.01	2.0 ± 0.05	24.9 ± 0.6
N. odorata	41.9 ± 0.23	2.5 ± 0.01	19.8 ± 0.2
T. latifolia	44.4 ± 0.14	2.1 ± 0.08	24.6 ± 0.8
P. cordata	41.2 ±0.07	1.3 ± 0.02	37.3 ± 0.6
L. oryzoides	43.1 ± 0.03	1.4 ± 0.13	35.9 ± 3.3
Belowground biomass			
Caged	35.9 ± 1.66	1.6 ± 0.07	26.0 ± 1.3
Uncaged	35.5 ± 2.00	1.6 ± 0.14	25.7 ± 1.7
Soil			
Caged	**6.5 ± 0.33**	0.5 ± 0.13	19.7 ± 0.7
Uncaged	**5.1 ± 0.36**	0.3 ± 0.02	20.8 ± 0.7

3.5. Decomposition Rates

Decomposition rates varied significantly across species; *N. odorata* decomposed faster than *T. latifolia*, *P. cordata* and *S. latifolia* ($F_{3,35}$ = 255, p < 0.001, Figure 3). Across all species measured, *N. odorata* was also the species with the lowest litter C:N (Table 2). Differences in these ratios were largely driven by species-specific differences in %N, with values ranging from a low of 1.3 ± 0.02% in *P. cordata* to 2.5 ± 0.01% in *N. odorata* (Table 2). Decomposition was greater in caged plots overall ($F_{1,35}$ = 7.3, p = 0.01), but the significant interaction between species and treatment showed that the increase in decomposition when grazers were excluded only occurred for *N. odorata* ($F_{3,35}$ = 6.5, p = 0.001, Figure 3).

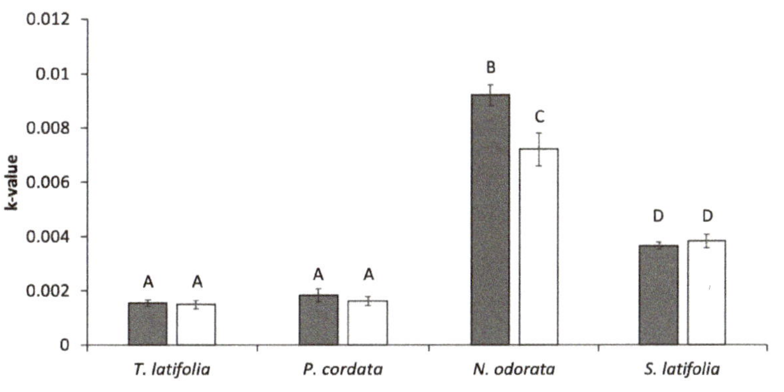

Figure 3. Decomposition rate of the four dominant macrophytes in caged (grey) and uncaged (white) plots. There was a significant interaction between species and treatment, and letters indicate statistical differences between average k-values, $p < 0.01$.

3.6. Gas Fluxes

Carbon dioxide fluxes differed between summer and fall and were strongly influenced by grazing (Table 1, Figure 4). Across grazer treatments, there was significantly higher NEE in the summer compared to the fall in both years (2017: $p = 0.0003$; 2018: $p = 0.003$). There was also a significant interaction in 2017 between season and treatment for NEE ($p = 0.01$), with summertime net uptake in caged plots more than double that of uncaged plots, whereas fall values were similar across treatments. Across years, ER was significantly higher in caged plots (2017: $p = 0.02$; 2018: $p = 0.03$); and in 2017, it was significantly higher in the summer compared to the fall ($p = 0.02$), but showed no interaction between treatment and season in either year. GPP showed a clear interaction between season and treatment, with a significant interaction in 2017 and a strong trend towards an interaction in 2018 (2017: $p = 0.02$; 2018: $p = 0.06$), with higher primary productivity in caged plots during the summer. In summer 2017, when the grazer effect was greatest, caged plots fixed 55% more carbon than uncaged plots. Across years and treatments there was also a significant linear relationship between aboveground biomass and GPP and ER (GPP: $R^2 = 0.3$, $F_{1,29} = 9.2$, $p = 0.005$, ER: $R^2 = 0.4$, $F_{1,29} = 17.6$, $p < 0.001$), with higher aboveground biomass correlated with higher GPP and ER (Figure 5).

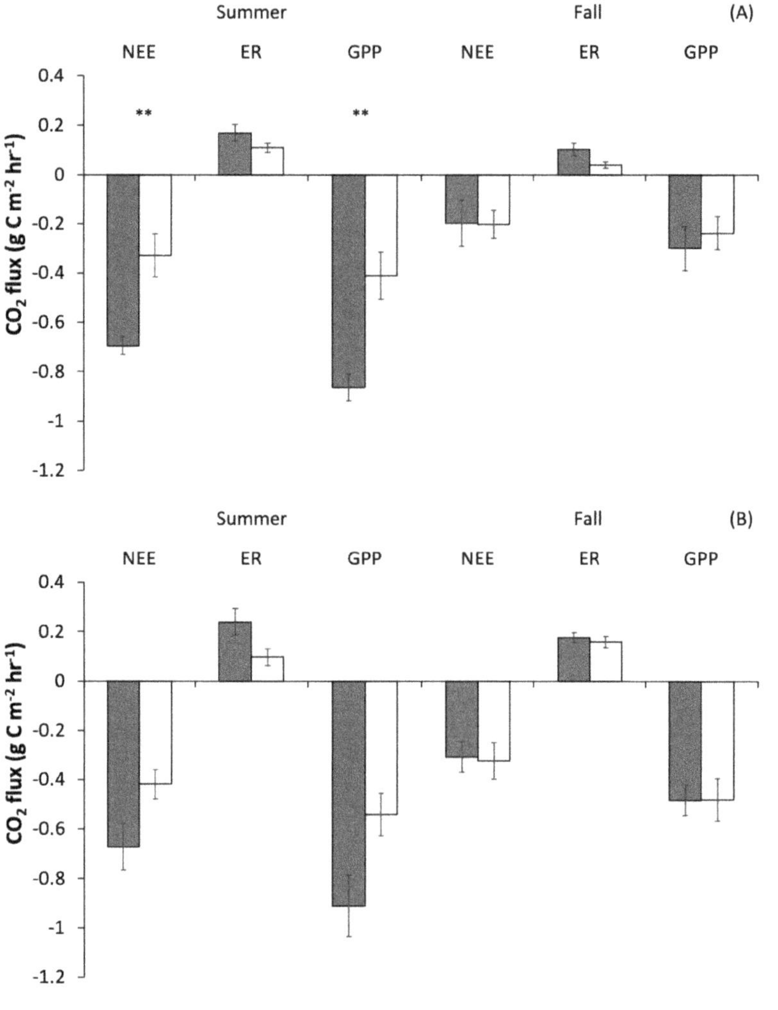

Figure 4. Net ecosystem exchange (NEE), ecosystem respiration (ER) and gross primary production (GPP) in 2017 (**A**) and 2018 (**B**) for caged (grey) and uncaged (white) treatments. Summer measurements were taken in the period June–July, fall measurements were taken in the period August–September. Stars indicate significant differences between caged and uncaged plots ($p < 0.05$) within a season, where there was a significant interaction between season and treatment.

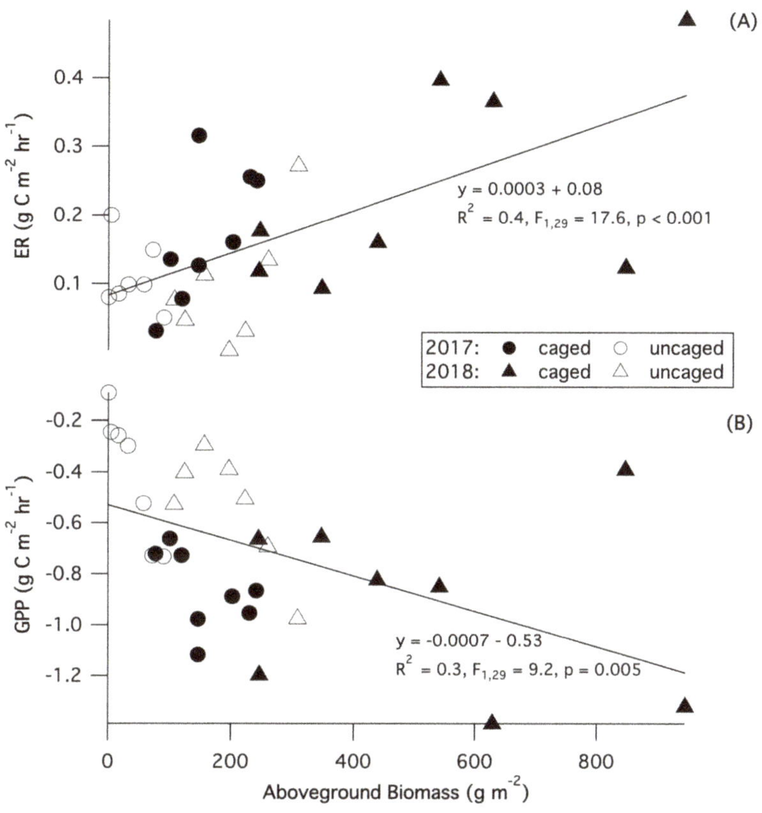

Figure 5. Relationship between peak growing season (July) aboveground biomass and (**A**) ecosystem respiration (ER) and (**B**) gross primary production (GPP). Regression analysis includes data from all plots and years in which both biomass and gas fluxes were measured (n = 31).

4. Discussion

Exclusion of herbivores had large impacts on the plant community and thereby carbon cycling and storage, with substantially lower biomass and gross productivity where grazers have access. While carbon losses in the form of ecosystem respiration and decomposition were also higher in the absence of grazers, higher rates of net ecosystem exchange and increased soil C content suggest that overall carbon sequestration is increased when herbivores are excluded. This points toward grazer management as a potential tool for establishing robust vegetation communities in created wetlands and for accelerating the accumulation of soil C, a characteristic that often differentiates created from natural wetlands [7,8,54].

Plant communities were impacted by grazers throughout the growing season, suggesting that exclusion of megaherbivores over multiple years has a persistent influence on the wetlands. Caged plots were more diverse and consistently had higher plant cover and aboveground biomass. When converted to units of carbon, using the average of 42.5% C in plant tissue measured at the site (Table 2), peak growing season aboveground biomass after five growing seasons of grazer exclusion was 175 g C m^{-2} compared to 75 g C m^{-2} in control plots (Figure 6). The two to three fold increase in aboveground biomass and vegetation cover in caged plots is greater than the 25–60% increase in aboveground biomass in riparian wetlands protected from grazers reported by Veen et al. [55] and the 50% increase in reed stem density observed by Dingemans et al. [33] when geese were excluded from wetland plots. Likewise, Mulder and Ruess [56] reported a significant decrease in aboveground biomass in their grazed treatments in salt marshes. The relative increase in cover in caged plots was greater than observed in earlier reported values from this experiment [21], suggesting that the impact of grazers is cumulative over time with continued access to young wetland sites. Further, the higher vegetation cover (but not biomass) in caged plots in spring 2018 suggests that the effects of grazer exclusion carry over year to year, and in created wetlands, which typically have young plant communities with lower stem density and belowground rhizomal development, limiting grazing may encourage the growth and establishment of stable and diverse plant communities [57].

Belowground biomass was higher than that measured by Lodge and Tyler [21], suggesting that as the wetland is maturing, belowground biomass is increasing and will likely continue to contribute carbon stocks to the soil. In contrast to aboveground biomass, belowground biomass was only slightly, and not significantly influenced by grazer access. Lodge and Tyler [21] previously observed significantly higher belowground biomass in these same caged plots, indicating that more intensive sampling may strengthen our observed trend. Unlike measurements of aboveground biomass, our measurements of belowground biomass only occurred once during the year (October), potentially missing some impacts of grazers on belowground processes. A more detailed sampling approach, including seasonal sampling and species-level identification of belowground biomass, would enable better linkages between aboveground responses to herbivory and belowground C storage.

Shifts in vegetation dominance when grazers were excluded were consistent with the grazing preferences of waterfowl at the site. In uncaged plots, semi-desirable species were frequently damaged by grazing, such as *S. latifolia*, resulting in lower biomass. In contrast, some plants in the caged plots were able to grow to maturity and may eventually become less palatable to grazers [58]. These species included *L. oryzoides* and *T. latifolia*, which were virtually absent in open plots. In 2018, we also observed the emergence of *N. odorata* in the open plots, suggesting that the removal of other palatable species allowed expansion of surface-covering lilies that may lead to further shifts in ecosystem structure. Bagchi and Richi [38] made similar observations in grasslands, reporting shifts in vegetation dominance consistent with grazer preferences, with a higher abundance of more palatable plants and higher diversity in ungrazed plots. In a review of the environmental impacts of expanding goose populations, a dominant grazer in our system, Buij et al. [59] found that grazers not only reduce overall plant biomass, but also reduce plant diversity and modify the habitat.

Growing season CO_2 fluxes in our system were comparable to other created/restored temperate emergent wetlands, with our estimated daily summer net CO_2 uptake for caged plots of -5.2 g C m^{-2} d^{-1} (Figure 6), falling within the range of August CO_2 uptake rates (-8.5 to -2.3 g C m^{-2} d^{-1}) (California; [60]) and our hourly uptake rates of -0.7 g C m^{-2} h^{-1} (caged) and -0.3 g C m^{-2} h^{-1} (uncaged) bracketing growing season static chamber flux values of -0.45–0.55 g C m^{-2} h^{-1} (Ohio; [61]). Changes in ecosystem CO_2 fluxes in response to grazing tracked patterns in aboveground biomass, with higher aboveground biomass correlated with higher rates of GPP and ER. Increases in growing season GPP in response to grazer exclusion was higher than that of ER, resulting in 25–100% higher summer net CO_2 uptake in caged plots. Higher temporal resolution of flux measurements would provide additional insights into the seasonality of grazer impacts on CO_2 fluxes and enable a more detailed assessment of shifts in carbon uptake in response to grazer exclusion. The relationship between CO_2 fluxes and plant cover confirms the strong influence of plant biomass on CO_2 uptake [62], suggesting that management that impacts the establishment and growth of vegetation communities can have large effects on carbon storage in these systems. Similar results were seen following grazer exclusion in alpine meadows [41] and high arctic wet meadows [40], where significantly lower carbon uptake in grazed plots was attributed to the substantial reduction in aboveground biomass.

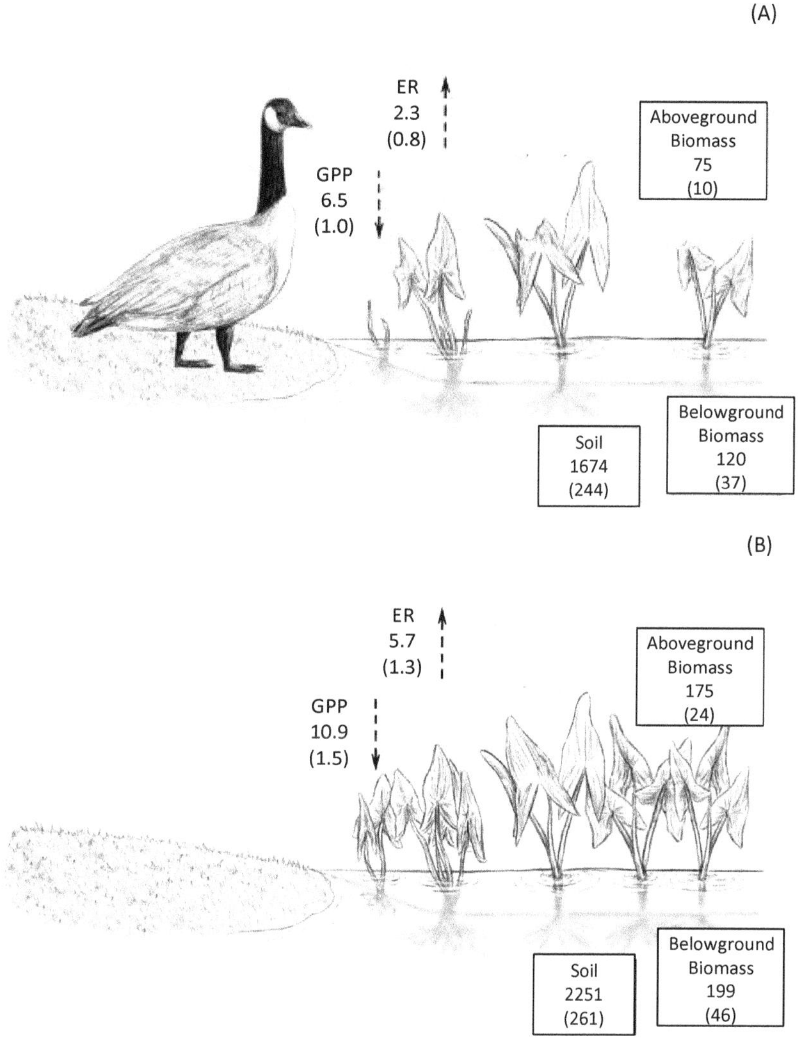

Figure 6. Estimated peak growing season carbon budget with (**A**) and after 4 years without (**B**) grazing. Average values of each variable are listed with standard error in parentheses, with values taken from caged and uncaged plots during the peak of the growing season (July) following five growing seasons of grazer exclusion (2018). Carbon pools are denoted by boxes, with units of g C m^{-2}; carbon fluxes are denoted by dashed arrows, with units of g C m^{-2} day^{-1}.

Decomposition rates varied strongly with plant litter type, suggesting that one of the most significant impacts of grazers on carbon cycling may be through the combined impact on species composition and total plant production. Variations in decomposition rates across species generally tracked differences in C:N, where species with the lowest C:N (*N. odorata*: 19.8 ± 0.2) decomposed fastest and the species with highest C:N (*P. cordata*: 37.3 ± 0.6) decomposed slowest. *T. latifolia* was the exception, with both low C:N and slow decomposition. In our system, shifts in species composition following grazer exclusion (e.g., Figure 2) contribute to slower decomposition rates. While the species with similar decomposition rates (*S. latifolia*, *P. cordata*) were more similar in biomass between treatments, the fast decomposing species, *N. odorata*, contributed <0.3% of the aboveground biomass in caged plots (<1 g m^{-2} in 2018) compared to 9% in uncaged plots at the end of the growing season (13 g m^{-2} in 2018). In contrast, the slower decomposing *T. latifolia* was only found in caged plots, with about 6% of the end of season biomass. While *L. oryzoides* (40% of end of season biomass in caged plots only) was not used in our decomposition study, its high C:N (35 .9 ± 3.3) suggests persistence within the system. *T. latifolia* and *L. oryzoides* together contributed an additional 160 g m^{-2} of relatively refractory material to caged plots in 2018. These species composition shifts, coupled with higher litter inputs associated with high aboveground biomass in caged plots will cascade over time to enhance the accumulation of soil carbon in the absence of grazers.

Our results support findings that changes in decomposition in response to grazing can arise from both changes in plant communities and alterations in the soil environment [42]. Grazer exclusion promoted decomposition of the most labile species in our study (*N. odorata*), suggesting that changes in soil environment due to grazing has the largest impact on more labile plant litter. In wetland ecosystems, high plant biomass can accelerate decomposition through two mechanisms [63], root oxygen loss creating aerobic sediment microsites [43,44] and the priming of the rhizosphere through the release of labile carbon from roots [64], thus facilitating the breakdown of plant litter in caged plots. It is important to note that *N. odorata* was not found in any caged plots, thus its higher decomposition rate in the absence of grazers did not impact overall decomposition rates. An additional impact of grazers on decomposition is changes in plant chemistry in response to grazing [42,65], which was not considered in our study. Plant responses to grazer can result in either increased [42] or decreased [66] C:N, which could exacerbate or reduce the patterns in decomposition observed in this study and bears further investigation.

A meta-analysis of the impacts of large grazers on carbon storage found that while data are more limited for wetlands, the response to grazers is similar to terrestrial ecosystems and experience reductions in soil carbon under grazing [67]. Overall carbon storage in the soil was increased >30% by grazer exclusion and carbon stored in belowground biomass was increased by >60% (Figure 6). While many wetland grazing studies focus on the impacts of mammalian grazers such as nutria (*Myocastor coypus*; e.g., [68]) or livestock (e.g., [69]), waterfowl grazing has been shown to reduce carbon storage in arctic tundra by a similar amount (35%; [70]). This pattern is consistent with grazing studies conducted across a range of ecosystems (e.g., North American, prairie [71]; Mongolian steppes [72]; Himalayan grasslands [38]). We measured belowground biomass in the top 20 cm of soil and soil C in the top 10 cm, and therefore may have underestimated overall belowground biomass and soil C, missing some impacts of grazing on belowground processes. However, grazer studies in other systems have found that the largest effects of grazer presence are observed in the soil surface (e.g., [73]), and therefore measurements in deeper soil layers are unlikely to significantly change our results.

While most prior studies have evaluated the impact of grazing on mature systems, with perhaps greater resilience, the results here illustrate the strongly divergent trajectories of ecosystem development that may occur when grazing pressure is applied soon after construction of emergent freshwater wetlands and extended over several years. In this nascent created wetland, where hydrology supports high waterfowl populations, both carbon pools and carbon fluxes were significantly impacted by grazing (Figure 6). The substantially high C fluxes were in the absence of grazers - approximately 4 g C m^{-2} d^{-1} greater intake of atmospheric carbon into the wetland through gross photosynthesis-cascaded to an additional 750 g C m^{-2} higher C in aboveground biomass, belowground biomass and soil pools after five growing seasons of grazer exclusion (Figure 6). Future trajectories may lead to slower development of emergent marsh structure, or transition to an alternate stable state dominated by submerged macrophytes. Thus, the impact of exclusion over longer time periods is worthy of future study to evaluate the persistence of these divergent trends in ecosystem development.

5. Conclusions

Grazing can significantly limit the carbon storage potential of created wetlands, underscoring the importance of managing waterfowl populations in newly created wetlands where vegetation communities are not fully established. As shifts in global climate alter behavior and migratory patterns of herbivorous waterfowl (e.g., [74]), a greater understanding of the interdependence between restoration planning and population dynamics of dominant species will be required. These results also highlight the need for frequent monitoring of not only plant populations and hydrology following wetland creation, but also the use (or overuse) of nascent systems by potentially damaging species. By limiting waterfowl numbers, and by extension grazing intensity, during the early stages of wetland development in created wetlands, a more stable and diverse plant community may form [35–37], maximizing potential C sequestration.

Supplementary Materials: The following are available online at https://www.mdpi.com/article/10.3390/land10080805/s1; Figure S1: Waterfowl observations June 2017 through November 2018; Figure S2: Water depth measured throughout the growing season (May–August) of 2017 to 2018; Figure S3: Vegetation cover during the 2017 and 2018 growing seasons; Figure S4: Grazer damage for each species normalized to species abundance; Table S1: Regression curves used to estimate aboveground biomass; Table S2: Results of analysis of variance on the effects and interactions of month (Mo), and treatment (Tr) for vegetation cover in 2017 and 2018.

Author Contributions: Conceptualization, A.C.T., C.K.M. and D.M.S.; methodology, A.C.T. and C.K.M.; formal analysis, A.C.T., C.K.M. and D.M.S.; investigation, D.M.S.; resources, A.C.T. and C.K.M.; writing—original draft preparation, D.M.S.; writing—review and editing, A.C.T. and C.K.M.; visualization, A.C.T., C.K.M. and D.M.S.; supervision, A.C.T. and C.K.M.; funding acquisition, A.C.T. and C.K.M. All authors have read and agreed to the published version of the manuscript.

Funding: This research was funded by Waste Management of New York and the Thomas H. Gosnell School of Life Sciences and the College of Science at the Rochester Institute of Technology.

Institutional Review Board Statement: Not applicable.

Informed Consent Statement: Not applicable.

Data Availability Statement: Not applicable.

Acknowledgments: We sincerely thank Waste Management of New York and the Thomas H. Gosnell School of Life Sciences and the College of Science at the Rochester Institute of Technology for funding this work. We are grateful to Bruce Cady for participating in grazer monitoring and to Kimberly Lodge for establishing the grazer experiment. We thank Ben Hamilton, Evan Squier, Michael McGowan, Briana Stringer, and Sydney VanWinkle for their assistance in the field and Nicole Fornof and Rebecca Zayatz for logistical support.

Conflicts of Interest: The authors declare no conflict of interest. The funders had no role in the design of the study; in the collection, analyses, or interpretation of data; in the writing of the manuscript, or in the decision to publish the results.

References

1. Zedler, J.B.; Kercher, S. WETLAND RESOURCES: Status, Trends, Ecosystem Services, and Restorability. *Annu. Rev. Environ. Resour.* **2005**, *30*, 39–74. [CrossRef]
2. Aerts, R.; Verhoeven, J.T.A.; Whigham, D.F. Plant-Mediated Controls On Nutrient Cycling in Temperate Fens and Bogs. *Ecology* **1999**, *70*, 2170–2181 [CrossRef]
3. DeAngelis, D.L.; Bartell, S.M.; Brenkert, A.L. Effects of Nutrient Recycling and Food-Chain Length on Resilience. *Am. Nat.* **1989**, *134*, 778–805. [CrossRef]
4. Costanza, R.; D'Arge, R.; de Groot, R.; Farber, S.; Grasso, M.; Hannon, B.; Limburg, K.; Naeem, S.; O'Neill, R.V.; Paruelo, J.; et al. The value of the world's ecosystem services and natural capital. *Nature* **1997**, *387*, 253–260. [CrossRef]
5. Chmura, G.L.; Anisfeld, S.C.; Cahoon, D.R.; Lynch, J.C. Global carbon sequestration in tidal, saline wetland soils. *Glob. Biogeochem. Cycles* **2003**, *17*, 1111. [CrossRef]
6. Kayranli, B.; Scholz, M.; Mustafa, A.; Hedmark, Å. Carbon Storage and Fluxes within Freshwater Wetlands: A Critical Review. *Wetlands* **2010**, *30*, 111–124. [CrossRef]

7. Campbell, D.A.; Cole, C.A.; Brooks, R.P. A comparison of created and natural wetlands in Pennsylvania, USA. *Wetl. Ecol. Manag.* **2002**, *10*, 41–49. [CrossRef]
8. Fennessy, M.S.; Rokosch, A.; Mack, J.J. Patterns of plant decomposition and nutrient cycling in natural and created wetlands. *Wetlands* **2008**, *28*, 300–310. [CrossRef]
9. Moreno-Mateos, D.; Power, M.E.; Comín, F.A.; Yockteng, R. Structural and functional loss in restored wetland ecosystems. *PLoS Biol.* **2012**, *10*, e1001247 [CrossRef] [PubMed]
10. Moreno-Mateos, D.; Alberdi, A.; Morriën, E.; van der Putten, W.H.; Rodríguez-Uña, A.; Montoya, D. The long-term restoration of ecosystem complexity. *Nat. Ecol. Evol.* **2020**, *4*, 676–685. [CrossRef] [PubMed]
11. Miller, R.L.; Fujii, R. Plant community, primary productivity, and environmental conditions following wetland re-establishment in the Sacramento-San Joaquin Delta, California. *Wetl. Ecol. Manag.* **2010**, *18*, 1–16. [CrossRef]
12. Rothman, E.; Bouchard, V. Regulation of carbon processes by macrophyte species in a Great Lakes coastal wetland. *Wetlands* **2007**, *27*, 1134–1143. [CrossRef]
13. Collins, M.E.; Kuehl, R. *Organic Matter Accumulation and Organic Soils*; Lewis Publishers: Boca Raton, FL, USA, 2000.
14. Mitsch, W.; Gosselink, J. *Wetlands*; Wiley: New York, NY, USA, 2007.
15. De Deyn, G.B.; Cornelissen, J.H.C.; Bardgett, R.D. Plant functional traits and soil carbon sequestration in contrasting biomes. *Ecol. Lett.* **2008**, *11*, 516–531. [CrossRef]
16. Mitsch, W.J.; Bernal, B.; Nahlik, A.M.; Mander, Ü.; Zhang, L.; Anderson, C.J.; Jørgensen, S.E.; Brix, H. Wetlands, carbon, and climate change. *Landsc. Ecol.* **2013**, *28*, 583–597. [CrossRef]
17. Yu, L.; Huang, Y.; Sun, F.; Sun, W. A synthesis of soil carbon and nitrogen recovery after wetland restoration and creation in the United States. *Sci. Rep.* **2017**, *7*, 1–9. [CrossRef] [PubMed]
18. Hossler, K.; Bouchard, V. Soil development and establishment of carbon-based properties in created freshwater marshes. *Ecol. Appl.* **2010**, *20*, 539–553. [CrossRef] [PubMed]
19. Murkin, H.R.; Murkin, E.J.; Ball, J.P. Avian Habitat Selection and Prairie Wetland Dynamics: A 10-Year Experiment. *Ecol. Appl.* **1997**, *7*, 1144–1159. [CrossRef]
20. Lor, S.; Malecki, R.A. Breeding Ecology and Nesting Habitat Associations of Five Marsh Bird Species in Western New York. *Waterbirds* **2006**, *29*, 427–436. [CrossRef]
21. Lodge, K.A.; Tyler, A.C. Divergent impact of grazing on plant communities of created wetlands with varying hydrology and antecedent land use. *Wetl. Ecol. Manag.* **2020**, *28*, 797–813. [CrossRef]
22. Isola, C.R.; Colwell, M.A.; Taft, O.W.; Safran, R.J. Interspecific Differences in Habitat Use of Shorebirds and Waterfowl Foraging in Managed Wetlands of California's San Joaquin Valley. *Waterbirds* **2000**, *23*, 196–203.
23. Ankney, C.D. An embarrassment of riches: Too many geese. *J. Wildl. Manag.* **1996**, *60*, 217–223. [CrossRef]
24. Lauridsen, T.L.; Jeppesen, E.; Andersen, F.Ø. Colonization of submerged macrophytes in shallow fish manipulated Lake Væng: Impact of sediment composition and waterfowl grazing. *Aquat. Bot.* **1993**, *46*, 1–15. [CrossRef]
25. Smith III, T.J.; Odum, W.E. The effects of grazing by snow geese on coastal salt marshes. *Ecology* **1981**, *62*, 98–106. [CrossRef]
26. Silliman, B.R.; Zieman, J.C. Top-down control of Spartina alterniflora production by periwinkle grazing in a Virginia salt marsh. *Ecology* **2001**, *82*, 2830–2845. [CrossRef]
27. Silliman, B.R.; Bertness, M.D. A trophic cascade regulates salt marsh primary production. *Proc. Natl. Acad. Sci. USA* **2002**, *99*, 10500–10505. [CrossRef]
28. Srivastava, D.S.; Jefferies, R. A positive feedback: Herbivory, plant growth, salinity, and the desertification of an Arctic salt-marsh. *J. Ecol.* **1996**, *84*, 31–42. [CrossRef]
29. Smith, L.M.; Kadlec, J.A. Fire and herbivory in a Great Salt Lake marsh. *Ecology* **1985**, *66*, 259–265. [CrossRef]
30. Jefferies, R.L.; Jano, A.P.; Abraham, K.F. A biotic agent promotes large-scale catastrophic change in the coastal marshes of Hudson Bay. *J. Ecol.* **2006**, *94*, 234–242. [CrossRef]
31. Jefferies, R.L.; Rockwell, R.F. Foraging geese, vegetation loss and soil degradation in an Arctic salt marsh. *Appl. Veg. Sci.* **2002**, *5*, 7–16. [CrossRef]
32. Ström, L.; Mastepanov, M.; Christensen, T.R. Species-specific Effects of Vascular Plants on Carbon Turnover and Methane Emissions from Wetlands. *Biogeochemistry* **2005**, *75*, 65–82. [CrossRef]
33. Dingemans, B.J.J.; Bakker, E.S.; Bodelier, P.L.E. Aquatic herbivores facilitate the emission of methane from wetlands. *Ecology* **2011**, *92*, 1166–1173. [CrossRef]
34. Winton, R.S.; Richardson, C.J. Top-down control of methane emission and nitrogen cycling by waterfowl. *Ecology* **2017**, *98*, 265–277. [CrossRef] [PubMed]
35. Lubchenco, J. Littornia and Fucus: Effects of Herbivores, Substratum Heterogeneity, and Plant Escapes During Succession. *Ecology* **1983**, *64*, 1116–1123. [CrossRef]
36. Elaine Evers, D.; Sasser, C.E.; Gosselink, J.G.; Fuller, D.A.; Visser, J.M. The impact of vertebrate herbivores on wetland vegetation in Atchafalaya Bay, Louisiana. *Estuaries* **1998**, *21*, 1–13. [CrossRef]
37. Kennedy, M.A.; Heck, K.L.; Michot, T.C. Impacts of wintering redhead ducks (Aythya americana) on seagrasses in the northern Gulf of Mexico. *J. Exp. Mar. Biol. Ecol.* **2018**, *506*, 42–48. [CrossRef]
38. Bagchi, S.; Ritchie, M.E. Introduced grazers can restrict potential soil carbon sequestration through impacts on plant community composition. *Ecol. Lett.* **2010**, *13*, 959–968. [CrossRef] [PubMed]

39. Speed, J.D.; Woodin, S.; Tømmervik, H.; Van der Wal, R. Extrapolating herbivore-induced carbon loss across an arctic landscape. *Polar Biol.* **2010**, *33*, 789–797. [CrossRef]
40. Sjögersten, S.; van der Wal, R.; Loonen, M.J.; Woodin, S.J. Recovery of ecosystem carbon fluxes and storage from herbivory. *Biogeochemistry* **2011**, *106*, 357–370. [CrossRef]
41. Hirota, M.; Tang, Y.; Hu, Q.; Kato, T.; Hirata, S.; Mo, W.; Cao, G.; Mariko, S. The potential importance of grazing to the fluxes of carbon dioxide and methane in an alpine wetland on the Qinghai-Tibetan Plateau. *Atmos. Environ.* **2005**, *39*, 5255–5259. [CrossRef]
42. Wang, Z.; Yuan, X.; Wang, D.; Zhang, Y.; Zhong, Z.; Guo, Q.; Feng, C. Large herbivores influence plant litter decomposition by altering soil properties and plant quality in a meadow steppe. *Sci. Rep.* **2018**, *8*, 1–12. [CrossRef]
43. Colmer, T. Long-distance transport of gases in plants: A perspective on internal aeration and radial oxygen loss from roots. *Plant Cell Environ.* **2003**, *26*, 17–36. [CrossRef]
44. Wolf, A.A.; Drake, B.G.; Erickson, J.E.; Megonigal, J.P. An oxygen-mediated positive feedback between elevated carbon dioxide and soil organic matter decomposition in a simulated anaerobic wetland. *Glob. Chang. Biol.* **2007**, *13*, 2036–2044. [CrossRef]
45. Lodge, K.A. Hydrology, Nutrient Availability, and Herbivory Interacting to Control Ecosystem Functions and Services in Created Emergent Freshwater Wetlands. Master's Thesis, Rochester Institute of Technology, ProQuest Dissertations Publishing, Rochester, NY, USA, 2017.
46. Bakker, J.P. The impact of grazing on plant communities, plant populations and soil conditions on salt marshes. *Vegetatio* **1985**, *62*, 391–398. [CrossRef]
47. Koh, H.S.; Ochs, C.A.; Yu, K. Hydrologic gradient and vegetation controls on CH_4 and CO_2 fluxes in a spring-fed forested wetland. *Hydrobiologia* **2009**, *630*, 271–286. [CrossRef]
48. Brinson, M.M.; Lugo, A.E.; Brown, S. Primary Productivity, Decomposition and Consumer Activity in Freshwater Wetlands. *Annu. Rev. Ecol. Syst.* **1981**, *12*, 123–161. [CrossRef]
49. Wang, H.; Chen, Z.X.; Zhang, X.Y.; Zhu, S.X.; Ge, Y.; Chang, S.X.; Zhang, C.B.; Huang, C.C.; Chang, J. Plant Species Richness Increased Belowground Plant Biomass and Substrate Nitrogen Removal in a Constructed Wetland. *CLEAN Soil Air Water* **2013**, *41*, 657–664. [CrossRef]
50. Chen, J.; Wang, Q.; Li, M.; Liu, F.; Li, W. Does the different photosynthetic pathway of plants affect soil respiration in a subtropical wetland? *Ecol. Evol.* **2016**, *6*, 8010–8017. [CrossRef] [PubMed]
51. Deghi, G.S.; Ewel, K.C.; Mitsch, W.J. Effects of Sewage Effluent Application on Litter Fall and Litter Decomposition in Cypress Swamps. *J. Appl. Ecol.* **1980**, *17*, 397–408. [CrossRef]
52. Moorhead, D.L.; Currie, W.S.; Rastetter, E.B.; Parton, W.J.; Harmon, M.E. Climate and litter quality controls on decomposition: An analysis of modeling approaches. *Glob. Biogeochem. Cycles* **1999**, *13*, 575–589. [CrossRef]
53. Carroll, P.; Crill, P. Carbon balance of a temperate poor fen. *Glob. Biogeochem. Cycles* **1997**, *11*, 349–356. [CrossRef]
54. Hossler, K.; Bouchard, V.; Fennessy, M.S.; Frey, S.D.; Anemaet, E.; Herbert, E. No-net-loss not met for nutrient function in freshwater marshes: Recommendations for wetland mitigation policies. *Ecosphere* **2011**, *2*, 1–36. [CrossRef]
55. Veen, G.; Sarneel, J.M.; Ravensbergen, L.; Huig, N.; van Paassen, J.; Rip, W.; Bakker, E.S. Aquatic grazers reduce the establishment and growth of riparian plants along an environmental gradient. *Freshw. Biol.* **2013**, *58*, 1794–1803. [CrossRef]
56. Mulder, C.P.H.; Ruess, R.W. Effects of Herbivory on Arrowgrass: Interactions Between Geese, Neighboring Plants, and Abiotic Factors. *Ecol. Monogr.* **1998**, *68*, 275–293. [CrossRef]
57. Myers, R.S.; Shaffer, G.P.; Llewellyn, D.W. Baldcypress (*Taxodium distichum* (L.) Rich.) restoration in southeast Louisiana: The relative effects of herbivory, flooding, competition, and macronutrients. *Wetlands* **1995**, *15*, 141–148. [CrossRef]
58. Goranson, C.E.; Ho, C.K.; Pennings, S.C. Environmental gradients and herbivore feeding preferences in coastal salt marshes. *Oecologia* **2004**, *140*, 591–600. [CrossRef]
59. Buij, R.; Melman, T.C.; Loonen, M.J.; Fox, A.D. Balancing ecosystem function, services and disservices resulting from expanding goose populations. *Ambio* **2017**, *46*, 301–318. [CrossRef] [PubMed]
60. Anderson, F.E.; Bergamaschi, B.; Sturtevant, C.; Knox, S.; Hastings, L.; Windham-Myers, L.; Detto, M.; Hestir, E.L.; Drexler, J.; Miller, R.L.; et al. Variation of energy and carbon fluxes from a restored temperate freshwater wetland and implications for carbon market verification protocols. *J. Geophys. Res. Biogeosci.* **2016**, *121*, 777–795. [CrossRef]
61. Altor, A.E.; Mitsch, W.J. Pulsing hydrology, methane emissions and carbon dioxide fluxes in created marshes: A 2-year ecosystem study. *Wetlands* **2008**, *28*, 423–438. [CrossRef]
62. Valach, A.C.; Kasak, K.; Hemes, K.S.; Anthony, T.L.; Dronova, I.; Taddeo, S.; Silver, W.L.; Szutu, D.; Verfaillie, J.; Baldocchi, D.D. Productive wetlands restored for carbon sequestration quickly become net CO_2 sinks with site-level factors driving uptake variability. *PLoS ONE* **2021**, *16*, e0248398. [CrossRef] [PubMed]
63. Mueller, P.; Jensen, K.; Megonigal, J.P. Plants mediate soil organic matter decomposition in response to sea level rise. *Glob. Chang. Biol.* **2016**, *22*, 404–414. [CrossRef] [PubMed]
64. Blagodatskaya, E.; Kuzyakov, Y. Mechanisms of real and apparent priming effects and their dependence on soil microbial biomass and community structure: Critical review. *Biol. Fertil. Soils* **2008**, *45*, 115–131. [CrossRef]
65. Bardgett, R.D.; Wardle, D.A.; Yeates, G.W. Linking above-ground and below-ground interactions: How plant responses to foliar herbivory influence soil organisms. *Soil Biol. Biochem.* **1998**, *30*, 1867–1878. [CrossRef]

66. He, M.; Zhou, G.; Yuan, T.; van Groenigen, K.J.; Shao, J.; Zhou, X. Grazing intensity significantly changes the C: N: P stoichiometry in grassland ecosystems. *Glob. Ecol. Biogeogr.* **2020**, *29*, 355–369. [CrossRef]
67. Tanentzap, A.J.; Coomes, D.A. Carbon storage in terrestrial ecosystems: Do browsing and grazing herbivores matter? *Biol. Rev.* **2012**, *87*, 72–94. [CrossRef] [PubMed]
68. Sasser, C.E.; Holm, G.O.; Evers-Hebert, E.; Shaffer, G.P. The nutria in Louisiana: A current and historical perspective. In *Mississippi Delta Restoration*; Springer: Berlin, Germany, 2018; pp. 39–60.
69. Yu, O.; Chmura, G. Soil carbon may be maintained under grazing in a St Lawrence Estuary tidal marsh. *Environ. Conserv.* **2009**, *36*, 312–320. [CrossRef]
70. Van Der WAL, R.; Sjögersten, S.; Woodin, S.J.; Cooper, E.J.; Jónsdóttir, I.S.; Kuijper, D.; Fox, T.A.; Huiskes, A. Spring feeding by pink-footed geese reduces carbon stocks and sink strength in tundra ecosystems. *Glob. Chang. Biol.* **2007**, *13*, 539–545. [CrossRef]
71. Frank, A.B.; Tanaka, D.L.; Hofmann, L.; Follett, R.F. Soil carbon and nitrogen of Northern Great Plains grasslands as influenced by long-term grazing. *Rangel. Ecol. Manag. J. Range Manag. Arch.* **1995**, *48*, 470–474. [CrossRef]
72. Cui, X.; Wang, Y.; Niu, H.; Wu, J.; Wang, S.; Schnug, E.; Rogasik, J.; Fleckenstein, J.; Tang, Y. Effect of long-term grazing on soil organic carbon content in semiarid steppes in Inner Mongolia. *Ecol. Res.* **2005**, *20*, 519–527. [CrossRef]
73. Elschot, K.; Bakker, J.P.; Temmerman, S.; van de Koppel, J.; Bouma, T.J. Ecosystem engineering by large grazers enhances carbon stocks in a tidal salt marsh. *Mar. Ecol. Prog. Ser.* **2015**, *537*, 9–21. [CrossRef]
74. Clermont, J.; Réale, D.; Giroux, J.F. Plasticity in laying dates of Canada Geese in response to spring phenology. *Ibis* **2018**, *160*, 597–607. [CrossRef]

land

Communication

Microbial Respiration and Enzyme Activity Downstream from a Phosphorus Source in the Everglades, Florida, USA

Sanku Dattamudi [1,*], Saoli Chanda [1] and Leonard J. Scinto [1,2]

[1] Department of Earth and Environment, Florida International University, Miami, FL 33199, USA; schanda@fiu.edu (S.C.); scintol@fiu.edu (L.J.S.)
[2] Southeast Environmental Research Center, Institute of Environment, Florida International University, Miami, FL 33199, USA
* Correspondence: sdattamu@fiu.edu

Abstract: Northeast Shark River Slough (NESS), lying at the northeastern perimeter of Everglades National Park (ENP), Florida, USA, has been subjected to years of hydrologic modifications. Construction of the Tamiami Trail (US 41) in 1928 connected the east and west coasts of SE Florida and essentially created a hydrological barrier to southern sheet flow into ENP. Recently, a series of bridges were constructed to elevate a portion of Tamiami Trail, allow more water to flow under the bridges, and attempt to restore the ecological balance in the NESS and ENP. This project was conducted to determine aspects of soil physiochemistry and microbial dynamics in the NESS. We evaluated microbial respiration and enzyme assays as indicators of nutrient dynamics in NESS soils. Soil cores were collected from sites at certain distances from the inflow (near canal, NC (0–150 m); midway, M (150–600 m); and far from canal, FC (600–1200 m)). Soil slurries were incubated and assayed for CO_2 emission and β-glucoside (MUFC) or phosphatase (MUFP) activity in concert with physicochemical analysis. Significantly higher TP contents at NC (2.45 times) and M (1.52 times) sites than FC sites indicated an uneven P distribution downstream from the source canal. The highest soil organic matter content (84%) contents were observed at M sites, which was due to higher vegetation biomass observed at those sites. Consequently, CO_2 efflux was greater at M sites (average 2.72 μmoles g dw^{-1} h^{-1}) than the other two sites. We also found that amendments of glucose increased CO_2 efflux from all soils, whereas the addition of phosphorus did not. The results indicate that microbial respiration downstream of inflows in the NESS is not limited by P, but more so by the availability of labile C.

Keywords: wetland soil; microbial respiration; CO_2 efflux; Everglades; enzyme activity; phosphorus

Citation: Dattamudi, S.; Chanda, S.; Scinto, L.J. Microbial Respiration and Enzyme Activity Downstream from a Phosphorus Source in the Everglades, Florida, USA. *Land* **2021**, *10*, 696. https://doi.org/10.3390/land10070696

Academic Editor: Richard C. Smardon

Received: 17 May 2021
Accepted: 25 June 2021
Published: 1 July 2021

Publisher's Note: MDPI stays neutral with regard to jurisdictional claims in published maps and institutional affiliations.

1. Introduction

Wetlands are considered to be both sources and sinks of terrestrial C and play an important role in the global C cycle. Wetlands are estimated to contain approximately 20–30% of the global C pool [1]. The total amount of C stored in wetlands in the conterminous US is about 11.52 Pg [2]. The nutrient dynamics and biogeochemistry of soils in wetlands are more complex than in uplands, specifically because of the intermittent anaerobic and semi-aerobic conditions. Microbial respiration and enzyme assays can be considered as potential indicators of soil health, nutrient enrichment, availability, and mineralization in wetlands [3–5]. Induced enzyme activity (EA) links environmental nutrient availability and microbial biomass stoichiometry [6]. For instance, higher EA in soil indicates low availability of labile nutrient contents. Additionally, small changes in the decomposition rate and enzyme activity will be able to alter both recalcitrant and labile C pools, impacting the long-term soil C stocks. Extracellular enzymes such as glucosidase and phosphatase are majorly studied for their ability to mineralize C from plant litter and P from nucleic acids, respectively, whereas the lability of soil organic matter (SOM) can be determined through the production of CO_2 in aerobic incubations [7]. These indicators are also helpful in predicting nutrient status in soils under varying hydrologic conditions. This short-term research study was conducted in the Florida Everglades, the largest wetland in the USA, and is internationally recognized by several organizations including the United Nations Educational, Scientific, and Cultural Organization (UNESCO). Hydrologic modifications enabling anthropogenic development of South Florida have greatly reduced the water flow in Northeast Shark River Slough (NESS) at the northeastern border of Everglades National Park (ENP). The construction of the Tamiami Trail (US highway 41) in 1928 to connect the Florida Coasts created a significant hydrologic barrier (shorter hydroperiod) between NESS and the remaining Everglades to the north. Additionally, nutrient loading from the northern border of NESS resulted in a phosphorus gradient consistent with that seen in other areas of the Everglades in proximity to discharge canals [8]. Studies have suggested that native periphytons (calcareous, filamentous blue-green algal communities), primary producers of Everglades ecosystem, are highly sensitive to P concentrations and small changes in P loading can create rapid changes in the periphyton community [9]. Overall, the reduced hydroperiods in much of the Everglades as a result of water diversion have resulted in net soil C loss [10]. Altered hydroperiods are responsible for the replacement of native aquatic plant species with non-native species [11] and the shifting of native periphyton communities, consequently hampering the long-term ecological balance in the Everglades; however, limited infrastructure, including culverts under the Tamiami Trail, were constructed to allow water to flow southward into the NESS. Since the water flow through the culverts was not sufficient, two large bridges were constructed in 2012 and 2019 to increase water flow into the NESS and to partially restore the hydrology and ecological structure and function.

Historically, the Everglades has largely been a P-limited system [12]; however, nutrient runoff from agricultural and urban development areas may have changed the dynamics of nutrient availability in NESS soils, specifically in nutrient-rich pockets. We contend that using biological indicators in the NESS region will help identify potential ecological and biogeochemical imbalances. The results from this study will be able to provide insights about the nutrient dynamics of NESS specifically after the construction of new bridges. We hypothesized that microbial respiration and enzymatic assays in response to current hydrological changes will be able to indicate the lability of organic matter and P biogeochemistry at certain distances downstream in the NESS. The specific objectives of this study were: (a) to characterize total P (TP), total C (TC) contents and other physicochemical properties of NESS soils at locations with at certain distances from the inflow; (b) to evaluate the potential of microbial respiration (CO_2 efflux) and extracellular enzyme activity as predictors of nutrient status in the NESS.

2. Materials and Methods

2.1. Site Location and Sample Collection

Soils were collected from NESS sites at certain distances from the bridge (inflow) on the Tamiami Trail (Figure 1). Sample collection locations included near-canal (n = 28), midway (n = 24), and far from canal (n = 24) sites that were about 0–150, 150–600, and 600–1200 m downstream, respectively. Intact soil cores (0 to 10 cm) were collected by inserting plastic tubes of a known diameter (surface area) into the soil. Additional care was taken to minimize compaction during collection. Subsamples of known mass fresh soils were used in CO_2 efflux incubations and enzyme activity analysis. Additional subsamples were dried, ground, and used in physicochemical analysis. Water depths during each sampling were measured using a standard ruler (one meter stick) to the nearest cm as the depths between the surface of the water and the resistance of the soil.

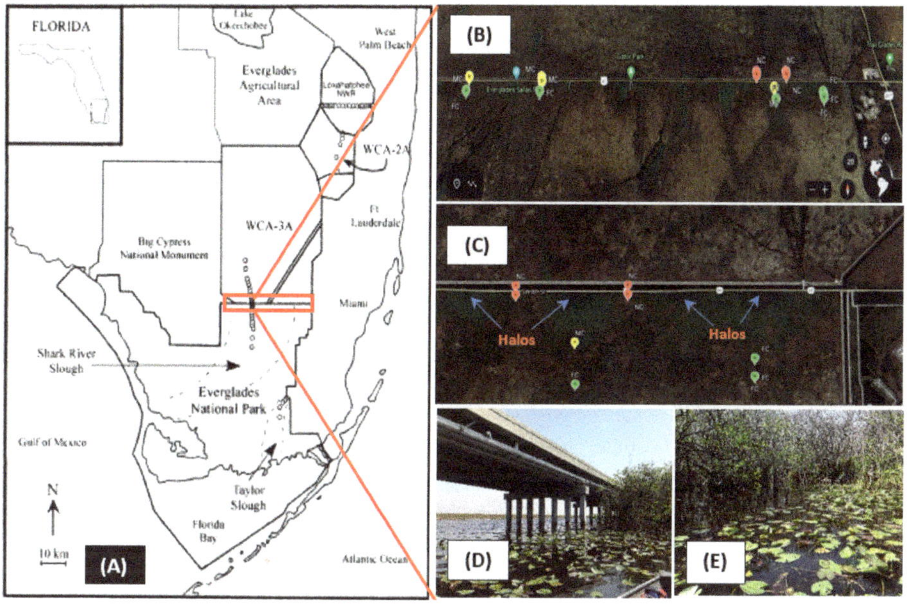

Figure 1. Locations of sampling sites in Northeast Shark Slough (NESS), Florida, USA (**A**), in proximity to a bridge constructed to partially restore southward water flow into Everglades National Park (ENP). Red tags are near-canal sites (0–150 m downstream of canal), yellow tags are midway (150–600 m), and green tags are far from the canal (600–1200 m) (**B**,**C**). Remnant vegetation "halos" (increased biomass and nutrients) are identified and show the locations of pre-bridge culverts (**C**). Google Earth images were used for context. Photographic images of the bridge (near-canal) (**D**) and sampling site (**E**).

2.2. Laboratory Analyses

The ability of soil microbes to generate CO_2 under controlled conditions was assessed by adding a known mass of fresh 1:1 soil slurries (soil/solution) to 20 mL CO_2-free air purged from headspace incubation vials (triplicate) and subjected to dark incubation at 25 °C for 72 h. Gas samples collected from headspace vials were then injected into a HP 5890 Series II gas chromatograph equipped with a flame ionization detector and a Shimadzu MTN-1 methanizer. The measured CO_2 efflux was expressed as µmoles per unit mass of dry soil following the methods described by [7,13].

Four treatments were evaluated, including a non-amended control (C) and amendments of 1.2 mmol L^{-1} of glucose (G), 0.4 mmol L^{-1} of phosphorus (P), and a combination of phosphorus and glucose (P + G). Treatment strengths were calculated following the method proposed [7] and our previous experiments in NESS soil.

Enzyme activities were determined for each of the four treatments (C, G, P, and P + G) by adding moieties of the fluorogenic substrate, 4-methylumbelliferyl (MUF; either -β-glucoside (MUF-C) or MUF–phosphatase (MUF-P)), to soil solutions made by diluting slurries to a final 10^{-3} dilution. Diluted samples (200 µL) were pipetted into 8 wells of a 96-well plate, to which 50 µL of either MUF-P or MUF-C substrates were added to obtain final concentrations of 10 µM. Plates were then incubated in the dark for 2 h (MUF-P) or 24 h (MUF-C). Substrates were added to replicate wells after incubation and immediately before fluorometric analysis to determine the initial fluorescence (i.e., t_0). No substrate was added to rows or columns containing standards or background blanks. A Synergy HT microplate reader was used for fluorometric determination with enzyme activity regressed from standard curves.

Total carbon (TC) and total nitrogen (TN) levels of oven dried soil samples were analyzed using a Carlo-Erba NA-1500 CNS analyzer [14]. Determination of Total P was done via oxidation (dry combustion) and hydrolysis to soluble forms (soluble reactive, ortho-P; SRP) using $MgSO_4/H_2SO_4$ and HCl [15], followed by standard colorimetric analysis using USEPA method 365.1 [16]. Dry bulk density (g cm^{-3}) and organic matter (%) levels of soil were measured following the standard methods of ASTM D4531-86 and ASTM D2974-87, respectively.

We analyzed and cleaned the data (using interquartile range calculation) outliers before performing statistical analyses. One-way analysis of variance (ANOVA) was carried out independently on treatments at certain distances and four treatments were carried out between distances using the SAS9.4 PROC MIXED procedure. Tukey–Kramer post-hoc tests were conducted to compare mean separation levels at $p < 0.05$ among treatments and between distances. Pearson correlation coefficient analysis was performed on soil characteristics at $p < 0.05$ (two-tailed).

3. Results and Discussion

The construction of the Tamiami Canal and Highway bisected the natural north-to-south sheet flow of water into the ENP. The nutrient-enriched canal water that flowed southward was funneled through a series of culverts and established the long-term P gradient evident in our analysis, whereby soil TP content decreased with distance (Table 1). The soil TP from near-canal (NC) and midway (M) sites were significantly higher ($p < 0.05$) by about 2.45 and 1.52 times, respectively, than far from canal (FC) sites. Additionally, certain samples with high TP contents (as high as 2630 μg g^{-1} dw) in NC sites were proximal to "halos" (enriched pockets), areas of increased water depth, nutrient content, and plant biomass at the outfall of culverts (Figure 1). Soil organic matter (SOM) was highest (average 87%) at M sites, which resulted in lowest bulk density values (ranged from 0.11 to 0.16 g dw cm^{-3}) in those soils. A possible explanation would be the presence of the higher vegetation cover in M sites as compared to NC and FC sites. The Pearson analysis (Table 2) of selected soil physicochemical parameters indicated that SOM was highly correlated with TC and TN, which was expected for high organic "peaty" wetland soils [17,18] where soil nutrients are largely in organic forms. Consequently, TN and TC contents were also highest in midway (M) soil samples. In unimpacted Everglades soils (those not influenced by allochthonous sources of P), TP is often correlated with TC [19]; however, this was not the case in our analysis, indicating the possibility of an influence of external P loading (causing a biogeochemical disconnect) from agricultural areas [20].

Table 1. Physiochemical parameters (mean ± SD), including water depth, soil bulk density (BD), organic matter (SOM), total P (TP), total N (TN), and total C (TC), at increasing distances from a source canal in the NESS. Different lowercase letters indicate significant differences ($p < 0.05$) between sites.

Sites	n	Water Depth	BD	SOM	TP	TN	TC
		cm	g dw cm^{-3}	%	ug g^{-1} dw	mg g^{-1} dw	
Near canal	28	19 ± 8 ^{a}	0.140 ± 0.035 ^{a}	54 ± 19 ^{b}	1143 ± 273 ^{a}	2.2 ± 0.9 ^{b}	32 ± 6 ^{c}
Midway	24	12 ± 4 ^{a}	0.129 ± 0.029 ^{a}	84 ± 17 ^{a}	712 ± 151 ^{b}	3.4 ± 0.3 ^{a}	46 ± 8 ^{a}
Far from canal	24	8 ± 4 ^{a}	0.149 ± 0.052 ^{a}	67 ± 11 ^{b}	467 ± 87 ^{c}	2.8 ± 0.4 ^{a}	40 ± 9 ^{b}

Table 2. Pearson's (two-tailed) correlation analysis between different physicochemical soil properties (bulk density, BD; total P, TP; total N, TN; total C, TC; soil organic matter, SOM) for all NESS sites combined.

Parameters	TP	Soil Moisture	TN	TC	SOM
BD	−0.33	−0.60 *	−0.35	−0.50	−0.51
TP		0.25	−0.28	−0.29	−0.21
Soil moisture			0.39 *	0.44 *	0.53 *
TN				0.95 *	0.95 *
TC					0.99 *

Note: * significant at $\alpha \leq 0.05$.

Soil organic matter accumulates in wetlands when C-fixation during photosynthesis exceeds decomposition through mineralization and respiration [21]. Several factors influence this balance, including those that affect fixation, decomposition, and the lability or stability of SOM [22,23]. Previous studies have suggested that decomposition of litter and belowground biomass in the oligotrophic Everglades is P-limited [24], as is the oft-cited P-limited productivity [12,25]. The hypothetical influence of P limitation and variations in SOM lability were studied in our CO_2 evolution experiment, where the CO_2 fluxes ranged from 0.18 in unamended controls to 5.62 (μmol CO_2 g^{-1} dw soil h^{-1}) under doubly-amended (P + G) treatments for all three locations along the gradient (Figure 2). The addition of P alone as an amendment did not significantly increase CO_2 emissions compared to the unamended controls at any location, suggesting that SOM respiration in these soils was not P-limited. Even in the FC sites that had the lowest TP contents (Table 1), where P addition might be suspected to have the greatest influence, P addition did not affect respiration. It should be noted that average TP contents at FC sites (467 mg g^{-1} dw) were much higher than the previously reported TP concentrations in unimpacted NESS soils (approximately 218 mg g^{-1} dw; 8). In unamended soils, CO_2 production was significantly greater ($p < 0.05$, as indicated by different capital letters) at the M sites than at the other sites. This could be reflective of the greater overall SOM content at this site rather than differences in the lability of the SOM. The addition of glucose (G) increased respiration relative to the unamended controls at the NC and M sites but not for FC sites (Figure 2). Adding a labile C source (glucose amendment) supports the hypothesis that the efflux of CO_2 from NESS soil is at least partially dependent on the SOM carbon quality. As might be expected, a major portion of the SOM present in the NESS soil is in more recalcitrant forms compared to labile glucose [7]. In a recent study conducted at Everglades National Park, [26] found that regardless of restoration activities, the decomposition and microbial respiration were limited by labile C rather than soil P. A potential ecological consequence of recalcitrant SOM under the influence of hydrologic restoration and concurrently longer hydroperiods would be an increase in peat accretion. An abundance of recalcitrant C indicates a healthy and productive wetland ecosystem [27]. In our study, across all treatments at all sites, the combination of phosphorus and glucose (P + G) produced significantly higher CO_2 than other treatments, suggesting readily available (labile) C and P increased microbial activity in the soils and caused a "priming effect", as reported in other studies [7].

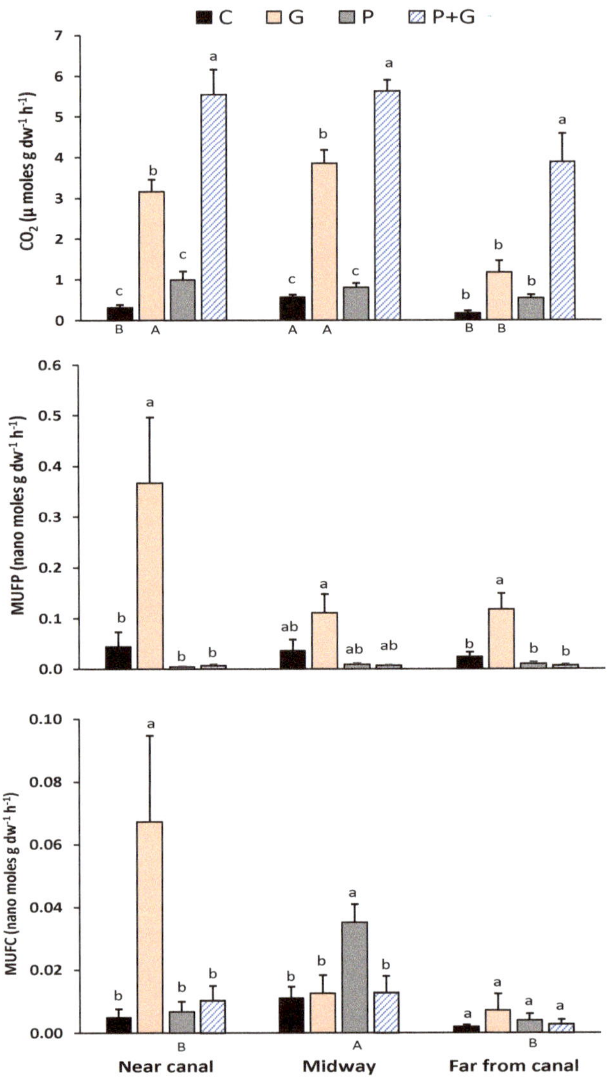

Figure 2. Carbon dioxide (CO_2) efflux and enzyme activity (phosphatase or MUFP and glucosidase or MUFC) analyses of incubated soil slurries collected at certain distances (near-canal, midway, and far from canal) downstream from the inflow in the Northeast Shark River Slough, ENP. Treatments included a control (C), glucose (G), phosphorus (P), and a combination of phosphorus and glucose (P + G). Lowercase letters indicate significant differences ($p < 0.05$) between treatments (within a distance), while uppercase letters indicate the treatment difference sacross distances (NC, M, and FC). Uppercase letters not used in this figure were for treatments with no significant difference between distances.

Enzyme assays were conducted to determine the activity levels of β-glucosidase and phosphatase enzymes in NESS soils. Soil microbes produce β-glucosidase or phosphatase enzymes to cleave the esterase bond between complex organic molecules and the glucose or phosphate, respectively, allowing these compounds to become available for uptake and degrade organic compounds in the process. Availability of nutrients in the soil can alter enzyme activity and soil respiration. In this study, the addition of G increased MUFP activity above the unamended controls at all sites (significantly so at NC and F sites), additionally indicating potential labile C limitation across all sites. Added labile C enhanced the microbial community to increase phosphatase production, likely as a response to a growing microbial population to take advantage of the added food source. In all sites, the MUFP responses were similar for each treatment, as they did not significantly differ for sites within treatments. Although not statistically significant, the addition of P appeared to reduce the phosphatase activity compared to controls at all sites, as would be expected for an inducible enzyme. The addition of glucose at the NC sites, where SOM was lowest and P content highest, significantly increased the MUFC activity over that of the controls, while P addition significantly increased this activity at the M sites, where SOM was highest relative to other locations. There were no other between-site differences regarding MUFC activity for any other treatments.

The biogeochemistry of P has major implications in controlling soil C dynamics in wetlands. The external addition of nutrients can potentially impact the enzyme activity and microbial respiration, specifically when those nutrients are not readily available in the soil; however, our soils had abundant P content (considering more than 450 mg g^{-1} dw TP contents in FC sites), so phosphatase enzyme activity was negligible when additional P sources were applied. We also observed that heterotrophic microbial activity was more responsive to additional food sources; therefore, the outcomes of this experiment indicated that respiration and extracellular enzyme activity, at least downstream of inflows in the NESS, are not limited by P, but rather by the availability of labile C.

Author Contributions: Conceptualization: S.D., S.C., L.J.S.; Methodology: S.D., S.C.; Formal analysis: S.D., S.C.; Resources: L.J.S.; Data curation: S.D.; Writing original draft: S.D.; Writing review and editing: S.D., S.C., L.J.S.; Funding acquisition: L.J.S. All authors have read and agreed to the published version of the manuscript.

Funding: This research was funded by Florida Coastal Everglades—National Science Foundation, grant number DEB-1832229.

Institutional Review Board Statement: Not applicable.

Informed Consent Statement: Not applicable.

Data Availability Statement: All data is available in the manuscript.

Acknowledgments: This research was partially supported by a task agreement between FIU and Everglades National Park (P14AC01639 and P16AC0032). The authors would like to thank Alex Crow and Diana Johnson for their assistance during laboratory analyses. This material was developed in collaboration with the Florida Coastal Everglades Long-Term Ecological Research Program under National Science Foundation Grant no. DEB-1832229. This is contribution number 1023 from the Southeast Environmental Research Center in the Institute of Environment at Florida International University.

Conflicts of Interest: The authors declare no conflict of interest.

References

1. Lal, R. Carbon sequestration. *Philos. Trans. R. Soc. B Biol. Sci.* **2008**, *363*, 815–830. [CrossRef]
2. Nahlik, A.M.; Fennessy, M.S. Carbon storage in US wetlands. *Nat. Commun.* **2016**, *7*, 13835. [CrossRef]
3. Nielsen, M.N.; Winding, A.; Binnerup, S. Microorganisms as Indicators of Soil Health. 2002. Available online: https://www.dmu.dk/1_Viden/2_Publikationer/3_Fagrapporter/rapporter/FR388.pdf (accessed on 30 June 2021).
4. Cabugao, K.G.; Timm, C.M.; Carrell, A.A.; Childs, J.; Lu, T.Y.S.; Pelletier, D.A.; Weston, D.J.; Norby, R.J. Root and rhizosphere bacterial phosphatase activity varies with tree species and soil phosphorus availability in Puerto Rico tropical forest. *Front. Plant. Sci.* **2017**, *8*, 1834. [CrossRef]
5. Margalef, O.; Sardans, J.; Fernández-Martínez, M.; Molowny-Horas, R.; Janssens, I.A.; Ciais, P.; Goll, D.; Richter, A.; Obersteiner, M.; Asensio, D.; et al. Global patterns of phosphatase activity in natural soils. *Sci. Rep.* **2017**, *7*, 1337. [CrossRef]
6. Sinsabaugh, R.L.; Lauber, C.L.; Weintraub, M.N.; Ahmed, B.; Allison, S.D.; Crenshaw, C.; Contosta, A.R.; Cusack, D.; Frey, S.; Gallo, M.E.; et al. Stoichiometry of soil enzyme activity at global scale. *Ecol. Lett.* **2008**, *11*, 1252–1264. [CrossRef] [PubMed]

7. Pisani, O.; Scinto, L.J.; Munyon, J.W.; Jaffé, R. The respiration of flocculent detrital organic matter (floc) is driven by phosphorus limitation and substrate quality in a subtropical wetland. *Geoderma* **2015**, *241*, 272–278. [CrossRef]
8. Childers, D.L.; Doren, R.F.; Jones, R.; Noe, G.B.; Rugge, M.; Scinto, L.J. Decadal change in vegetation and soil phosphorus pattern across the Everglades landscape. *J. Environ. Qual.* **2003**, *32*, 344–362. [CrossRef] [PubMed]
9. Gaiser, E. Periphyton as an indicator of restoration in the Florida Everglades. *Ecol. Indic.* **2009**, *9*, S37–S45. [CrossRef]
10. Malone, S.L.; Starr, G.; Staudhammer, C.L.; Ryan, M.G. Effects of simulated drought on the carbon balance of Everglades short-hydroperiod marsh. *Glob. Chang. Biol.* **2013**, *19*, 2511–2523. [CrossRef]
11. Brown, P.; Wright, A.L. The Role of Periphyton in the Everglades. (2013, February). Available online: http://edis.ifas.ufl.edu/pdffiles/SS/SS52200.pdf (accessed on 30 June 2021).
12. Noe, G.B.; Childers, D.L.; Jones, R.D. Phosphorus biogeochemistry and the impact of phosphorus enrichment: Why is the Everglades so unique? *Ecosystems* **2001**, *4*, 603–624. [CrossRef]
13. Dattamudi, S.; Wang, J.J.; Dodla, S.K.; Viator, H.P.; DeLaune, R.; Hiscox, A.; Darapuneni, M.; Jeong, C.; Colyer, P. Greenhouse gas emissions as influenced by nitrogen fertilization and harvest residue management in sugarcane production. *Agrosyst. Geosci. Environ.* **2019**, *2*, 1–10. [CrossRef]
14. Nelson, D.W.; Sommers, L.E. Total carbon, organic carbon, and organic matter. In *Methods of Soil Analysis: Part 3 Chemical Methods*; Sparks, D.L., Ed.; SSSA Book Series 5; SSSA: Madison, WI, USA, 1996; pp. 961–1010.
15. Solorzano, L.; Sharp, J.H. Determination of total dissolved phosphorus and particulate phosphorus in natural waters. *Limnol. Oceanogr.* **1980**, *25*, 754–758. [CrossRef]
16. O'Dell, J.W. *Method 365.1—Determination of Phosphorus by Semi-Automated Colorimetry*; USEPA, Environmental Monitoring Systems Laboratory, Office of Research and Development: Cincinnati, OH, USA, 1993.
17. Rivero, R.G.; Grunwald, S.; Osborne, T.Z.; Reddy, K.R.; Newman, S. Characterization of the spatial distribution of soil properties in Water Conservation Area 2A, Everglades, Florida. *Soil Sci.* **2007**, *172*, 149–166. [CrossRef]
18. Steinmuller, H.E.; Stoffella, S.L.; Vidales, R.; Ross, M.S.; Dattamudi, S.; Scinto, L.J. Characterizing hydrologic effects on soil physicochemical variation within tree islands and marshes in the coastal florida everglades. *Soil Sci. Soc. Am. J.* **2021**. [CrossRef]
19. Irick, D.L.; Li, Y.C.; Inglett, P.W.; Harris, W.G.; Gu, B.; Ross, M.S.; Wright, A.L.; Migliaccio, K.W. Characteristics of soil phosphorus in tree island hardwood hammocks of the southern Florida Everglades. *Soil Sci. Soc. Am. J.* **2013**, *77*, 1048–1056. [CrossRef]
20. Schade-Poole, K.; Möller, G. Impact and mitigation of nutrient pollution and overland water flow change on the Florida Everglades, USA. *Sustainability* **2016**, *8*, 940. [CrossRef]
21. DeBusk, W.F.; Reddy, K.R. Nutrient and hydrology effects on soil respiration in a northern Everglades marsh. *J. Environ. Qual.* **2003**, *32*, 702–710. [CrossRef]
22. DeBusk, W.F.; Reddy, K.R. Litter decomposition and nutrient dynamics in a phosphorus enriched everglades marsh. *Biogeochemistry* **2005**, *75*, 217–240. [CrossRef]
23. Pisani, O.; Gao, M.; Maie, N.; Miyoshi, T.; Childers, D.L.; Jaffé, R. Compositional aspects of herbaceous litter decomposition in the freshwater marshes of the Florida Everglades. *Plant. Soil* **2018**, *423*, 87–98. [CrossRef]
24. Amador, J.A.; Jones, R.D. Response of carbon mineralization to combined changes in soil moisture and carbon-phosphorus ratio in a low phosphorus histosol. *Soil Sci.* **1997**, *162*, 275–282. [CrossRef]
25. Wetzel, P.R.; Van Der Valk, A.G.; Newman, S.; Coronado, C.A.; Troxler-Gann, T.G.; Childers, D.L.; Orem, W.H.; Sklar, F.H. Heterogeneity of phosphorus distribution in a patterned landscape, the Florida Everglades. *Plant. Ecol.* **2009**, *200*, 83–90. [CrossRef]
26. Medvedeff, C.A.; Inglett, K.S.; Inglett, P.W. Patterns and controls of anaerobic soil respiration and methanogenesis following extreme restoration of calcareous subtropical wetlands. *Geoderma* **2015**, *245*, 74–82. [CrossRef]
27. Lane, R.R.; Mack, S.K.; Day, J.W.; DeLaune, R.D.; Madison, M.J.; Precht, P.R. Fate of soil organic carbon during wetland loss. *Wetlands* **2016**, *36*, 1167–1181. [CrossRef]

land

Article

Surface Water Quality Differs between Functionally Similar Restored and Natural Wetlands of the Saint Lawrence River Valley in New York

Brendan Carberry, Tom A. Langen and Michael R. Twiss *

Department of Biology, Clarkson University, Potsdam, New York, NY 13699, USA; brendanjcarberry@gmail.com (B.C.); tlangen@clarkson.edu (T.A.L.)
* Correspondence: mtwiss@clarkson.edu; Tel.: +1-315-268-2359

Abstract: We tested the hypothesis that upland wetland restorations provide the same quality of wetland, in terms of ecosystem services and biodiversity, as natural wetlands in the St. Lawrence River Valley. Water quality (pH, alkalinity, colored dissolved organic matter, phytoplankton community composition, chlorophyll-a, fecal coliform, total phosphorus, dissolved nitrate, turbidity, specific conductivity) in 17 natural and 45 restored wetlands was compared to determine whether wetland restoration provided similar physicochemical conditions as natural wetlands in the Saint Lawrence River Valley of northeastern New York State. Natural wetlands were more acidic, which was hypothesized to result from the avoidance of naturally acidic regions by farmers seeking to drain wetlands for crop and pasture use. Natural wetlands had significantly greater fecal coliform concentrations. Restored wetlands had significantly greater specific conductivity and related ions, and this is attributed to the creation of wetlands upon marine clay deposits. Other water quality indicators did not differ between restored and natural wetlands. These findings confirm other research at these same wetlands showing no substantial differences between restored and natural wetlands in major biotic indicators. Thus, we conclude that wetland restoration does result in wetlands that are functionally the same as the natural wetlands they were designed to replicate.

Keywords: cyanobacteria; phosphorus; restoration ecology; water quality; wetland

Citation: Carberry, B.; Langen, T.A.; Twiss, M.R. Surface Water Quality Differs between Functionally Similar Restored and Natural Wetlands of the Saint Lawrence River Valley in New York. *Land* **2021**, *10*, 676. https://doi.org/10.3390/land10070676

Academic Editor: Richard C. Smardon

Received: 30 May 2021
Accepted: 25 June 2021
Published: 27 June 2021

1. Introduction

Wetland restoration is a form of ecological engineering wherein these valued ecosystems in the landscape are reestablished for communal (human, ecosystem) good [1]. Public-private partnerships for wetland restoration are a mechanism by which private landowners and government agencies work together to improve the environmental quality of a human-modified landscape. US federal wetland restoration programs such as those administered by the US Department of Agriculture's Natural Resources Conservation Service (NRCS) or US Fish and Wildlife Service (FWS) are intended to restore wetlands and the ecosystem services wetlands provide on agricultural landscapes where wetlands have been drained or degraded in the past [2,3]. Wetlands are important features in the Upper St. Lawrence Valley landscape that provide numerous ecosystem services such as fish and wildlife habitats, natural water quality improvement, flood protection, opportunities for recreation, and aesthetics. NRCS and FWS collaborate with private landowners to restore or enhance wetlands on former or currently productive agricultural lands. In the St. Lawrence River Valley of New York, over 200 landowners have had wetlands restored on their property via these programs. Evaluating the success of wetland restorations is essential to program expansion and the design of the best approach to achieve communal benefits. A key question to evaluate programs success is "Do these restorations provide the same quality of wetland, in terms of ecosystem services and biodiversity, as natural wetlands?"

Regional assessments of wetland restoration programs indicate that aggregating over restoration projects, wetland restoration programs do augment ecosystem services in agricultural landscapes [4,5], though not necessarily the same quality as the former natural wetlands that had been lost on the landscape due to drainage or other hydrological alterations. There is a lack of studies, however, that evaluate restored wetlands on a project level to natural wetlands in the same landscape [6,7].

We conducted biotic surveys and informational surveys of landowners at a large set of restored and similar natural wetlands within the St. Lawrence River Valley of New York. Reference natural wetlands were similar in landscape context and size to wetlands restorations, and in proximity to them. Landowners had voluntarily enrolled in wetland restoration programs because they want to improve the environmental quality of their property by establishing and protecting well-functioning wetlands [8]. Restored wetlands were similar to natural wetlands in terms of birds, amphibians, reptiles, fish, and vegetation [6,8,9]. By various ecological indicator metrics, restored wetlands were qualitatively similar to natural wetlands, albeit quantitatively most indices scores, on average, were a little lower (i.e., lower environmental quality) than natural wetlands [10]. In comparison, these restored wetlands scored much higher (i.e., better environmental quality) that wetlands in a nearby Great Lakes Area of Concern.

One wetland ecosystem service is improvement of water quality, and water quality is used as one indicator of wetland state. Thus, one way to evaluate the success of wetland restoration programs at restoring well-functioning wetlands is to compare water quality between restorations and natural wetlands in the same landscape. We used a modified water quality index developed for coastal marshes in the Laurentian Great Lakes [11]; this index used water quality parameters that are significantly related to Great Lakes basin-wide land use stressors and sensitive to road density [12]. Our implementation of the index incorporated water turbidity, pH, temperature, conductivity, total nitrogen, total phosphate, and chlorophyll-a. We found that the water quality index was, on average, quantitatively slightly lower (indicating poorer water quality) than natural wetlands. However, the water quality index averaged much higher (i.e., better water quality) than wetlands in the nearby Great Lakes Area of Concern at Massena/Akwesasne, where significant anthropogenic stressors are known to be present [10]. However, this water quality index, surprisingly, was not correlated with other biotic and landscape indices of wetland quality. We surmised in [10] that water quality was a poor indicator of wetland habitat quality for wetland-associated plants and animals, and this may be because water quality parameters in shallow wetlands are highly variable at short timescales and within short distances.

Water quality properties were compared between sets of restored and natural wetlands to determine if adverse effects are a result of wetland restorations, e.g., nutrient enrichment of the aquatic environment leading to eutrophication, and the related increase in potentially toxigenic cyanobacteria (commonly referred to as blue-green algae). If adverse effects were detected in water quality, then social impacts could reduce program acceptance and efficiency. Here we examine water quality attributes between a set of restored and natural wetlands in the St. Lawrence River Valley of New York, including chemical, physical, and microbial parameters that are indicators of nutrient runoff-associated eutrophication and other anthropogenic stressors.

2. Materials and Methods

Wetland sampling for water quality was conducted on 17 natural wetlands and 45 restored wetlands (Figure 1) over a four-week period (25-July-2014 to 25-August-2014); this set of wetlands underwent extensive biotic assessment in 2009–2011 and 2014 [6]. Wetland restoration techniques included removal of drainage tiles, blocking drainage ditches, excavation of potholes, creation of dikes and berms, and installation of water control structures on outflow streams. Reference natural wetlands were selected to match restorations in terms of size and landscape context (see [6] for details on site selection and geographic dispersion). The wetlands were shallow (under 2 m maximum depth) and small (1–3 ha surface area), with bordering upland vegetation that varied from old-field to hardwood and coniferous forest. Introduced and invasive wetlands plants were present at most wetlands [9,13]. Landowners rarely managed water levels using the water control structures [8]. Most wetlands with adjacent forest had signs of beaver (*Castor canadensis*) activity, and all likely had muskrat (*Ondatra zibethicus*) present [6].

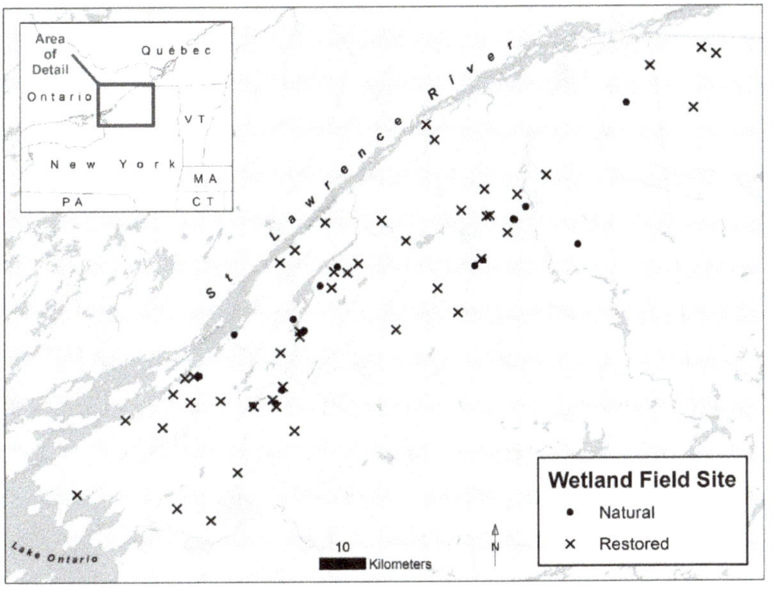

Figure 1. Location of sampled natural (n = 17) and restored (n = 45) wetlands in the Saint Lawrence River Valley, northern New York.

Grab samples (one liter) were collected in acid-clean polycarbonate bottles from the surface water present in each wetland, stored cool in the dark, and processed that day. Wetlands were sampled on a schedule determined by the logistics of travel; sampling was avoided after heavy rainfall by going into the field minimally three days after a thunderstorm in the area.

We measured 12 physicochemical and biological state variables: total chlorophyll-a (acetone extraction and quantification by fluorometry [14]; phytoplankton community (pigment-specific fluorometry); fecal coliform bacteria (Petri-Film; 3M Corp.); dissolved nitrate, sulfate, and chloride by ion exchange chromatography; colored dissolved organic matter (CDOM) by fluorometry (TD-700 using Suwanee River fulvic acid reference material (International Humic Substances Society); turbidity by absorbance at 500 nm in a 5 cm path length cuvette; pH by potentiometry, and alkalinity by Gran titration using HCl; specific conductivity was measured using an electronic meter (YSI model 600XL); and total phosphorus (TP) by colorimetry following persulfate digestion at 121 °C [15]. Dissolved solutes were measured after filtration through a 0.2-µm polyether sulfone membrane syringe filter (Whatman). All measurements were made using standard limnological and analytical methods. Water temperature and dissolved oxygen were not measured due to their inherent high magnitude of diel variation in wetlands.

The phytoplankton community composition was assessed using the FluoroProbe (bbe Moldaenke, GmbH), an instrument capable of classifying the community into four major phytoplankton groupings based on pigment content [16]. Each sample was corrected for background fluorescence using water filtered through 0.2-µm pore-size syringe filters prior to evaluating the non-filtered sample.

Statistical hypothesis tests of specific water quality parameters between restored and natural wetlands were done using Student's t-test for unequal variances on non-transformed data, with two-tailed distributions.

3. Results and Discussion

No significant difference (*t*-test; $p < 0.05$) were observed between natural and restored wetlands for CDOM, turbidity, phytoplankton community composition, nitrate, total chlorophyll-a and total phosphorus (Figure 2). Significant differences were observed for chloride ($p = 0.010$) and sulfate ($p = 0.014$) concentrations, alkalinity ($p = 0.006$), specific conductivity ($p = 0.002$), pH ($p = 0.001$), and fecal coliform concentrations ($p = 0.005$). The complete data set is an electronic appendix at https://data.mendeley.com/datasets/m7dycy7gt6/2, accessed on 30 May 2021. Although the analysis in this study is based on a measurement campaign in a single season, we acknowledge that there might be seasonal differences within one waterbody. Repeated measurements of water quality parameters over a summer season in a smaller subset of these wetlands showed that water quality parameters were consistent between sampling dates over five months [13]. Thus, we believe that single visit to this larger set of wetlands over a short duration (one month) was adequate for assessing differences.

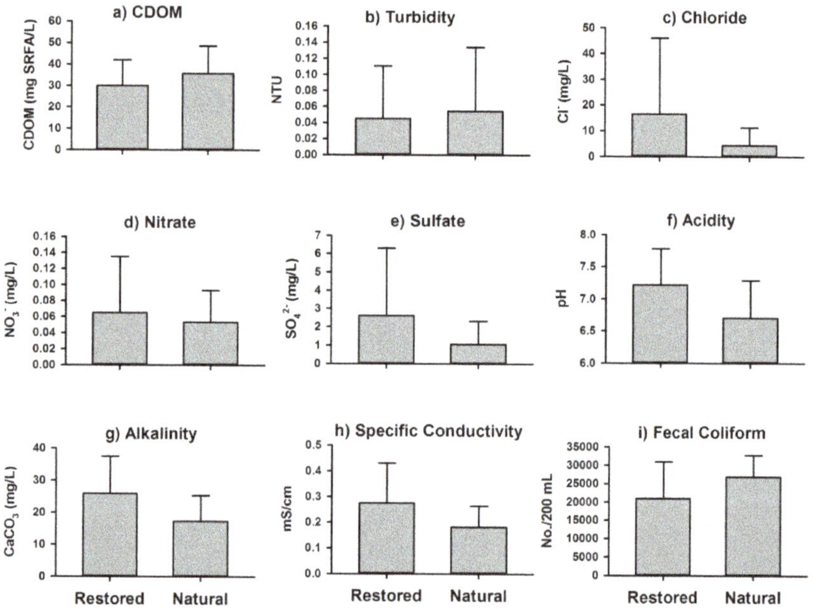

Figure 2. Observed water quality in a set of restored ($n = 45$) and natural ($n = 17$) wetlands found in the Saint Lawrence River Valley in northern New York. Significant difference between wetland types ($p < 0.05$) were present for: (**a**) CDOM, (**b**) turbidity, (**c**) chloride, (**d**), nitrate, (**e**) sulfate, (**f**) acidity, (**g**) alkalinity, (**h**) specific conductivity, and (**i**) fecal coliform. Values are mean ± SD.

Total mercury and methylmercury was determined (see [13] for details) in surface waters from four natural and 16 restored wetlands from among the set described here. The wetlands were sampled three to five times at approximately monthly intervals (over the period of May to October 2015). There was no significant difference between mercury concentration and mercury speciation between the two types of wetlands [13]: total mercury and percentage mercury in natural and restored wetlands was 1.0 ± 0.4 ng/L (37 ± 17%) and 1.1 ± 0.5 ng/L (46 ± 15%), respectively (values are mean ± standard deviation; SD).

Both natural and restored wetlands had similar phytoplankton community composition (Table 1). We were most interested to determine if restored or natural wetlands contained more phycocyanin-rich cyanobacteria, a group of phytoplankton that contain species capable of producing potent toxins such as microcystins and anatoxins [17]. There was no significant difference in the proportion of phycocyanin-rich cyanobacteria between the wetland types nor the absolute amount of potentially toxigenic phytoplankton between wetland types (Table 1).

Table 1. Phytoplankton community composition observed in vivo in restored (n = 45) and natural (n = 17) wetlands in the Saint Lawrence River Valley of northern New York, measured using spectrofluorometry. Total chlorophyll-a (Chl-a) was measured following solvent extraction. PC = phycocyanin; PE = phycoerythrin. Values are mean ± SD.

	Phytoplankton Groups (% Total)				
Wetland Type	**Chlorophyta and Euglenophyta**	**PC-Rich Cyanobacteria**	**Pyrrophyta and Heterokontophyta**	**PE-Rich Cyanobacteria and Cryptophyta**	**Total Chl-*a* (μg/L)**
Restored (n = 45)	38 ± 24	19 ± 19	35 ± 22	7.2 ± 13	29 ± 50
Natural (n = 17)	36 ± 19	24 ± 17	30 ± 18	10 ± 17	18 ± 15

Chlorophyll-a content within wetlands was highly variable and both wetland types were highly productive on average, as seen by mean chlorophyll-a at the 20 μg/L threshold for eutrophy (Table 1); this is not surprising given wetlands are shallow aquatic systems and highly productive during summer. There was no significant difference between phytoplankton groupings in natural or restored wetlands, based on pigment-based groupings of the phytoplankton community.

Significant differences (t-test; $p < 0.05$) between restored and natural surface water quality parameters were detected for fecal coliform concentrations, pH, alkalinity, and specific conductivity. Natural wetlands had 30% greater fecal coliform concentrations. Natural wetlands (pH 6.70) were 3.3 times more acidic than restored wetlands (pH 7.22), calculated by comparing {H+} derived from lab pH. Natural wetlands had 1.5-times lower alkalinity and specific conductivity than restored wetlands.

Other potential sources of dissolved ions in wetland surface waters were sewage and proximity to roads. Although there are significantly more fecal coliforms in natural wetlands, there is no indication of sewage input from human sources or manure run off from livestock farming, based on visits to these sites. Road salt applications for winter road management can have profound impacts on roadside waterways and groundwater [18]. Reference natural wetlands were more distant to roads than restorations on average (natural: 352 ± SD 305 m, restoration: 152 ± 135 m), but the average wetland was distant enough that it was unlikely that elevated chloride was from deicing road salt. Moreover, for restorations there was no correlation between distance to a road and chloride or conductivity (distance-chloride r = −0.01, one tailed p = 0.5, distance-conductivity r = −0.13, p = 0.2); the two highest chloride concentrations were at wetlands 100 m from a potential source road.

Although TP was positively associated with specific conductivity in natural wetlands (r = 0.48, one-tailed $p < 0.03$) with a similar trend in restored (r = 0.23, p = 0.06) and there was significantly greater fecal coliform densities in natural wetlands (Figure 2), there was no visible evidence during site visits that natural wetlands were impacted by human sewage or manure spreading. Thus, it appears that restored wetlands had greater concentrations of solutes in them; however, this did not affect the trophic status of the restored wetlands, as could be inferred from differences in total phosphorus concentrations ($TP_{Restored}$ = 32 ± 22 μg/L, $TP_{Natural}$ = 32 ± 24 μg/L).

We hypothesize that the significantly greater salt content (as indicated by specific conductivity and chloride) in restored wetlands is due to the creation of these wetlands, which often occurs by simply scraping off top soil until an impervious clay layer is reached or creating a dike in a region that has impervious soil layers (e.g., clay). Clays in the St. Lawrence River Valley are remnants of glacial activity and the Champlain Sea that existed in this area as late as 6000 years ago when these marine clays deposited under briny conditions [19]. There is a significant difference in the molar ratio (SO_4^{2-}:Cl^-) between restored (0.20 ± 0.32) and natural (0.43 ± 0.54) wetlands. The molar ratio of sulfate to chloride is exceeded in all wetlands relative to seawater (SO_4^{2-}:Cl^-) ≈ 0.05), which suggest that there has been more chloride flux from the clays in to overlying fresh waters. Chloride would be more mobile from clays, and more so with exposed clays as in restored wetlands. In support of these observations, a long-term study of restored wetlands in a comparable wet landscape in central New York concluded that establishment of soil conditions critical for water quality in restored wetlands can require decades to centuries to reach reference conditions [20].

Natural wetlands were more acidic but this was not solely due to higher concentrations dissolved weak organic acids (humic and fulvic acids) as indicated by CDOM concentration (Figure 1). Natural wetlands frequently had mature conifer tree stands growing along the margins; the acidifying effects of conifer litter may have caused the lower pH. There is a history of high and sustained levels of atmospheric sulfate deposition in this region that would contribute to sulfate content, reduction in alkalinity, and increased acidity in the ground and surface water [21], but this would affect restored and natural wetlands alike.

We suspect that wetlands that were in naturally acidic areas had adjacent acidic (and hence low quality) soils and were thus avoided by farmers due to poor soil condition and their collective traditional ecological knowledge of crop production in such soil. Hence, this set of natural wetlands were those selected to remain in a natural state. Further study (sensu [22]) would be required to examine whether soil infertility explains the more acidic natural wetlands in this region.

4. Conclusions

We conclude that water quality is similar between natural and restored wetlands in the St Lawrence River Valley, and any differences are minor and may be a result of how the restoration projects were done (site selection for wetland restoration) or the result of site characteristics of remnant natural wetlands (agricultural bias against removing a wetland from the landscape). From a water quality perspective, some differences exist between wetland types (salinity, pH, fecal coliform content) yet they do not have an impact on criteria of interest, such as trophic status (as indicated by the concentration of phosphorus) or the abundance of potentially toxigenic cyanobacteria. Overall, wetland restoration programs do meet their objectives of providing wetlands that are functionally similar to natural wetlands on the landscape.

Author Contributions: B.C.: investigation; T.A.L.: funding acquisition, project administration, writing—review and editing; M.R.T.: conceptualization, investigation, writing—original draft, review and editing. All authors have read and agreed to the published version of the manuscript.

Funding: This research was funded by a grant from the University of Michigan Water Center and the Erb Family Foundation.

Data Availability Statement: The datasets generated and analyzed during the current study are available in the Mendeley Data repository: https://data.mendeley.com/datasets/m7dycy7gt6/2, accessed on 30 May 2021.

Conflicts of Interest: The authors declare that they have no known competing financial interests or personal relationships that could have appeared to influence the work reported in this paper.

References

1. Mitsch, W. What is ecological engineering? *Ecol. Eng.* **2012**, *45*, 5–12. [CrossRef]
2. Filsinger, M.; Milmoe, J. *Restore and Enhance: The 25th Anniversary of the Service's Partners for Fish and Wildlife Program*; Supplemental Material, Reference S4; US Fish and Wildlife Service: Washington, DC, USA, 2012.
3. US Department of Agriculture Natural Resource Conservation Service. Agricultural Conservation Easement Program. 2021. Available online: https://www.nrcs.usda.gov/wps/portal/nrcs/main/national/programs/easements/acep/ (accessed on 30 May 2021).
4. Brinson, M.M.; Eckles, S.D. U.S. Department of Agriculture conservation program and practice effects on wetland ecosystem services: A synthesis. *Ecol. Appl.* **2011**, *21*, S116–S127. [CrossRef]

5. Cheng, F.Y.; Van Meter, K.J.; Byrnes, D.K.; Basu, N.B. Maximizing US nitrate removal through wetland protection and restoration. *Nat. Cell Biol.* **2020**, *588*, 625–630. [CrossRef]
6. Benson, C.E.; Carberry, B.; Langen, T.A. Public–private partnership wetland restoration programs benefit Species of Greatest Conservation Need and other wetland-associated wildlife. *Wetl. Ecol. Manag.* **2017**, *26*, 195–211. [CrossRef]
7. Lewis, K.E.; Rota, C.T.; Anderson, J.T. A comparison of wetland characteristics between Agricultural Conservation Easement Program and public lands wetlands in West Virginia, USA. *Ecol. Evol.* **2020**, *10*, 3017–3031. [CrossRef] [PubMed]
8. Welsh, R.; Webb, M.E.; Langen, T.A. Factors affecting landowner enrollment in wetland restoration in northeastern New York State. *Land Use Policy* **2018**, *76*, 679–685. [CrossRef]
9. Benson, C.E.; Carberry, B.; Langen, T.A. Public–Private Partnership Wetland Restorations Provide Quality Forage for Waterfowl in Northern New York. *J. Fish Wildl. Manag.* **2019**, *10*, 323–335. [CrossRef]
10. Stryszowska-Hill, K.M.; Benson, C.E.; Carberry, B.; Twiss, M.R.; Langen, T.A. Performance of wetland environmental quality assessment indicators at evaluating palustrine wetlands in northeastern New York State. *Ecol. Indic.* **2019**, *98*, 743–752. [CrossRef]
11. Chow-Fraser, P. Development of the water quality index (WQI) to assess effects of basin-wide land-use alteration on coastal marshes of the Laurentian Great Lakes. In *Coastal Wetlands of the Laurentian Great Lakes: Health, Habitat and Indicators*; Simon, T.P., Stewart, P.M., Eds.; AuthorHouse: Bloomington, IN, USA, 2006; pp. 137–166.
12. Decatanzaro, R.; Cvetković, M.; Chow-Fraser, P. The Relative Importance of Road Density and Physical Watershed Features in Determining Coastal Marsh Water Quality in Georgian Bay. *Environ. Manag.* **2009**, *44*, 456–467. [CrossRef] [PubMed]
13. Wang, T.; Driscoll, C.T.; Hwang, K.; Chandler, D.; Montesdeoca, M. Total and methylmercury concentrations in ground and surface waters in natural and restored freshwater wetlands in northern New York. *Ecotoxicology* **2020**, *29*, 1602–1613. [CrossRef] [PubMed]
14. Welschmeyer, N.A. Fluorometric analysis of chlorophyll a in the presence of chlorophyll b and pheopigments. *Limnol. Oceanogr.* **1994**, *39*, 1985–1992. [CrossRef]
15. Wetzel, R.G.; Likens, G.E. *Limnological Analyses*, 3rd ed.; Springer: New York, NY, USA, 2000; pp. 85–111.
16. Kring, S.A.; Figary, S.E.; Boyer, G.L.; Watson, S.B.; Twiss, M.R. Rapid in situ measures of phytoplankton communities using the bbe FluoroProbe: Evaluation of spectral calibration, instrument intercompatibility, and performance range. *Can. J. Fish. Aquat. Sci.* **2014**, *71*, 1087–1095. [CrossRef]
17. Carmichael, W.W. Freshwater Blue-Green Algae (Cyanobacteria) Toxins—A Review. In *The Water Environment*; Carmichael, W.W., Ed.; Springer: Boston, MA, USA, 1981; pp. 1–13.
18. Regalado, S.A.; Kelting, D.L. Landscape level estimate of lands and waters impacted by road runoff in the Adirondack Park of New York State. *Environ. Monit. Assess.* **2015**, *187*, 1–15. [CrossRef] [PubMed]
19. Desaulniers, D.E.; Cherry, J.A. Origin and movement of groundwater and major ions in a thick deposit of Champlain Sea clay near Montréal. *Can. Geotech. J.* **1989**, *26*, 80–89. [CrossRef]
20. Ballantine, K.; Schneider, R. Fifty-five years of soil development in restored freshwater depressional wetlands. *Ecol. Appl.* **2009**, *19*, 1467–1480. [CrossRef] [PubMed]
21. Driscoll, C.T.; Lawrence, G.B.; Bulger, A.J.; Butler, T.J.; Cronan, C.S.; Eagar, C.; Lambert, K.F.; Likens, G.E.; Stoddard, J.L.; Weathers, K.C. Acidic Deposition in the Northeastern United States: Sources and Inputs, Ecosystem Effects, and Management Strategies: The effects of acidic deposition in the northeastern United States include the acidification of soil and water, which stresses terrestrial and aquatic biota. *AIBS Bull.* **2001**, *51*, 180–198.
22. Oudwater, N.; Martin, A. Methods and issues in exploring local knowledge of soils. *Geoderma* **2003**, *111*, 387–401. [CrossRef]

land

MDPI

Article

Typing Colonial Perceptions of Carrum Carrum Swamp: The Expected and the Surprising

Meredith Frances Dobbie

Monash Art Design and Architecture, Monash University, Caulfield East, Melbourne, VIC 3145, Australia; meredith.dobbie@monash.edu

Abstract: Carrum Carrum Swamp was a vast wetland to the south-east of Melbourne, Victoria, Australia, at the time that it was first sighted by white colonists in 1803. By 1878, the colonists had commenced converting the swamp to dry land for agricultural and horticultural pursuits, and 100 years later it was predominantly residential land. Shifting values in the 1970s led to environmental concerns about water quality in local creeks and Port Phillip Bay and subsequent residential development on the former swamp included the construction of stormwater treatment wetlands. Perceptions of wetlands are now diverse, including positive perceptions that support their presence in urban settings. In contrast, traditionally, wetlands have been perceived negatively, as waste lands, leading to their drainage. Nevertheless, alternative, perhaps positive, perceptions could have existed, only to be overwhelmed by the negative perceptions driving drainage. Understanding the full range of past perceptions is important to ensure that the historical record is correct and to provide historical context to contemporary perceptions of wetlands. It will better equip natural resource managers and designers and managers of constructed wetlands in urban locations to ensure that wetlands are healthy, functioning and appreciated by their local and wider communities. Thus, the perceptions of Carrum Carrum Swamp by colonists from 1803 to 1878 were examined through qualitative content analysis of historical documents, and a typology was developed. Seven different perceptions were identified: scientific, premodern, exploitative, romantic, aesthetic, medico-mythic and ecological. Most could be traced to the colonists' predominantly British heritage, but one perception arose in the colony in response to the specific environmental conditions that the colonists encountered. This ecological perception valued wetlands as places of predictable water supply in a land of unpredictable rainfall. It recognised wetlands as part of a broader hydrological system, with influences on the local climate. Its proponents promoted the need for a different approach to the management of wetlands than in Britain and Europe. Nevertheless, a dominant exploitative perception prevailed, leading to the drainage of Carrum Carrum Swamp. The typology developed in this study will be useful for exploring perceptions of other wetlands, both colonial and contemporary.

Keywords: wetlands; colonization; nature-culture relationship; perceptual typology

Citation: Dobbie, M.F. Typing Colonial Perceptions of Carrum Carrum Swamp: The Expected and the Surprising. *Land* **2022**, *11*, 311. https://doi.org/10.3390/land11020311

Academic Editor: Richard C. Smardon

Received: 3 January 2022
Accepted: 11 February 2022
Published: 18 February 2022

1. Introduction

Traditionally, wetlands have been perceived as waste lands—dangerous, smelly, and unsightly places, breeding grounds of mosquitoes, snakes, and other pests, producers of dangerous gases, and physical obstacles to progress. Their value has lain in their potential as agricultural land, which could only be realized by drainage [1–4]. Draining wetlands has been described as "one of the oldest and commonest forms of land modification in American history" ([5], p. x); it is likely that this is the case across the developed world.

The draining of swamps has been explored in many publications, over decades. (e.g., [5–9], yet there are few studies of the past perceptions of specific wetlands, leading to their drainage. Meindl (2000) explores perceptions of the Floridian Everglades in the early 20th century [10]. He highlights the importance of beliefs and values in forming perceptions of the wetland, in the absence of direct knowledge of the place. Giblett (1996) describes different perceptions of wetlands during the establishment of Perth, Western Australia in the early 1800s [2]. Sornig (2018) provides insights into a range of perceptions of Dudley Flats and West Melbourne Swamp, which were the site of a shanty town well into the 20th century on the western edge of Melbourne, the capital of Victoria, Australia [11]. These studies show that past perceptions of swamps were more nuanced than convention allows.

Shifting values in the 1970s have led to environmental concerns about water quality in local creeks and waterbodies receiving stormwater discharge. Consequently, constructed wetlands are often part of stormwater management systems to harvest, treat, and reuse stormwater. Throughout Australia, sustainable stormwater management is being implemented, including the construction of treatment wetlands. Perceptions of wetlands are now diverse, including positive perceptions that support their presence in urban settings [12,13].

Many residential developments with constructed stormwater wetlands are being established on sites of former wetlands that have been drained. This is the case at Carrum Carrum Swamp.

Carrum Carrum Swamp was a vast wetland about 25 km to the south-east of what was to become Melbourne at the time it was first sighted by white colonists in 1803. The shallow trough that later became the swamp had been formed at the same time as Port Phillip Bay, after a period of glaciation 25 to 35 million years ago. The shoreline of the bay had since retreated a little, leaving sand dunes stranded inland behind which lay the swamp. Its area was more than 5000 ha, fed by what were to be named Dandenong and Eumemmerring Creeks, draining a total catchment of 735 km^2. Two creeks flowed through the sand dunes to Port Phillip Bay, one at the northern edge of the swamp, a small inlet now known as Mordialloc Creek, and another towards its southern edge, Kananook Creek [14,15]. Much of the swamp contained permanent water, which remained stagnant for long periods. Water was only slowly released to the bay. Any heavy rainfall or sudden storm rapidly flooded the entire swamp and often extended for a large area beyond.

By 1878, the colonists had commenced converting the swamp to dry land for agricultural and horticultural pursuits. Always marginal land at best, most of the swamp had been drained within 100 years.

Since Australia was first colonized in 1788, and Victoria in the 1830s, the landscape has been transformed, for "a settler society, whether or not numerically dominant, was an invading, investing, **transforming** society with an internal frontier, both natural and cultural" ([16], p. 10). Transformation was a result of the perceptions, beliefs, and values that the colonists held of the new environment, which were translated into action. Perception of Carrum Carrum Swamp by settlers, neighbouring residents, citizens, government authorities, and the media has influenced this landscape change. These perceptions and underlying beliefs and values attached to the swamp have directed action towards it.

In this study, these perceptions are explored to determine if they were as narrow as popular opinion would have us believe or broader, akin to current perceptions towards wetlands. The recent studies of perception of contemporary Victorian wetlands [12,13] implemented an empirical landscape assessment methodology, based on a transactional human-landscape model [17,18]. The method involved interviews and questionnaires to reveal perceptions directly. In contrast, this study adopts an historical methodology, in which literature is reviewed, archival material consulted, and colonial newspapers read, to infer perceptions of Carrum Carrum Swamp during the early period of its colonization. An understanding of past perceptions is important to complement knowledge about current perceptions of wetlands and to ensure that the historical record is correct.

Landscape perception is an outcome of a transaction between a human observer and the landscape [19]. There are multiple nested scales at which this transaction can occur, but the perceptible realm comprises visible landscape patterns, evoking perceptual processes and affective reactions. The observer's beliefs, values, knowledge, experience, and sociocultural context, amongst other personal attributes, will influence perception, as will attributes of the landscape, such as land type and use, spatial extent, ownership, etc. An aesthetic perceptual response is an obvious and immediate reaction to a landscape, but this is not the only way of perceiving a landscape. Meinig (1999) describes how one particular landscape can be perceived in ten different ways: as nature, habitat, artifact, system, problem, wealth, ideology, history, place, or aesthetic [20].

Different values that an observer holds for nature can inform the different ways of perceiving a landscape. Kellert (2009, 2012, 2018) proposes biophilic values held by humans towards nature, which can be expected to influence their perception of a natural landscape [21–23]. Originally there were nine values: humanistic, aesthetic, negativistic, dominionistic, utilitarian, ecologistic/scientific, naturalistic, symbolic, and moralistic [21]. More recently, Kellert has condensed them into eight: affection, attraction, aversion, control, exploitation, intellect, symbolism, and spirituality [22,23]. They are defined in Table 1. As values inherent in all humans [23], these might be expected to exist in colonial times, informing colonial perceptions of Carrum Carrum Swamp.

Table 1. Biophilic values of nature from Kellert [21,23]. Classification of values has changed slightly as Kellert developed the concept, decreasing the number from nine to eight and attaching different descriptors.

Value	Definition [22,23]	Earlier Descriptor [21]
Affection	Strong emotional attachment and love for natural world	Humanistic
Attraction	Aesthetic appeal of nature, from superficial sense of the pretty to profound realization of beauty	Aesthetic
Aversion	Antipathy toward and sometimes fearful avoidance of nature	Negativistic
Control	Tendency to master, dominate, subjugate nature	Dominionistic
Exploitation	Desire to utilize and materially exploit the natural world as source of materials and resources	Utilitarian
Intellect/Reason	Desire to know and intellectually comprehend the world, from basic facts to more complex understanding	Ecologistic/scientific and naturalistic
Symbolism	Symbolic representation of nature through image, language, and design	Symbolic
Spirituality	Pursuit of meaning and purpose through connection to the world beyond ourselves	Moralistic

Differing landscape perceptions, beliefs and values shaped the five visions of the environment that Heathcote (1972) identified since colonization of Australia—scientific, romantic, colonial, national, and ecological [24]. A scientific vision drove the exploration of the southern Pacific in the late 18th century and developed after colonization into a curiosity into natural phenomena for their own sake and an interest in the environmental obstacles to successful colonization. There was great interest in Australia's landscape, environment, and unique flora and fauna, leading to scientific descriptions and analysis. These involved both pure and applied scientific approaches, identifying local Australian patterns to help understand general global patterns, and collecting data as potential resources. A romantic vision existed concurrently with the scientific vision. In this vision, colonists were sympathetic to the Aborigines and regarded the countryside, known colloquially as bush, as almost a paradise. An important factor was the apparent lack of human influence, "a wilderness apparently unmodified by the hand of man" ([24], p. 87). The advance of civilization was unwelcomed. A colonial vision focused on the potential resources of the land, with both financial and aesthetic dimensions. The landscape was perceived as large, empty, unattractive expanses of wilderness. Its value lay in its potential as productive land, which would require the introduction of European plants and animals. Transformation involved land clearing, which was often rapid, widespread, and indiscriminate, with the goal to establish familiar and tidy agricultural landscapes. A national vision appeared from the late 1800s, with the celebration of the production of "a unique, Australian landscape" ([24], p. 91) from the original wilderness. This national vision reflected the establishment of cities as commercial and cultural centres. As cities became more urbane, a romantic dimension developed in this nationalism, which celebrated the bush in art and literature. Within this vision was a belief that the potential for national development was limited only by capital and labour. In the 20th century, an ecological vision emerged. This vision has characteristics of the scientific, romantic, and national visions. It emphasized the need to preserve the flora and fauna for their pure and applied scientific value. It implied a romantic conception of the natural environment as antidote to the built environment and the need for its appreciation by the public. There were also national overtones, in that preservation of the natural environment and its flora and fauna would generate pride and interest.

Heathcote [24] attributes a linear chronology to these visions, commencing with the scientific vision expressed by the explorers that 'discovered' and surveyed Australia and culminating in the ecological vision in the late 20th century. However, he cautions that one vision does not displace another, but each remains to create a dynamic complex of visions, some more conspicuous than others, changing over time.

This study draws on Kellert's typology of nature values [21] and Heathcote's typology of environmental visions [24] to interpret historical data to answer the following research question:

> What were the perceptions of the colonists who first settled in Victoria, towards Carrum Carrum Swamp, between their first sighting in 1803 and settlement and drainage, up to 1878?

In answering this question, a typology is developed for wetland perceptions to guide future studies.

2. Methods

Historical material extending in time from the first observation of Carrum Carrum Swamp by white colonists in 1803 to its settlement and drainage, up to 1878, was consulted. Searches using the keywords 'Carrum Carrum', 'Carrum Carrum Swamp', 'Carrum Swamp', 'Mordialloc Creek', 'Mordialloc Common' located material in the State Library of Victoria, the Public Records Office of Victoria, National Library of Australia, Dandenong Library, City of Greater Dandenong council records, and Historical Society of Victoria. Sources included Victorian Parliamentary Papers with records of parliamentary debates in the Legislative Assembly and reports of parliamentary committees and commissions; Parliamentary Acts of Victoria; Victorian Government Gazettes; contemporary newspapers, e.g., *The Argus, The Dandenong Journal, Mordialloc-Chelsea News*; applications for land selection and letters from selectors to the Victorian Commissioner of Lands; Dandenong District Road Board and Dandenong Shire rate books and minute books. In addition, published personal memoirs and local histories were found. Letters to the Editor in Melbourne newspapers contained commentary about other swamps near Melbourne. This was included in the study to augment material directly relating to Carrum Carrum Swamp as it revealed perceptions towards wetlands more generally that material specifically relating to Carrum Carrum Swamp might not.

This material was content analysed qualitatively to infer perceptions of the first white colonists to see Carrum Carrum Swamp, settlers of the swamp, other citizens of Melbourne, state and local government politicians, colonial scientists appointed by the state government to the Swamp Commission, and the media. Quotations are given in the text to illustrate perceptions. Newspaper quotations are taken from *The Argus*. Public commentary on local issues was very active and often reported similarly in the various newspapers. To provide continuity of commentary, where necessary, quotations are drawn from this single newspaper.

Additional references were consulted to aid interpretation of perceptions and provide context to them.

3. Results

Colonial perceptions of Carrum Carrum Swamp between 1803 and 1878 can be characterized as scientific, premodern, exploitative, romantic, aesthetic, medico-mythic, or ecological. Six of these relate to Kellert's nature values [21–23] and Heathcote's environmental visions [24]. However, the premodern perception is not in either typology.

3.1. Scientific Perception

A scientific perception was revealed, which reflects Heathcote's scientific vision in which there is a scientific curiosity about the natural environment and a desire to identify opportunities for settlement and development [24]. At Carrum Carrum Swamp, this is evident in diary entries and annotated maps of the first explorers to sight the swamp. Subsequent surveys described the land and its potential for agricultural development.

The first written account of Carrum Carrum Swamp, in the journal of the exploration of Port Phillip in 1803 by Charles Grimes, Acting Surveyor-General of New South Wales, reveals both pure and applied scientific interest. The journal was kept by James Flemming, a gardener, assigned to the expedition to assess soil quality [25]. The published journal includes the daily entries by Flemming and explanatory footnotes by the editor, J. J. Shillinglaw [26]. The swamp was described and thereby classified for its potential value as a resource. On Sunday 30 January 1803, Flemming noted that

> *"I ascended a hill (footnote—back of Frankston) where I could see eight or ten miles, hills without trees, narrow valleys with scrubby brush. The soil black, g[r]avelly sand; at a mile-and-a-half from the beach a run of fresh water to a lagoon. Came to a river* (footnote—Cananook Creek)*; it was salt; traced it to the beach; crossed it up to the knees about a mile farther; went in about a quarter of a mile found a fine fresh water river about 30 feet wide, and deep enough for a boat; Mr. Grimes took the bearings of it; traced it six or eight miles; it runs in a parallel line with the sea."*

The following day, they

> *"crossed a neck of land about half a mile over* (footnote—referring to Long Beach and Carrum Swamp)*; went along the beach a little way and ascended a hill; the country appearing very barren."*

On Wednesday, 2 February 1803, they came to

> *"a large swamp, with three lagoons in it, all dry. The land appears covered with water in wet seasons".*

On the map produced from Grimes' survey [27], the area of Carrum Carrum Swamp is described as "Open and barren country" (Figure 1, indicated by red arrow).

On subsequent charts and maps produced of Port Phillip Bay and its shoreline, Carrum Carrum Swamp is always indicated, reflecting its potential as a resource for the colonists. At first its description is general, such as "Large swamp overgrown by reeds" in 1804 [28], "Swampy country" in 1827 [29] or "Swampy land" in 1836 [30]. With time, the swamp was named and its description on maps became more detailed, indicating the diversity of the swamplands and suggesting its potential for colonization and agriculture, e.g., "Sandy ridge", "Open plain liable to winter floods black soil well grassed", "Tea tree scrub", "Fine agricultural soil lightly timbered with gum, cherry & lightwood", "Fine agricultural land—partly subject to flood", "Dense tea tree scrub water plentiful" [31].

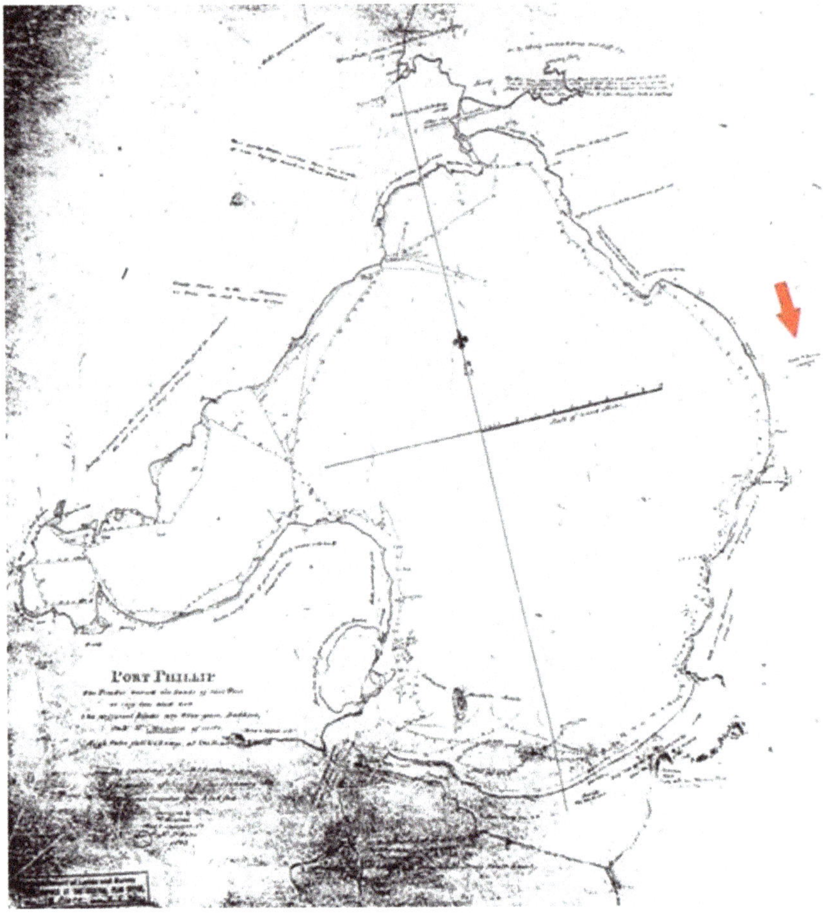

Figure 1. Survey of Port Phillip Bay, by Grimes, 1803. Red arrow indicates the description of Carrum Carrum Swamp. Source: Crown Lands and Survey Historical Plan CS26(1), Port Phillip, C. Grimes, 1803, Public Records Office of Victoria.

3.2. Premodern Perception

A premodern perception of swamps is likely to have accompanied those colonists who first settled on Carrum Carrum Swamp as squatters. Their successful colonization would have depended on it. Sluyter (2002) contrasts the Modern West, those developed Western societies in which the natural and the social had become separate dichotomies, with the Premodern Rests, in which the natural and the social were intertwined [32]. Many colonists were from countries that had long adapted over centuries to living with coastal wetlands; their premodern perception of Carrum Carrum Swamp revealed in this study would have been shaped by their experience of similar coastal wetlands in England and would have, in turn, shaped their attitudes and actions towards the swamp. This premodern perception was expressed by the colonists utilising the swamp within its biogeophysical constraints, in which the social and the natural were not binaries but enmeshed. There is no equivalent to this premodern perception in either Kellert's nature values [21–23] or Heathcote's environmental visions [24].

By the time of colonization of Australia, many of the coastal wetlands of England had been transformed from their prehistoric mosaics of intertidal mudflats and more elevated vegetated saltmarshes to landscapes of arable land, pasture, and meadows [33]. Darby (1956) describes the transformation process in The Fens, an area of marsh in East Anglia on the east coast of England, edging The Wash, a large embayment of the North Sea [7]. It was a large flooded plain, with an area of 330,000 ha, with peat on the landward side and silt on the seaward, fed by rivers. Before large-scale drainage in the 17th century, The Fens consisted of open pastures and meadows, with reedy swamps and pools, connected by a network of channels. In winter, the entire area was covered with water, punctuated by islands. From medieval times, clusters of villages had been established on the high land, e.g., Isle of Ely, sharing common land for grazing and watering stock. Residents lived on the resources that The Fens provided: fishing, fowling, gathering reeds and rushes, and making salt within the marshes; making hay, grazing livestock, and cutting turves on the land less frequently inundated; and farming on the islands or higher arable land. The local economy was bound with the cycles in The Fens, with land practices based on tradition. Seasonal variation was accepted as inevitable, maybe even beneficial. For example, winter floods made the pasture richer; summer floods caused little damage. Although much of the southern part of The Fens had been drained by the 18th century, with a regular pattern of channels and dykes superimposed on the older natural streams, there were still areas of deep water and patches of swamp into the 19th century.

This was the inheritance of the most successful of the squatters who first settled Carrum Carrum Swamp in 1837. The northern half of the swamp was described in a survey in 1868 [34] as marshland, elevated 1–3 m above the level of the highest spring tides at the mouth of Mordialloc Creek. It was unwooded except for a few patches covered with tea tree or with red gum and honeysuckle (banksia). It flooded in wet seasons [35]. Morass, overgrown with reeds and sedge, covered half the rest of the swamp, less than 1 m above the highest observed spring tides at the mouth of Mordialloc Creek or about 1 m above the level of ordinary high water. This area had been described as swamp with permanent water in an earlier survey in 1866 [36], with loose peaty soil over sand. The rest of the swamp was sandy hillocks and flats, covered with scrub, fern or spear grass and lightly wooded with stunted eucalypts, banksia, and casuarina.

The attraction of land on Carrum Carrum Swamp to colonial settlers was plentiful water [15], either in Mordialloc or Eumemmerring Creek or the permanent water holes and lagoons along the creeks or in the swamp itself. In time, four runs were established by squatters on the swamp, vast holdings to graze cattle and sheep. That much of the swamp might be too wet at times to be grazed fully was less important than access to water. The land that was not permanently inundated was valued highly, "extensive plains of rich black loam covered in rib grass, one of the most nourishing grasses in Australia" (J. Hawdon, quoted in [15], p. 22). To be successful, the squatters had to work within the physical constraints of the swamp. Many of the squatters failed and, by 1847, there was a single large run in the southern part of the swamp operated by the Wedge brothers, known as Banyan Waterholes (Figure 2). The Wedges had come to Australia from East Anglia [15], so it is highly likely that they were familiar with a watery landscape such as Carrum Carrum Swamp and had the skills to manage it.

As British tradition dictated, a farmers' common was also established on Carrum Carrum Swamp. An Aboriginal Reserve had been established at the mouth of Mordialloc Creek on the northern edge of the swamp in the early 1850's [15]. With the colonial occupation of Carrum Carrum Swamp limited to the southern half, 2007 ha in the northern half of the swamp was declared a Farmers' Common, under the *Nicholson Land Act 1860*, in February 1861 [37]. This farmers' common at Mordialloc included the Aboriginal Reserve. It was extended twice so that by late 1866 it occupied the entire swamp from Mordialloc Creek, along the coast of Port Phillip Bay to the boundaries of the squatters' holdings in the south. The farmers' common allowed the benefits of grazing cattle (sheep were not allowed) to be shared by purchasers of land within 5 miles of the common [37–44]. Access to the common was particularly important during times of drought when the swamp, although drying out, would have supported more vegetation than elsewhere. The farmers using the common valued the land as their own, wishing to use the revenue from its rental for its improvement [45].

Figure 2. Carrum Carrum Swamp, extending south from Mordialloc to Frankston and east towards Dandenong, showing the Wedge's run, Banyan Waterholes. Source: Map of Victoria; constructed and engraved at the Surveyor General's Office; G.A. Windsor, draughtsman; William Slight, engraver. Melbourne: Published by authority of Government under direction of A.J. Skene, M.A. Surveyor General, The Hon. J.J. Casey, President, Board of Land & Works & Comr. Of Lands & Survey, 1872, National Library of Australia.

3.3. Exploitative Perception

An increasing concern for the potential resources of the new colony and a weariness with the unfamiliar landscape found expression in an exploitative perception of the landscape. This encompasses Heathcote's concept of a colonial vision [24] and is associated with a utilitarian value of nature [21]. A broad-scale process of landscape change began, in which the natural landscape was modified to meet utilitarian needs of the colonists or their aesthetic preference. Landscape change in wetlands involved their drainage. The colonists' aesthetic preferences were expressed in the conversion of wet land to dry land, as were the values that drove the change. Inevitably, this involved transformation of the colonial landscape to a predominantly British landscape, reflecting the preferences of the colonists.

Attitudes towards Carrum Carrum Swamp would have been influenced by the process of landscape change in wetland regions of England, from where many colonists came. In prehistoric times, the coastal wetlands in England were mosaics of intertidal mudflats and more elevated vegetated saltmarshes [33]. Freshwater peatlands developed on the inland areas, which were settled in the Roman period to exploit their natural resources. Some reclamation efforts were made late in this period, but most coastal wetlands were abandoned and reverted to their natural state by the end of the Roman period. Human settlement resumed during the Middle Ages when drainage efforts commenced [33]. The drainage process in The Fens marshland is well documented [7]. By the Middle Ages, local and individual effort had reclaimed the edges of the marshland. Legislation for comprehensive draining of The Fens was first passed in 1600, in response to a treatise by Humphrey Bradley in 1589. He proposed drainage of The Fens to increase the population of the area, to increase productivity for local, regional, national, and international trade, to provide employment and in "many other ways redound to the great advantage and strengthening of the nation" ([7], p. 68). Little drainage was achieved, however, until after passage of the *Act for the draining the Great Level of the Fens, extending itself into the Counties of Northampton, Norfolk, Suffolk, Lincoln, Cambridge and Huntingdon, and the Isle of Ely, or some of them* in 1649 [7]. By the 18th century, much of the southern part of The Fens had been drained, to take advantage of the fertility of the peat and silt lands for agriculture and to open the area for more settlement. By the 19th century, technological advance through the Industrial Revolution and improved agricultural practices had contributed to effective drainage of much of The Fens [7].

The colonization of Victoria at the time of the Industrial Revolution equipped the colonists with the technical skills to effect dramatic landscape change readily [8,16], bringing the colony into the Modern West [32]. They had access to technology that would enable them to implement changes consistent with their desire to exploit the land occupied by wetlands, their familiarity with transformed waterscapes in their home countries [46], and a legislative system to enforce them.

In Victoria, colonial attitudes that valued wetlands as a potential resource were enshrined in legislation, replicating the practice in Britain. A series of Land Acts was proclaimed, regulating the survey, selection, and sale of land in the colony. Acts relevant to Carrum Carrum Swamp were the *Land Act 1862* [47], the *Amending Land Act 1865* [48], and the *Land Act 1869* [49].

Section 38 of the *Amending Land Act 1865* legislated for the granting of a lease of Crown Land to anyone "willing to make and construct canals or to undertake works for the drainage or reclamation of any swamp or morass". This did not apply to land simply subject to flood [50]. A Professional Board was appointed to inquire into applications in October 1865 [51]. It received several applications for schemes involving Carrum Carrum Swamp. Although much of the swamp was under permanent water, the thick layer of decaying organic matter and silt trapped in the swamp promised very fertile soil. In October 1865, Lockhart Morton applied for a lease of the entire Carrum Carrum Swamp. A lease for his scheme, details of which were not given, was recommended by the board, with the explanation that the land, currently drawing a revenue of £9 2 s. per annum, would be worth at least £4 or £5 per acre. Subsequently, objections were submitted on the grounds that "a portion of the so-called swamp was good grazing land" [52] and that construction of a canal was being considered [53]. In 1869, another proposal was canvassed, to lease a portion of the Carrum Carrum Swamp for sugar beet cultivation. The scheme was promoted to benefit the district and the colony, by the employment the lease would provide (anticipated to be 1500 [54]), and "the conversion of what is a barren waste into a luxuriant agricultural district" [55]. The issue of the lease was supported by 258 farmers and other residents of Mordialloc near the swamp. They believed that the proposal to drain the swamp and use it for growing sugar beet would open up several hundred hectares for cultivation and be the means of supplying the district with fresh water in the area and greatly preferred it to Morton's proposal [54].

As *The Argus* anticipated on 30 April 1869 [56], soon the government believed that the value of Carrum Carrum Swamp lay in it being drained for agriculture: "in its undrained state it would be perfectly useless" ([57], p. 1079). The swamp had been surveyed in 1868 in preparation of its sale and schemes for its drainage were debated in Parliament. Ultimately, the cost of drainage was considered too great for either the government or an individual to bear ([57], p. 1144). Considering sale by auction inadvisable, the land was made available for selection.

Land on Carrum Carrum Swamp was selected under the *Land Act 1869* [49] (Figure 3). Strict conditions were specified in Section 20, which the selector was obliged to meet.

These included the requirement to enclose the land "with a good and substantial fence", to cultivate at least 1 acre out of every 10 acres, to occupy the allotment within 6 months of the issue of a licence for the duration of the licence (minimum residency of 2 1/2 years), and to undertake substantial and permanent improvements to the value of £1 per acre by the end of the third year of the licence.

Most of the selectors of allotments on Carrum Carrum Swamp were seeking land from which they could make a living, as graziers, farmers, and market gardeners. To do so for many required drainage of the land; drainage was also necessary to meet the conditions of Section 20. Only allotments on the sandy hillocks and flats in the south of the swamp might have been able to support cultivation and allow fencing and construction of a residence.

A typical selector was Alfred Bishop, who wrote to the Secretary of Lands in 1878 that "the land in question is of no use to anyone unless drained" [58]. Norman McSwain wrote to the President of the Board of Lands and Works in 1876 that "Through the whole year three fourths of the land is under water and during the past winter there was not one acre dry in the whole of it ... the land is not returning anything. I had five head of cattle and two horses on it but when the first flood came down last winter I found the cows over their knees in water where they had been for two days without any food and it was with no small difficulty I got them off so that it is very hard for a poor man to spend money on such land. When the proposed drainage is finished it will alter the case" [59].

Edgar Pettit may have had different expectations. He was a basket maker and hoped to be able to grow willows on his selection [60,61], an undertaking that would require wet land, but his land was so wet "I cannot reside there to grow willows, and it is no good taking up ground to grow willows upon a mountain, and it is wet even now; I cannot go on with the fencing" ([61], p. 3).

3.4. Romantic Perception

Some colonists had a romantic perception of the landscape, as described by Heathcote, delighting in its "uncivilised" appearance, and dreading the inevitable changes of civilization [24]. They perceived the landscape as pristine, untouched by human activity. This is related to Kellert's humanistic value [21]. Such an attitude was evident in the perception of Carrum Carrum Swamp by William Bruton, born in 1854. In his memoirs [62], he recalls the vegetation and bird life of the swamp of his childhood:

> *"When I first visited Carrum ... .the foreshore was a growth of honeysuckle ferns and wild currants, and when these trees were flowering, a large number of birds were seen. Magpies and crows preferred the other side of the swamp. The call of the Kookaburra was heard everywhere, and amongst the trees were wattle birds, leatherheads, woodpeckers, thrushes, kingfishers, robins and many different kinds of parrots, and as we camped near the swamp, we heard plovers chattering and chanting the whole night through. As we cannot go through the swamp we go around a large clump of swamp ti-tree, when—oh! wild turkeys—they know the human beings, and are up and off quickly.*
>
> *But what are those tall things over yonder? A flock of native companions. Rare as they are, they are still birds, but they are far more conceited than any other bird. Note their stately stride; their very conceit of themselves constitutes the joy of life.*
>
> *Here the gum trees lay prone where they have lain for hundreds of years, and others in the full glory of life send their spreading limbs and luxuriant foliage out, displaying their pride of life.*
>
> *Here also are the possums in plenty, disporting themselves amongst the branches.*
>
> *"The rich man has many goods, but here we have all we require. We eat, drink, and are merry, and the rich man has no grand busy tail as we have." "Oh! That one out there and in the daytime, too. Look out pussy! If that gunman comes around the corner he may want you for dog's food.""* ([62], pp. 4–5).

Figure 3. Lot plan of Carrum Carrum Swamp, 1868. Source: Crown Lands and Survey Historical Plan, Roll plan 17A, Public Records Office of Victoria.

Bruton regretted the loss of the swamp, concluding

"And so it seems that the energy of the axeman, the drainer, the builder have turned the heavenly paradise of thousands of years, into a joy somewhat like unto the dog of old, racing with the jam tin, which rascally boys have attached to his tail" ([62] p. 5).

In this romantic perception of swamps, changes inevitably associated with colonization were unwelcome.

3.5. Aesthetic Perception

An aesthetic perception, albeit negative, is evident amongst Victorian colonists, who regarded wetlands as unsightly as well as unproductive land. Heathcote describes aesthetic responses within the romantic, colonial, and national visions [24], whereas Kellert identifies aesthetic values of landscapes as a distinct category [21]. As perception is often primarily visual, a fundamental perceptual response when viewing a landscape is aesthetic [19]. Gobster et al. (2007) define a landscape aesthetic experience as "a feeling of pleasure attributable to directly perceivable characteristics of spatially and/or temporally arrayed landscape patterns" ([19], p. 964). However, aesthetic appreciation of wetlands is difficult because it does not usually fit the scenic canon [2,63]. Appreciation often requires an understanding of their complexity and ecological functioning [64–68] and perhaps some imagination [69] and arousal [70]. It is cognitive as well as affective. Consequently, aesthetic appreciation of wetlands is often unfavourable; wetlands are often perceived as unattractive [19].

In colonial Victoria, negative aesthetic perceptions of wetlands such as Carrum Carrum Swamp encouraged their drainage, in the process creating the more familiar and favoured landscapes of Britain. A correspondent to *The Argus* in 1866 suggested Batman's Swamp near Melbourne should be drained using trenches, to create "fine grassy emerald meads, dotted with clean, sleek, well-fed cows; not, as now, with dirty, raw-boned, wretched animals, wading up to their bellies in semi-fluid, black putrid mud, to crop the rank, innutritious garbage growing on its surface" [71]. It is expected that a similar aesthetic perception applied to Carrum Carrum Swamp.

3.6. Medico-Mythic Perception

Colonial attitudes towards wetlands were also influenced by folklore that had developed over centuries. Although The Fens had been inhabited, albeit sparsely, since Roman occupation, most people regarded swamps and marshes with fear and horror. The Fenlanders were regarded as a breed apart. Swamps were perceived as unhealthy places, to be avoided. As well as a source of disease, they were believed to be occupied by evil spirits and devils, a belief that passed into folklore [5]. This can be interpreted as a medico-mythic perception. It has no parallel in Heathcotes's typology of visions [24], but it falls within Kellert's negativistic value [21].

By the 18th century, scientific knowledge in Europe and Britain supported the miasma theory [72]. The colonists brought this theory to Victoria. Miasmas were accumulations of infectious particles floating in the air. It was believed that air held a frightening mixture of gases in suspension [73]. In particular, the air around wetlands was full of dangerous emissions from it, including the decomposition products of plants, animals, and insects. The smell associated with a wetland was regarded to be evidence of these dangerous particles, contaminating the air around it.

The miasma theory supported the perception of swamps as unhealthy places, requiring drainage to protect the health of the colonists. Such colonial perceptions were expressed in Letters to the Editor, published in the daily newspaper. An interesting exchange of letters was published in *The Argus* [71,74–78], debating the proposal to drain Batman's Swamp. The swamp was a common, used by residents to graze their stock throughout the year. A proposal had been submitted under Section 38 of the *Amending Land Act 1865* [48] to drain it for cultivation of a market garden. There were conflicting opinions about the value of the swamp as public or private land ([74], p. 5). Prominent were concerns about the health risks of the swamp, although it is difficult to separate concerns about the swamp itself from those of the drainage of the adjoining manure depot and the neighbourhood's sewage draining into it. One medical correspondent wrote that during the hot season, the "atmosphere" around the swamp was filled with "the most deadly malaria" [75]. He continued, "Ague is not a common disease in this country, except in some parts of Queensland, but on the northern part of this swamp I have attended case of the very worst form, with great danger to life ... the stench arising from it (the swamp), sufficient to poison a whole neighbourhood". The writer concluded with his hope that the government would convert "a deleterious swamp into a healthy and salubrious garden". Another letter on 3 March 1866 expressed the same concern about the "poisonous gases" produced by the swamp [77]. The writer believed that it was a "pestilential miasma arising from Batman's Swamp and the manure depot". Similar perceptions no doubt applied also to Carrum Carrum Swamp.

3.7. Ecological Perception

An ecological perception of wetlands, extending to Carrum Carrum Swamp, is evident amongst Victoria's colonial scientists, supported by the local newspapers. This perception pre-empts the ecological vision identified by Heathcote, which he dates to the late 1970s [24]. This earlier ecological perception recognized the importance of the local flora and fauna and the ecosystems of which they were a part, although the discipline of ecology was not named until 1869 by the German scientist Ernst Haekel and defined as "the study of the natural environment including the relations of organisms to one another and to their surroundings" ([79], p. 3). Importantly, this perception acknowledged that the environment of Australia differed markedly from that of Europe and the United Kingdom and thus needed to be managed differently. It falls within Kellert's ecologistic/scientific and naturalistic values [21].

Under Section 38 of the *Amending Land Act 1865* [48], applications were made for leasing and drainage of many swamps on Victorian Crown Land, in addition to Carrum Carrum Swamp. Applications proposing grazing, farm and garden production, pastoral and agricultural pursuits, cultivation of cereals and English grasses, and cultivation of willows were assessed by a Professional Board appointed by the government, forming a Swamp Commission. Objections raised by the Professional Board highlighted such issues as the value of the swamps for watering of stock (e.g., [80]), for general access to water (e.g., [80]), and for the availability of pasture during droughts [81].

With remarkable early insight into the environmental constraints of the colony, the Professional Board wrote that many swamps in the Western District of Victoria "are the recipients of water derived from large areas of drainage. They arrest the rapid conduction of that water to the sea, and render less unequal the summer and winter discharges of the streams which are fed by these swamps. Many of the swamps are also very useful as watering places for the stock of owners and occupiers of adjacent lands. In nearly every case the general interests of the public would be more promoted by raising the levels of the water in these swamps by dams than by lowering those levels by drains." [82].

They also highlighted the ameliorating effect of the swamps on the local climate. "We could not help observing on our journey that the surface of the country in the vicinity of large swamps and well-wooded hills showed no signs of the drought which has so seriously affected many districts. It is true that we were traveling through a part of the colony which has a greater inland water surface than any other. Everywhere swamps lie at the base of the volcanic hills; and lakes of fresh, and brackish, and salt water are numerous, and some of them very large. No doubt these extensive sheets of water serve to modify the local climate; and it is certain that the observant traveler cannot fail to perceive a change when he approaches one of the larger lakes. The cool moist breeze, the rich colours of the landscape, the character of the foliage, and the luxuriance of the grasses, all indicate that the local conditions are different from those which obtain in other parts where a water surface is absent." [82].

The members of this Professional Board were Charles Whybrow Ligar, the Surveyor-General (formerly Surveyor-General of New Zealand), Clement Hodgkinson, Assistant

Commissioner of Crown Lands and Survey, and Robert Brough Smyth, Secretary of Mines and formerly Government Meteorologist (Figure 4). Each was a prominent member of the colonial scientific community, both professionally and personally, and contributed to the colonial scientific discourse in their areas of training and interest. All were active members of the Royal Society of Victoria, which had the aim to embrace "the whole field of science, with a special reference to the cultivation of those departments that are calculated to develop the natural resources of the country" [83]. In their contributions to the Royal Society of Victoria, their training was complemented by their interest in natural history. Each presented papers to the Society in his area of interest and expertise and served as an office bearer or councilor.

Figure 4. Professional Board appointed to assess swamp reclamation applications, 1865: Charles Whybrow Ligar, Surveyor-General (**left**), Robert Brough Smyth, Secretary of Mines (**centre**) and Clement Hodgkinson, Assistant Commissioner of Crown Lands and Survey (**right**). Sources: Charles Whybrow Ligar, Surveyor-General, ca. 1859, State Library of Victoria; Robert Brough Smyth, 188–?, by George Gordon McCrae, 1833–1927, National Library of Australia; Clement Hodgkinson, Johnstone, O'Shannessy & Co. photographers, Melbourne; David Syme & Co., 2 October 1893, State Library of Victoria.

As members of the Royal Society of Victoria, the members of the Professional Board would have been well informed about contemporary scientific issues. The Society played an important role in the scientific development of the colony and contributed to a detailed knowledge of its natural resources [83]. They would also have had access to scientific periodicals from overseas, exchanged for the publications of the Royal Society of Victoria. They were part of an imperial scientific collaboration, whereby British imperial officers and scientists exchanged environmental data and insights [72,84].

Much had been written in this scientific literature about the negative effects of deforestation in the colonies [85]. Clement Hodgkinson was clearly aware of such literature when he commented that "to drain all the swamps in the country would be as great a calamity as to denude it of all timber" [80]. He established a program of reservation, regulation, administration and education to control the use of Victoria's forests, which became a model for the future forestry profession (www.asap.unimelb.edu.au/bsparcs/biogs/P002057b.htm; accessed 25 November 2021). This system had its origins in the forest-conservation system set up in colonial India [72,85].

This scientific exchange cast doubt on imported environmental theories and associated management practices [72]. The importance of the specific environmental conditions of each colony was acknowledged, as was the need to develop local environmental management strategies. Water management was critical to the survival of the colony. The Professional Board's recommendations reflected this.

The media supported the conclusions of the Professional Board. *The Argus* published weekly accounts of the Board's meetings [50,80,86–89]. On 19 December 1865, the editor wrote that "the general tenor of the report (of the Professional Board, in which many swamps were exempted from reclamation) confirms the impression that any extensive drainage of our "waste lands" would be productive of very dangerous consequences. The belief that any general or extensive drainage of bogs and fens ought to prove a public benefit is one of the many old country ideas which is still inapplicable here. In the United Kingdom, with its constantly dropping skies—in Holland, where so much of the soil has been wrested from the sea—there is, as a rule, only too much water on the earth and in the air; and in Western Europe generally, the man who undertakes to remove what has been termed those "blurs on the fair face of nature", justly deserves to be regarded as a public benefactor. But the case in Australia, with its strong sun and with its perpetual liability to drought, where water is the great physical want, and where every expanse of water, however unsightly or seemingly wasteful to old country eyes, is really precious from its influence in lending the atmosphere humidity . . . Our water-covered lands are not now waste lands; they contribute to a purpose, and no one will deny that it is the most important of all others here." [90].

4. Discussion

4.1. The Expected and the Surprising in Colonial Perceptions of Carrum Carrum Swamp

From their homelands, the colonists brought with them to Carrum Carrum Swamp scientific, premodern, exploitative, romantic, aesthetic, and medico-mythic perceptions towards wetlands. The presence of these perceptions is to be expected. They represent two different perspectives of wetlands: those that accept the wetland as wet land and those that support drainage and conversion of the wetland to dry land. A premodern perception of wetlands had enabled the first colonists, the squatters, to settle on the swamp before its drainage and to manage it as a resource. This perception enabled occupants of The Fens and other coastal wetlands in England and elsewhere to survive for centuries [7] but has not been revealed before in a study of colonial perceptions in Australia. The romantic perception also accepted the wetland in its natural state, to be enjoyed as a wild refuge from everyday life. In contrast, the scientific, exploitative, aesthetic, and medico-mythic perceptions of wetlands all contributed to the drainage of wetlands, both in Australia and elsewhere (1–10). Within 41 years of its settlement by colonists, the wet land of Carrum Carrum Swamp was converted to dry land. A mosaic of swamp, morass, permanent water, and higher sandy land—waste land—became a landscape of market gardens, farmland, and grazing land, with some areas of water too deep to be drained.

The presence of an ecological perception of swamps, expressed by colonial scientists and promoted by the local media, was a surprising result of this study. This ecological perception attached a different value to wetlands and encouraged their retention as wet land. It arose in response to the unique environmental conditions of the new colony and predates the ecological vision identified by Heathcote [24], albeit for utilitarian reasons, and before the discipline of ecology had been named and defined in 1869 [79]. The members of the Professional Board of the Swamp Commission understood that the swamps were elements of hydrological systems, connected to creeks and rivers, and that they moderated the local climate. They recognised that swamps had value beyond utilitarian purposes.

This ecological perception of wetlands in colonial Victoria is consistent with the rise of environmentalism in the colonies on the periphery of the modern world [85]. There were several aspects to this environmental consciousness, related to the practical concern for the physical wellbeing and survival of colonists; a new valuing of the environment as an expression of the "other" represented by the newly colonized land; a growing awareness of extinction processes; and an understanding of the dynamics of species change and the origin of humans. Scientific members of the colonies were aware of the potential for harmful environmental change with colonization. Such changes could jeopardize the long-term survival of the colony. In such circumstances, conservationism had economic advantages, particularly in ensuring forest protection and water supply. The perception, beliefs and values attached to Victorian wetlands by the Professional Board is a local example of this environmentalism in colonies.

Early colonial conservation policies were almost always perceived as being the legitimate concern of the state rather than of the individual [85]. The unfamiliar environment of the colony presented risks to its survival. To ensure the future of the colony, some understanding of those risks was necessary, as was the implementation of some controls. The advice of scientists was critical to the colonial government in identifying and assessing the risks. The Victorian colonial government recognised this in establishing the Professional Board to assess applications for leases of swamps on Crown Land. Consequently, controls were imposed on the drainage of selected swamps, the value of which for the colony lay in their retention as water sources in a country of unreliable rainfall.

Water management was critical to the survival of the colony. The environmental conditions in the colony, however, differed dramatically from those of Britain. The premodern and exploitative perceptions, values, and attitudes towards wetlands had developed in a stable and known relationship between the people and the land, between society and nature, over generations in Britain [85]. The establishment of the Professional Board acknowledged the uncertainty of the relevance of these inherited conventions to the new colony. The involvement of scientists in the assessment of wetlands officially introduced empiricism into management of wetlands in the colony, although some colonists were already aware of the need for empirical testing in their management of the new land [72]. Exploitative perceptions and practices might not be appropriate. An empirical approach was necessary, whereby new land management practices could be developed and their efficacy tested in the new and unfamiliar environment.

This ecological perception must be distinguished from an ecological aesthetic perception [19], which focuses on the pleasure experienced from the appearance of a landscape. An ecological aesthetic perception has been defined in contrast to a scenic aesthetic perception, which is problematic when appreciating many types of landscapes, particularly wetlands [2,19,66,67]. An ecological aesthetic perception requires knowledge and involves cognition [65]. It is multimodal, involving senses other than vision and does not involve a "frame". It accepts that the landscape is dynamic, living and changing and often messy. It is not possible to discount an ecological aesthetic perception by the colonial scientists in this study, but the data did not reveal it specifically.

An ecological perception has also been described by Sewall (1995) [91] but it, too, differs from the colonial ecological perception. This ecological perception is specifically visual, focusing on dynamic relationships within a landscape. It has not been identified empirically but is proposed as a way of "seeing" to bring humans closer to the natural world. It demands five practices: paying attention; seeing relationships between things rather than objects independent of context; being flexible in perceiving, by seeing familiar things in a new way; seeing from a position within the biosphere; and encouraging visual imagination. Ecological knowledge is not included. This is not equivalent, either, to the ecological aesthetic perception, which is cognitive as well as affective.

4.2. Typology for Perception of Wetlands

Landscape perceptions, beliefs, and values shaped the five visions of the environment that Heathcote [24] identified since colonization of Australia—scientific, romantic, colonial, national, and ecological. The colonial perceptions of Carrum Carrum Swamp lie within four of these visions. The scientific, romantic, and colonial (exploitative) visions would be expected. However, the fourth, the ecological, has generally been regarded as a phenomenon of the 20th century. The fifth—a national vision—was absent from Carrum Carrum Swamp. Yet there were other perceptions that are evident amongst the colonists. One, a premodern perception, which was a heritage of their homeland, led to adaptive land management when colonizing the swamp. In addition, there were aesthetic and medico-mythic perceptions.

These seven perceptions bear some relationship to the nine values of nature proposed by Kellert [21]. However, utilitarian values underlie both premodern and exploitative perceptions of Carrum Carrum Swamp; naturalistic, symbolic, humanistic, and moralistic values underlie romantic perceptions; and negativistic values can underlie medico-mythic perception. Utilitarian values can also underlie the colonial ecological perception of wetlands. Kellert's typology with nine values [21] is preferred to his later typology with eight values [22,23]. It offers more scope, with separate ecologistic/scientific and naturalistic values and uses adjectives as descriptors of values instead of nouns.

From Heathcote's and Kellert's typologies, a typology for wetland perception has been developed (Table 2). An earlier study of contemporary perceptions of wetlands identified ecological/scientific, aesthetic, negativistic, utilitarian, naturalistic and symbolic perceptions [12]. Ecological/scientific and aesthetic perceptions were dominant. This classification used Kellert's earlier terminology [21]. Interpreted in terms of the typology presented in Table 2, these perceptions would be described as scientific, ecological, aesthetic, medico-mythic, exploitative, and romantic. Thus, present perceptions essentially follow the same typology as past colonial perceptions. However, in contemporary perceptions, the predominant were ecologistic/scientific and aesthetic, and aesthetic values related to the perception of the wetland as habitat.

Table 2. Typology for perceptions of wetlands.

Perception of Wetland	Description
Scientific	Wetland as repository of scientific information
Premodern	Wetland as resource, pre-Industrial Revolution
Exploitative	Wetland as potential resource, post-Industrial Revolution
Romantic	Wetland as uncivilized landscape, appreciated for its wildness rather than as physical resource
Aesthetic	Wetland as object to be aesthetically appreciated
Medico-mythic	Wetland as dangerous and unhealthy landscape understood through myths and legends
Ecological	Wetland as system of plants, animals, soils and climate

A premodern perception was not revealed amongst contemporary perceptions [12]. This can be the result of different personal attributes of the 21st century participants from those of the 19th century or the specific landscape attributes of the wetlands compared with Carrum Carrum Swamp. Certainly, a premodern perception requires familiarity with pre-Industrial Revolution landscapes.

This typology also includes negativistic perceptions of wetlands associated with insect-borne disease, within the medico-mythic perception. It does not accommodate perception of insect pests themselves, e.g., mosquitoes, identified with contemporary wetlands. This might be specific to the colonists of Carrum Carrum Swamp as mosquitoes in the Perth swamps were a concern [2].

Herein lies a limitation of this study. Perceptions have been inferred from historical documents located using selected keywords. Searches with additional keywords might have located other documents, analysis of which might have revealed additional perceptions. Further studies of colonial perceptions of other wetlands are needed to clarify this issue and to test this typology. The typology is likely to have broader application for understanding contemporary perceptions of wetlands. This, too, should be explored in future studies.

5. Conclusions

In contrast to popular opinion that all wetlands were regarded in the past as waste land, dangerous, smelly, and unsightly places, this study has shown that colonial perceptions of Carrum Carrum Swamp were not homogeneous and uniformly negative. Certainly, the overriding perception was exploitative, which resulted in its drainage. However, in the mix of seven perceptions identified, the premodern, romantic, and ecological perceptions did not demand drainage of the swamp. An ecological perception of colonial wetlands was surprising, as ecological awareness is associated with the 20th century. Colonial scientists, through their ecological perception of wetlands, prevented the drainage of many swamps in Victoria.

Colonial perceptions of Carrum Carrum Swamp were similar in type to contemporary perceptions of Victorian wetlands, differing only in the presence of a premodern perception. A premodern perception is likely to apply only to those with experience of pre-Industrial Revolution landscapes. However, perceptions in the two periods differed in the relative importance of each, with consequences for wetland management. In colonial times, an exploitative perception predominated, resulting in wetland drainage. In contrast, ecologistic/scientific and aesthetic perceptions prevail in recent times, supporting the use of wetlands in sustainable stormwater management in urban locations.

The typology for perceptions of wetlands was developed using a suite of nature values believed to be inherent in all humans. Thus, this typology should be suitable to classify wetland perceptions regardless of period.

Funding: This research received no external funding other than an Australian Postgraduate Scholarship to support the author during her PhD candidature, undertaken at the University of Melbourne.

Institutional Review Board Statement: Not applicable.

Informed Consent Statement: Not applicable.

Data Availability Statement: Not applicable.

Conflicts of Interest: The author declares no conflict of interest.

References

1. Fritzell, P.A. American wetlands as cultural symbol: Places of wetlands in American culture. In *Wetlands Functions Values and the State of Our Understanding*; Gleeson, P.E., Clark, J.R., Clark, J.E., Eds.; American Water Resources Association: Minneapolis, MN, USA, 1978; pp. 523–534.
2. Giblett, R.J. *Postmodern Wetlands: Culture, History, Ecology*; Edinburgh University Press: Edinburgh, Scotland, 1996.
3. Prince, H. *Wetlands of the American Midwest: A Historical Geography of Changing Attitudes*; University of Chicago Press: Chicago, IL, USA, 1997.
4. Smardon, R.C. Human perception of utilization of wetlands for waste assimilation, or how do you make a silk purse out of a sow's ear. In *Constructed Wetlands for Wastewater Treatment: Municipal, Industrial, Agricultural*; Hammer, D.A., Ed.; CRC Press: Boca Raton, FL, USA, 1989.
5. Carlson, A.E. "Drain the Swamps for Health and Home": Wetlands Drainage, Land Conservation and National Water Policy, 1850–1917. Ph.D. Thesis, University of Oklahoma, Norman, OK, USA, 2010.
6. Darby, H.C. *The Medieval Fenland*; Cambridge University Press: Cambridge, UK, 1940.
7. Darby, H.C. *The Draining of the Fens*, 2nd ed.; Cambridge University Press: Cambridge, UK, 1956.
8. Grasteyer, S.P.; Flora, C.B. Modernizing the savage: Colonization and perceptions of landscape and lifescapes. *Sociol. Rural.* **2000**, *40*, 128–149. [CrossRef]
9. Maxwell, W. The Back Swamp drainage project, Robeson County, North Carolina: Biopolitical intervention in the lives of Indian farmers. *Water Hist.* **2017**, *9*, 9–28. [CrossRef]
10. Meindl, C.F. Past perceptions of the Great American Wetland: Florida's Everglades through the early twentieth century. *Environ. Hist.* **2000**, *5*, 378–395. [CrossRef]
11. Sornig, D. *Blue Lake: Finding Dudley Flats and the West Melbourne Swamp*; Scribe Publications: Melbourne, Australia, 2018.
12. Dobbie, M.F.; Green, R.J. Public perceptions of freshwater wetlands in Victoria, Australia. *Landsc. Urban Plan.* **2013**, *110*, 143–154. [CrossRef]
13. Dobbie, M.F. Public aesthetic preferences to inform sustainable wetland management in Victoria, Australia. *Landsc. Urban Plan.* **2013**, *120*, 178–189. [CrossRef]

14. Griffiths, T. Ecology and empire: Towards an Australian history of the world. In *Ecology and Empire: Environmental History of Settler Societies*; Griffiths, T., Robin, L., Eds.; Melbourne University Press: Carlton, VIC, Australia, 1997.
15. Brennan, N. *Chronicles of Dandenong*; The Hawthorn Press: Melbourne, VIC, Australia, 1973.
16. Hibbins, G.M. *A History of the City of Springvale: Constellation of Communities*; City of Springvale in Conjunction with Lothian Pub.: Port Melbourne, VIC, Australia, 1984.
17. Zube, E.H.; Sell, J.L.; Taylor, J.G. Landscape perception: Research, application and theory. *Landsc. Plan.* **1982**, *9*, 1–33. [CrossRef]
18. Zube, E.H. Themes in landscape perception theory. *Landsc. J.* **1984**, *3*, 104–110. [CrossRef]
19. Gobster, P.H.; Nassauer, J.I.; Daniel, T.C.; Fry, G. The shared landscape: What does aesthetics have to do with ecology? *Landsc. Ecol.* **2007**, *22*, 959–972. [CrossRef]
20. Meinig, D. The beholding eye. In *The Interpretation of Ordinary Landscapes*; Meinig, D., Ed.; Oxford University Press: New York, NY, USA, 1979; pp. 33–48.
21. Kellert, S.R. A biocultural basis for an environmental ethic. In *The Coming Transformation: Values to Sustain Human and Natural Communities*; Kellert, S.R., Speth, J.G., Eds.; Yale School of Forestry and Environmental Studies: New Haven, CT, USA, 2009; pp. 21–38.
22. Kellert, S.R. *Birthright: People and Nature in the Modern World*; Yale University Press: New Haven, CT, USA, 2012.
23. Kellert, S.R. *Nature by Design and the Practice of Biophilic Design*; Yale University Press: New Haven, CT, USA, 2018.
24. Heathcote, R.L. The visions of Australia 1770–1970. In *Australia as Human Setting*; Rapoport, A., Ed.; Angus and Robertson: Melbourne, VIC, Australia; Sydney, NSW, Australia, 1972.
25. East, R.; Lee, D. *Some Notes on the Dandenong Creek*; Royal Historical Society of Victoria: Melbourne, Australia, 1973.
26. Shillinglaw, J.J. *Historical Records of Port Phillip*; Heinemann: Melbourne, Australia, 1972.
27. Crown Lands and Survey Historical Plan CS26(1), Port Phillip, C. Grimes. 1803; Land Victoria, Australia.
28. Crown Lands and Survey Historical Plan CS113, Port Phillip and Bass Strait, Lt Tuckey. 1804; Land Victoria, Australia.
29. Crown Lands and Survey Historical Plan MCS 24, Chart of Part of New South Wales, J. Cross. 1827; Land Victoria, Australia.
30. Map of Port Phillip, Lts Symonds and Henry. 1836; State Library of Victoria, Australia.
31. Crown Lands and Survey, Historical Plan P/A E78, County Lands in the Parishes of Eumemmerring and Lyndhurst, County of Mornington, Callanan. 1858; Land Victoria, Australia.
32. Sluyter, A. *Colonialism and Landscape: Postcolonial Theory and Applications*; Rowman and Littlefield Publishers: Lanham, MD, USA, 2002.
33. Rippon, S. 'Uncommonly rich; fertile' or 'not very salubrious? The perception; value of wetland landscapes. *Landscapes* **2009**, *10*, 39–60. [CrossRef]
34. Crown Lands and Survey Historical Plan, Roll plan 17A, Plan of the lowlands comprising the Carrum Swamp and situated between the sold lands of the Parish of Lyndhurst, C. Hodgkinson; T. Couchman. 1868; Land Victoria, Australia.
35. *The Argus*; 9 December 1869; p. 5. Available online: https://trove.nla.gov.au/newspaper/page/224463 (accessed on 25 November 2021).
36. Crown Lands and Survey Historical Plan, Roll plan 17, Survey of the Carrum Swamp in the County of Mornington, T. Rawlinson. 1866; Land Victoria, Australia.
37. Ferres, J. *Votes and Proceedings of the Legislative Assembly*; Government Printer: Melbourne, Australia, 1861–1862; Volume 1, p. 759.
38. *Victoria Government Gazette*; No. 27; 22 February 1861; p. 380. Available online: http://gazette.slv.vic.gov.au/view.cgi?year=1861&class=general&page_num=379&state=V&classNum=G27&id= (accessed on 25 November 2021).
39. *Victoria Government Gazette*; No. 49; 28 March 1861; p. 656. Available online: http://gazette.slv.vic.gov.au/view.cgi?year=1861&class=general&page_num=655&state=V&classNum=G49&id= (accessed on 25 November 2021).
40. *Victoria Government Gazette*; No. 50; 3 April 1861; p. 775. Available online: http://gazette.slv.vic.gov.au/view.cgi?year=1861&class=general&page_num=671&state=V&classNum=G50&id= (accessed on 25 November 2021).
41. *Victoria Government Gazette*; No. 133; 30 August 1861; p. 1659. Available online: http://gazette.slv.vic.gov.au/view.cgi?year=1861&class=general&page_num=1659&state=V&classNum=G133&id= (accessed on 25 November 2021).
42. *Victoria Government Gazette*; No. 122; 30 October 1866; p. 2354. Available online: http://gazette.slv.vic.gov.au/view.cgi?year=1866&class=general&page_num=2353&state=V&classNum=G122&id= (accessed on 25 November 2021).
43. *Victoria Government Gazette*; No. 82; 10 July 1868; p. 1257. Available online: http://gazette.slv.vic.gov.au/view.cgi?year=1868&class=general&page_num=1257&state=V&classNum=G82&id= (accessed on 25 November 2021).
44. Ferres, J. *Votes and Proceedings of the Legislative Assembly*; Government Printer: Melbourne, VIC, Australia, 1870; Volume 1, p. 205.
45. *The Argus*; 25 December 1869; p. 5. Available online: https://trove.nla.gov.au/newspaper/page/224711 (accessed on 25 November 2021).
46. Carman-Brown, K. *Environmental History and the Hydrological Cycle in Colonial Gippsland, Australia, 1838–1900*; ANU Press: Canberra, ACT, Australia, 2019.
47. Ferres, J. The Land Act 1862. In *Parliamentary Acts of Victoria, 1851–1900*; Government Printer: Melbourne, Australia, 1988.
48. Ferres, J. The Amending Land Act 1865. In *Parliamentary Acts of Victoria, 1851–1900*; Government Printer: Melbourne, Australia, 1988.
49. Ferres, J. The Land Act 1869. In *Parliamentary Acts of Victoria, 1851–1900*; Government Printer: Melbourne, Australia, 1988.
50. *The Argus*; 2 November 1865; pp. 4–5. Available online: https://trove.nla.gov.au/newspaper/page/210622 (accessed on 25 November 2021).
51. *Victorian Government Gazette*; No. 144; 13 October 1865; p. 2365. Available online: http://gazette.slv.vic.gov.au/view.cgi?year=1865&class=general&page_num=2365&state=V&classNum=G144&id= (accessed on 25 November 2021).

52. *The Argus*; 3 February 1866; Reclamation of swamps; p. 5. Available online: https://trove.nla.gov.au/newspaper/page/215261 (accessed on 25 November 2021).
53. Greater Dandenong Council. Dandenong District Road Board Minutes.
54. *The Argus*; 29 January 1869; Lease of the Carrum-Carrum Swamp; p. 7. Available online: https://trove.nla.gov.au/newspaper/page/222389 (accessed on 25 November 2021).
55. *The Argus*; 28 January 1869; p. 5. Available online: https://trove.nla.gov.au/newspaper/page/222368 (accessed on 25 November 2021).
56. *The Argus*; 30 April 1869; p. 5. Available online: https://trove.nla.gov.au/newspaper/page/223799 (accessed on 25 November 2021).
57. *Parliamentary Debates, Session 1871*, Legislative Council & Legislative Assembly: Melbourne, VIC, Australia; Volume 13.
58. Commissioner of Crown Lands, Selection File, VPRS 625/P, Unit 1694, file 10340/19.20 Lyndhurst. Public Records Office: Victoria.
59. Commissioner of Crown Lands, Selection File, VPRS 625/P, Unit 268, file 17920/19,20 Lyndhurst. Public Records Office: Victoria.
60. Greater Dandenong Council. Rate Book, District of Dandenong. 1873–1875.
61. Ferres, J. *Report from the Select Committee on the Carrum Carrum Swamp Selectors*; Government Printer: Melbourne, Australia, 1876.
62. Bruton, W. *Local History: Carrum to Cheltenham*; Self-Published by W. Bruton: Melbourne, VIC, Australia, 1977.
63. Carlson, A. Admiring the mirelands: The difficult beauty of wetlands. In *Suo on Kornis*; Hakala, K., Ed.; Maakenki Oy: Helsinki, Finland, 1999; pp. 173–181.
64. Leopold, A. *A Sand County Almanac*; Oxford University Press: New York, NY, USA, 1968.
65. Gobster, P.H. An ecological aesthetic for forest landscape management. *Landsc. J.* **1999**, *18*, 54–64. [CrossRef]
66. Rolston, H. Aesthetics in the swamp. *Perspect. Biol. Med.* **2000**, *43*, 584–597. [CrossRef] [PubMed]
67. Carlson, A. Aesthetic preferences for sustainable landscapes: Seeing and knowing. In *Forests and Landscapes: Linking Ecology, Sustainability and Aesthetics*; Sheppard, S.R.J., Harshaw, H.W., Eds.; CABI Publishing: Wallingford, UK, 2001; pp. 31–41.
68. Eaton, M.M. Fact and fiction in the aesthetic appreciation of nature. *J. Aesthet. Art Crit.* **1998**, *56*, 149–156. [CrossRef]
69. Brady, E. Imagination and the aesthetic appreciation of nature. *J. Aesthet. Art Crit.* **1998**, *56*, 139–147. [CrossRef]
70. Carroll, N. Emotion, appreciation, nature. In *Beyond Aesthetics: Philosophical Essays*; Carroll, N., Ed.; Cambridge University Press: Cambridge, UK, 2001; pp. 384–394.
71. *The Argus*; 10 February 1866; The Batman's Swamp Question, To the Editor of the Argus; p. 6. Available online: https://trove.nla.gov.au/newspaper/page/215314 (accessed on 25 November 2021).
72. Powell, J.M. Enterprise; dependency: Water management in Australia. In *Ecology and Empire: Environmental History of Settler Societies*; Griffiths, T., Robin, L., Eds.; Melbourne University Press: Carlton, VIC, Australia, 1997.
73. Corbin, A. *The Foul and the Fragrant: Odour and the Social Imagination*; Picador: London, UK, 1994.
74. *The Argus*; 9 February 1866; Lease of Batman's Swamp for Market Gardens; p. 5. Available online: https://trove.nla.gov.au/newspaper/page/215303 (accessed on 25 November 2021).
75. *Supplement to The Argus*; 17 February 1866; Batman's Swamp as a Source of Disease. To the Editor of The Argus; p. 1. Available online: https://trove.nla.gov.au/newspaper/page/215373 (accessed on 25 November 2021).
76. *The Argus*; 20 February 1866; Batman's Swamp, To the Editor of The Argus; p. 6. Available online: https://trove.nla.gov.au/newspaper/page/215388 (accessed on 25 November 2021).
77. *The Argus*; 3 March 1866; Batman's Swamp, To the editor of The Argus; p. 5. Available online: https://trove.nla.gov.au/newspaper/page/215477 (accessed on 25 November 2021).
78. *The Argus*; 5 March 1866; Melbourne Swamps v. Cabbage Gardens, To the Editor of The Argus; p. 6. Available online: https://trove.nla.gov.au/newspaper/page/215488 (accessed on 25 November 2021).
79. Odum, E.P.; Barrett, G.W. *Fundamentals of Ecology*, 5th ed.; Thomson Brooks/Cole: Boston, MA, USA, 2004.
80. *The Argus*; 27 October 1865; p. 5. Available online: https://trove.nla.gov.au/newspaper/page/210580 (accessed on 25 November 2021).
81. *The Argus*; 7 April 1866; p. 4. Available online: https://trove.nla.gov.au/newspaper/page/210580 (accessed on 25 November 2021).
82. Ferres, J. *Reclamation of Swamps. Report of the Professional Board*; Government Printer: Melbourne, VIC, Australia, 1866.
83. Pescott, R.T.M. *The Royal Society of Victoria from Then, 1854 to Now, 1959*; Royal Society of Victoria: Melbourne, Australia, 1961.
84. Lowenthal, D. Empires and ecologies: Reflections on environmental history. In *Ecology and Empire: Environmental History of Settler Societies*; Griffiths, T., Robin, L., Eds.; Melbourne University Press: Carlton, VIC, Australia, 1997.
85. Grove, R. *Green Imperialism: Colonial Expansion, Tropical Island Edens and the Origins of Environmentalism, 1600–1860*; Cambridge University Press: Cambridge, UK, 1995.
86. *The Argus*; 19 October 1865; p. 5. Available online: https://trove.nla.gov.au/newspaper/page/210520 (accessed on 25 November 2021).
87. *The Argus*; 21 October 1865; pp. 4–5. Available online: https://trove.nla.gov.au/newspaper/page/210534 (accessed on 25 November 2021).
88. *The Argus*; 6 December 1865; p. 5. Available online: https://trove.nla.gov.au/newspaper/page/210868 (accessed on 25 November 2021).
89. *The Argus*; 13 January 1866; p. 5. Available online: https://trove.nla.gov.au/newspaper/page/215107 (accessed on 25 November 2021).
90. *The Argus*; 19 December 1865; p. 5. Available online: https://trove.nla.gov.au/newspaper/page/214929 (accessed on 25 November 2021).
91. Sewall, L. The skill of ecological perception. In *Ecopsychology: Restoring the Earth, Healing the Mind*; Roszak, T., Gomes, M.E., Kanner, A.D., Eds.; Sierra Club Books: San Francisco, CA, USA, 1995; pp. 201–215.

land

Article

Understanding Farmers' Intention towards the Management and Conservation of Wetlands

Naser Valizadeh [1,*], Samira Esfandiyari Bayat [2], Masoud Bijani [2], Dariush Hayati [1], Ants-Hannes Viira [3], Vjekoslav Tanaskovik [4], Alishir Kurban [5,6,7,8] and Hossein Azadi [5,9,10]

1 Department of Agricultural Extension and Education, School of Agriculture, Shiraz University, Shiraz 7144165186, Iran; hayati@shirazu.ac.ir
2 Department of Agricultural Extension and Education, College of Agriculture, Tarbiat Modares University (TMU), Tehran 1497713111, Iran; s.esfandiyarbayar@modares.ac.ir (S.E.B.); mbijani@modares.ac.ir (M.B.)
3 Institute of Economics and Social Sciences, Estonian University of Life Sciences, 51006 Tartu, Estonia; Ants-Hannes.Viira@emu.ee
4 Faculty of Agricultural Sciences and Food-Skopje, Ss. Cyril and Methodius University in Skopje, 1000 Skopje, North Macedonia; vtanaskovik@fznh.ukim.edu.mk
5 Xinjiang Institute of Ecology and Geography, Chinese Academy of Sciences, 818 South Beijing Road, Urumqi 830011, Xinjiang, China; alishir@ms.xjb.ac.cn (A.K.); hossein.azadi@ugent.be (H.A.)
6 Research Center for Ecology and Environment of Central Asia, Chinese Academy of Sciences, 818 South Beijing Road, Urumqi 830011, Xinjiang, China
7 Sino-Belgian Joint Laboratory for Geo-Information, Urumqi 830011, Xinjiang, China
8 University of Chinese Academy of Sciences, Beijing 100049, China
9 Department of Geography, Ghent University, 9000 Ghent, Belgium
10 Faculty of Environmental Sciences, Czech University of Life Sciences Prague, 165 00 Prague, Czech Republic
* Correspondence: n.valizadeh@shirazu.ac.ir; Tel.: +98-713-228-6300; Fax: +98-713-228-6072

Citation: Valizadeh, N.; Esfandiyari Bayat, S.; Bijani, M.; Hayati, D.; Viira, A.-H.; Tanaskovik, V.; Kurban, A.; Azadi, H. Understanding Farmers' Intention towards the Management and Conservation of Wetlands. *Land* **2021**, *10*, 860. https://doi.org/10.3390/land10080860

Academic Editor: Richard C. Smardon

Received: 17 July 2021
Accepted: 8 August 2021
Published: 16 August 2021

Abstract: The aim of the present research was to analyze the farmers' intention towards participation in the management and conservation of wetlands through the lens of the extended theory of planned behavior (TPB). To do this, a cross-sectional survey of Iranian farmers was carried out. To select the samples, a multi-stage random sampling process with a proportional assignment was employed. The research instrument was a researcher-made questionnaire whose validity and reliability were verified using various quantitative and qualitative indicators. The results of the extended TPB using structural equation modeling showed that four variables, namely moral norms of participation in management and conservation (MNPMC), attitude towards participation in management and conservation (APMC), subjective norms towards participation in management and conservation (SNPMC), and self-concept about participation in management and conservation (SCPMC) had positive and significant impacts on intention towards participation in management and conservation (IPMC). The results also revealed that that entering MNPMC and SCPMC into TPB could increase its explanatory power. Also, the fit indicators supported the extended TPB. From a practical point of view, the present study provides justifications and insights for the use of MNPMC, APMC, SNPMC, and SCPMC in policies and programs intended to encourage farmers and local communities to participate in wetlands management and conservation.

Keywords: wetlands ecosystems; moral norms; self-identity; theory of planned behavior; sustainability

1. Introduction

Wetlands are among the most valuable ecosystems on Earth [1]. There are about 1280 million hectares of wetlands around the world, which include inland and coastal wetlands such as lakes, rivers, swamps, and constructed wetlands such as rice fields and reservoirs [2]. Wetlands are actually water-saturated soils [3] that include flowing, fresh, brackish, and saline water bodies. In some cases, wetlands contain a part with marine water that has a depth not exceeding six meters at low tides [4,5]. Wetlands, with their numerous plants, animals, and microorganisms, play a crucial role in conserving global biodiversity. Their sustainable conservation and management can also play a key role in achieving 17 goals of the Sustainable Development Agenda of the United Nations and directly or indirectly contribute to the sustainability of 75 SDG indicators (out of 230) [6–8]. This means that the importance of wetlands has been emphasized in 75 indicators of the 2030 Sustainable Development Agenda. Wetlands are of great importance from ecological, economic, and socio-cultural perspectives. In the ecological dimension, with the conservation of aquatic wildlife, mammals, native and migratory bird species, resident amphibians, reptiles, and various species of insects [2–9], wetlands are safe habitats for plants and animals [10,11] and thus help to preserve and develop biodiversity. Wetlands are natural filtration systems for runoffs and can improve water quality in an area [3]. Scientific evidence [12–14] shows that wetlands play a crucial role in reducing the risks of natural disasters such as storms and floods and preventing soil erosion. Ecosystem services provided by wetlands include the reduction of climate change impacts, pollutants treatment, nutrients and human waste recycling [14,15], the rehabilitation of degraded groundwater aquifers and increasing water use potential [2], and the conservation of watersheds, carbon sequestration, and storage [14].

The wetlands are economically important because they provide many essential ecosystem services for the welfare of human communities [15]. Wetlands are turned into pivotal sources in eradicating poverty via the provision of ecosystem services [16]. In this regard, the role of wetlands resources is especially important in the livelihood of the poor in developing countries [17]. Tourists' access to these places and the development of the tourism industry can be a good source of income for local communities. Normally, the economic conditions of an area are closely associated with the health and stability of wetlands. In general, wetlands have numerous commodities and services that have economic value not only for the local population but also for the people living outside the wetlands [2]. Major economic services of wetlands ecosystems are human habitation, water supply for various uses including plantation and seasonal agriculture, the production of fishery products, the grazing source of local livestock, the growth of wild and medicinal plants, energy production, the procurement of building raw materials, the protection of genetic resources, the supply of constructing raw materials, and industrial uses for individuals [2,17].

Rodrigues et al. [18] consider socio-cultural services of wetlands ecosystems to be very important. The researchers state that wetlands are a place for recreation that can be very inspiring due to their visual and aesthetic qualities. In addition, wetlands can represent cultural heritage and identity that have high scientific and educational potential [14]. The potential role of aquatic environments (wetlands) in improving health and reducing stress has rarely been investigated so far. However, there is evidence that natural habitat and biodiversity are effective therapies to reduce mental problems such as post-disaster stress and other psychological diseases [19]. In addition, wetlands reduce intensive and widespread population movements, as the loss of livelihood opportunities and migration of people living around wetlands who depend on the existence of wetlands is one of the main social outcomes of wetlands degradation [20]. This increase in migration has caused many issues and problems including cultural shocks, slum development, illegal jobs, hidden unemployment, and other similar things in migration destinations [21].

Despite the great importance of wetlands from an ecological, economic, and socio-cultural perspective, unfortunately, many of them are being destroyed. Such disasters can have various consequences such as reducing the capacity of ecosystems to provide services [14], increasing soil salinity, increasing salt storms, the occurrence of the dust phenomenon, population mobility, international conflicts, famine, desertification, etc. There are many reasons for wetlands degradation. Examples include high population density and urbanization pressure [22], drought, storm, and sea level changes [23]. However, it should be noted that human-induced factors are the main reasons for wetlands' degradation [17]. Failure to assess the impacts of industrial and tourism development, rapid urbanization and population growth, dam construction, and agriculture are among these factors [4,23,24].

Although many factors such as climate change and the depletion of groundwater resources play a critical role in wetlands degradation around the world, the overdevelopment of agricultural activities is one of the most important factors in the destruction of most wetlands in some countries such as Iran. In this regard, paying more attention to the role of farmers as one of the influential stakeholders in the process of wetlands management, conservation, and rehabilitation is crucial [17]. Farmers can contribute to wetlands management and conservation in a variety of ways. Reducing the intensity of agricultural activities around the wetlands, launching non-agricultural businesses to prevent the destruction of wetlands ecosystems, using modern irrigation methods to reduce water consumption, and low use of chemicals in agricultural crops' production are among the contributions of farmers to the management and protection of the wetlands. Furthermore, respect for the rights of native and migratory animals of the wetlands, payment for wetlands ecosystem services, and participation in programs and projects can also be considered as the other roles and contributions for farmers in the field of wetlands management and conservation [25].

Despite the great importance of the participation of local communities (such as farmers living around wetlands) in the management and conservation of wetlands in Iran, managers, decision makers, and policy makers have not paid much attention to it. There are many wetlands that are drying for various reasons such as droughts, unbalanced agricultural development, population growth, climate change, and mismanagement in Iran. For example, Ghara Gheshlagh wetland is one of the large wetlands located in northwestern Iran near the shores of Lake Urmia. As the socio-behavioral dimensions are one of the main dimensions of sustainable wetlands management, many social intervention programs have been implemented for the rehabilitation and sustainable management of Ghara Gheshlagh wetland in this area. However, according to the Iranian Department of Environment [25], many of these social interventions have not been very successful. One of the main reasons for this is the low awareness of planners and intervention organizations about the factors determining farmers' participation and socio-psychological mechanisms of the intention to participate in the management and conservation programs of Ghara Gheshlagh wetland. Thus, identifying and analyzing the farmers' intentions to participate in the conservation and management of Ghara Gheshlagh wetland was selected as the main purpose of the current research. The main objectives included developing an appropriate theoretical framework based on the theory of planned behavior (TPB) to analyze farmers' intention to participate in the management and conservation of the wetlands, test the original version of TPB, test the extended version of TPB, and provide practical recommendations for improving the wetlands management and conservation. The present study is novel from four perspectives. First, no similar study has been conducted in the study area. Therefore, this study can provide a basis for further research on socio-psychological dimensions of wetlands conservation and management. The application of the theory of planned behavior (TPB) to analyze the farmers' intention to participate in the management and conservation of wetlands is the second novelty of the study. Furthermore, the present study extends the original version of TPB by adding the self-concept about participation in management and conservation (SCPMC) and moral norms of participation in management and conservation (MNPMC) to the analysis. This can be considered as the most important theoretical contribution of the study. The last novelty is that it provides the readers and end-users of the study with innovative results and insights about the mechanisms of the relationships among socio-psychological variables.

2. Theoretical Framework: TPB

TPB is a significant and well-known social-cognitive theory aiming at explaining the variance in volitional behaviors [26]. Fishbein and Ajzen [26] first conceptualized and presented the Theory of Reasoned Action (TRA) to describe individuals' thinking and predict their future intentions and behaviors. Attitude towards a specific behavior and subjective norm are the two main structures of TRA [27]. In 1988, Ajzen developed the TRA to overcome its limitations [28], and finally, TPB was first articulated and introduced by

Ajzen in 1991 [29]. Unlike TRA, TPB states that not all behaviors are completely voluntary and under one's control. A person may have a strong intention to perform a behavior, but in certain circumstances, this behavior is prevented. Thus, perceived or actual behavioral control may affect intention or the relationship between intention and behavior [28]. There are three main variables in TPB, including the attitude towards behavior, the subjective norm about behavior, and the perceived behavioral control [29]. Attitude is defined as the favorable or unfavorable appraisal towards the actual representation of a particular behavior [30,31]. In the present study, attitude refers to the degree of farmers' favorable or unfavorable evaluation towards participation in the management and conservation of wetlands. The subjective norm denotes "the perception of a person (farmer) that most people think that he/she must or must not perform a particular type of behavior", i.e., participating in wetlands conservation and management [26–31]. Perceived behavioral control indicates the perception of individuals about their control over the target behavior, which in this study is participation in the conservation and management of wetlands. Kolvereid [32] explains that in TPB, "attitudes or beliefs do not directly predict behaviors. Instead, these factors are completely or partially absorbed into intention" [33,34].

TPB is one of the most widely used behavior prediction theories due to its global recognition [35]. Despite the evolution of TPB over the past decades, there are still concerns about its comprehensiveness and efficiency. One of the limitations of this theory is the inadequate accuracy in prediction. Fishbein and Ajzen [26] stated that comparative importance may vary depending on the type of behavior and population. However, TPB has been criticized for being too focused on the individuals' behavior and paying insufficient attention to the respondent's identity [36,37].

It is also claimed that TPB underestimates the influence and role of social norms. Based on a study by Ajzen [30], in TPB, social-oriented norms are usually regarded as subjective-base norms [29]. Another critique of this theory is that TPB is based on individualism and all variables in it are rational predictors [29]. Some researchers [38] believe that environmentally-friendly behaviors are influenced by public-sphere and altruistic motives. Therefore, they argue that moral norms are important in perceiving individuals' behaviors and should be included in the TPB model [29]. Ajzen and Fishbein [39] also showed that when encountering behaviors with moral dimensions, moral criteria should be included in TPB to determine whether it helps to explain the intention and behavior or not. High personal or moral norms motivate individuals to follow social behaviors, while low moral norms prevent social and altruistic behaviors [29]. This theoretical approach is very useful for explaining the behaviors of farmers and other stakeholders in agriculture and natural resource management [38]. As the goal of TPB is to explain global behaviors (and not exclusively rational behaviors) [40], a variable of moral norm should be added to the extended version of TPB to improve its predictive power. Moral norms are internal moral rules or values that are motivated by predicted self-developed rewards or punishments [29–40]. In the present study, moral norms of participation in management and conservation of wetlands refer to the personal or moral justification of participation by farmers (farmers' personal beliefs about what it is right or wrong to do).

Some studies [41] have also demonstrated that in addition to moral norms, personal self-image (self-concept) can also be useful in increasing the TPB's predictive power [42], but this variable is ignored in the original version of TPB. The self-image includes all the individual roles. Individual choices may also be determined by the degree to which a particular intention is compatible with his/her own feelings [43,44]. The concept of self can encompass a broader social context for the individuals and connect intention and action to certain personal characteristics. In the present study, self-image (self-concept) is defined as the label that farmers use to describe themselves in the field of participation in the management and conservation of wetlands. The idea of self-concept integration in TPB is not new, and many previous studies [43] have shown that self-concept is an important predictor and improves the standard TPB model. Even some researchers such as Carfora et al. [37] present self-concept as the strongest predictor of pro-environmental behavioral intentions. Therefore, in order to eliminate the weaknesses of the original form of the theory, two variables of moral norms and self-concept were included in TPB, and the theoretical framework of the study was articulated as Figure 1. This method of extending a theory is not new and is one of the most widely used and effective methods for developing behavioral theories. There is strong support for this methodology in the literature see [43,45–47], and many researchers have used this method to develop the theory of planned behavior. In order to achieve the main purpose, five hypotheses were defined:

All the studied variables including the attitude towards participation in management and conservation (APMC) (hypothesis 1), subjective norms towards participation in management and conservation (SNPMC) (hypothesis 2), perceived behavioral control on participation management and conservation (PBCPMC) (hypothesis 3), self-concept about participation in management and conservation (SCPMC) (hypothesis 4), and moral norms of participation in management and conservation (MNPMC) (hypothesis 5) will have positive and significant effects on the intention towards participation in management and conservation (IPMC).

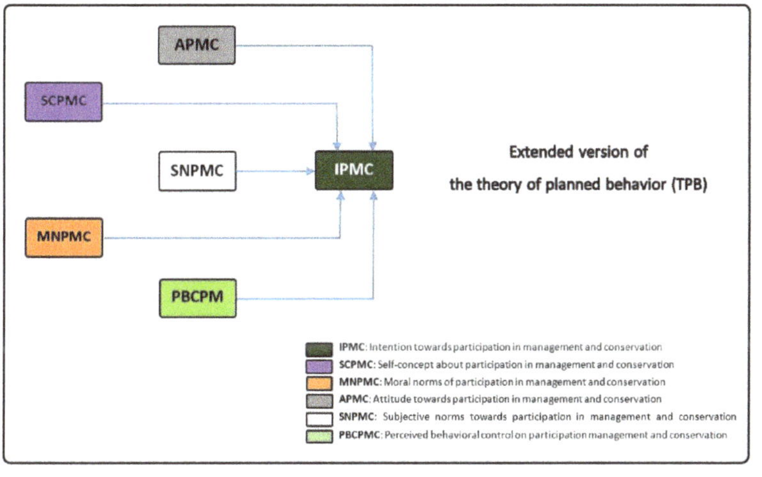

Figure 1. Theoretical framework of the research.

3. Materials and Methods

3.1. Research Design and Research Type

In terms of paradigm, this study is a quantitative one in which a non-experimental research design was used, and in terms of purpose, it is an applied one that was conducted using a cross-sectional survey among farmers.

3.2. Study Area

Ghara Gheshlagh wetland is located in the northwest of Iran and in the southeastern part of Lake Urmia (Figure 2). This wetland is located in the geographical area of the two provinces of West and East Azerbaijan and has an area of about 48 square kilometers. This means that the wetland is located on the border between the two provinces of West and East Azerbaijan in Iran. Its average height above the sea level is 1275.9 m. The water depth of this wetland is about 15 cm. It is noteworthy that its water depth varies in different seasons of the year due to many factors such as rainfall in wet seasons, evaporation in hot seasons, the consumption of water sources by different users, the extent of exploitation of underground resources in the wetland, and the decrease in the water level of Lake Urmia. The main livelihood and income sources of the villagers are agriculture and mobile/fixed livestock husbandry. The agricultural products of the farmers of the region are alfalfa, watermelon, wheat, barley, melon, tomato, beet, and orchards (apple and grape), among which a significant proportion is allocated to the first four crops. Today, about three-quarters of the region's agricultural land is used for alfalfa, watermelon, wheat, and then barley. Only a quarter of the rest is allocated to other products. Alfalfa, meanwhile, is a crop that needs plenty of water, and wheat and barley are irrigated rather than rainfed [25]. Therefore, it can be said that the agriculture sector in this region is impressively water consuming. Handicrafts such as weaving kilims and carpets, hunting waterfowl, and fishing are other sources of income for local people. The presence of 185 species of native and migratory birds, numerous mammals, and rare plant and aquatic species indicates the richness of this habitat as one of the main and ecological reserves. Ghara Gheshlagh's biodiversity has made it one of the important habitats for natural vegetation and wildlife. This wetland is home to thousands of migratory birds (especially flamingos) that migrate to this area from distant areas every year during the cold season [25].

Figure 2. The study site.

3.3. Population and Sampling

The population in the present cross-sectional survey was farmers around Ghara Gheshlagh wetland (Figure 2) (N = 9536). As this wetland is located in the West and East Azerbaijan provinces, we attempted to select a sample from farmers in both provinces. In order to obtain the required sample size of the population around the wetland, the Krejcie and Morgan table was used (n = 373). Once the population size is figured out, the Krejcie and Morgan table is a simple way to estimate the sample size. This sampling table is one of the most widely used methods for calculating a statistical sample size [44]. For logical distribution of the sample in the population and selection of a representative random sample, a multi-stage stratified method with proportional assignment was used. In this sampling method, the study area should be divided into several categories based on special criteria. Therefore, first, the area was divided into two categories/counties. In this classification, based on the Statistics Center of Iran, Bonab and Miandoab counties were considered as separate classes. In the second stage, the name and characteristics of the villages around Ghara Gheshlagh wetland were determined. To do this, the data and information provided by the Forests and Rangelands Organization of West Azerbaijan and East Azerbaijan provinces were used. Wetlands may directly and indirectly influence agricultural, livelihood, and industrial activities at a distance of 5 km. Therefore, the villages located within a 5 km radius of the wetland were included in the population (23 villages). Some other studies have used this approach in the past [48]. However, it should be noted that in some cases, the direct and indirect effects of wetlands may be felt at distances of more than 5 km [25]. In the third stage, the number of farmers in each village was estimated. For this purpose, the demographic statistics presented in the agricultural reports of the Agricultural Jihad Administrations of Bonab and Miandoab counties were used. In the fourth step, the required sample size for each village was estimated by comparing its agricultural population with the total agricultural population and the selected sample. This made it possible to distribute the whole sample among different villages. Finally, farmers (samples) were randomly selected from the farmers of each of these villages in proportion to the volume.

3.4. Research Instrument

The study instrument was a researcher-made and close-ended questionnaire (Table S1). Different methods and indicators were used to verify the validity and reliability of the study device. In order to confirm the face and content validity, the opinions of the expert group were employed. These experts were either faculty members or field practitioners with extensive experience in socio-ecological interventions based on sustainable wetlands management.

The pilot study was conducted among farmers around Parishan wetland in Fars province in the Southeast of Iran. This area was selected for the pilot study because it was similar to Ghara Gheshlagh wetland in terms of its conditions. Alpha coefficients were used to verify the reliability. Table 1 shows the calculated reliability for the various variables. After the cross-sectional survey, other complementary methods were used to confirm the validity and reliability. At this stage, two indicators of composite reliability (CR) and loading factors of items in the first-order confirmatory factor analysis were used to appraise and examine the reliability of the research instrument.

Three other indicators (composite reliability, convergent validity (CV) or average variance extracted (AVE), and divergent validity) were also applied to confirm the validity indices of the research instrument. Divergent validity of the questionnaire was evaluated by the average shared squared variance (ASV) and the maximum shared squared variance (MSV). Finally, all the questionnaire items (see Table 1 for the complete list of items) were scored by means of a five-point Likert scale (1 = Strongly disagree to 5 = Strongly agree). In order to calculate the scores of the variables in the conceptual framework, the scores of the constituent items were added together.

Table 1. Items employed to measure the variables and alpha coefficients.

Var.	No.	Items
		Attitude towards participation in management and conservation (APMC): (α = 0. 78)
APMC	1	Usefulness of participation in wetlands conservation and management
	2	Prioritization of personal interests such as increasing products over wetlands conservation and management
	3	Necessity of wetlands conservation and management in water shortage conditions
	4	Necessity of cooperation and participation of all stakeholders in wetlands conservation and management
	5	The desirability of engagement in wetlands conservation and management
	6	Considering participation in wetland conservation and management as a wise behavior
	7	Trusting the wetlands conservation and management plans
		Perceived behavioral control on participation management and conservation (PBCPMC): (α = 0.75)
PBCPMC	1	Having the time and skills needed to participate in wetlands management and conservation
	2	Having the necessary economic capability to participate in wetlands conservation and management activities
	3	Ease of participation in wetlands conservation and management
	4	Having the knowledge needed to participate in wetlands conservation and management
		Moral norms of participation in management and conservation (MNPMC): (α = 0.84)
MNPMC	1	I feel that I am a good person if I help protect and revitalize the wetlands.
	2	I consider it a moral duty to participate in the better conservation and management of the wetlands.
	3	I feel that I have to do useful work for better conservation and management of wetlands.
	4	I have a sense of commitment to participate in wetlands conservation and management.

Table 1. *Cont.*

Var.	No.	Items
		Intention towards participation in management and conservation (IPMC): (α = 0.82)
IPMC	1	I want to pay for the rehabilitation and conservation of Ghara Gheshlagh wetland.
	2	I want to learn the necessary skills for the conservation and management of Ghara Gheshlagh wetland.
	3	I would like to participate in the management and conservation of Ghara Gheshlagh wetland.
	4	I would like to cooperate with the government, experts, and stakeholders involved in the rehabilitation of Ghara Gheshlagh wetland.
		Subjective norms towards participation in management and conservation (SNPMC): (α = 0.73)
SNPMC	1	Commitment to participate in the management and conservation of the wetlands will lead to my approval by those around me.
	2	I think my acquaintances and friends expect me to be as committed as I can to participate in the management and conservation of the wetlands.
	3	My acquaintances and friends think that I should be committed to participating in the management and conservation of the wetlands.
		Self-concept about participation in management and conservation (SCPMC): (α = 0.77)
SCPMC	1	I believe that I am a person with environmental concerns in Ghara Gheshlagh wetland.
	2	My self-image is a person committed to participating in the management and conservation of the wetlands.
	3	I think I am a person who participates in the management and conservation of the wetlands (water conservationist).

3.5. Data Collection and Data Analysis

The researchers used face-to-face interviews with farmers for data collection. Data collection was performed by the first author. To collect data, the first author used an experienced data collecting team. The number of members of the data collection team was 4 and they were engaged in data collection for two months. According to the estimated sample size, 373 farmers in different villages were selected and interviewed, but 33 of them refused to answer. Therefore, 340 questionnaires were collected. Then, 25 questionnaires were excluded due to high missing data. Finally, 315 questionnaires were analyzed. Data analysis was performed using SPSS24 and AMOS20 software. Data normality was tested using Mardia's skewness and kurtosis coefficients. The values of these coefficients were at the acceptable level (less than ± 1.96). First-order confirmatory factor analysis models were applied to calculate the measuring models and analyze the structure of the extended TPB variables. Confirmatory factor analysis is a method to determine the power of a predefined factor model that has a set of observational data. In other words, this type of factor analysis tests the degree of conformity between the theoretical model and the experimental model of the research. In this method, first, the relevant variables and indicators are selected based on the initial theory and then, factor analysis is used to determine whether these variables and indicators are loaded on the predicted factors or not [49,50]. The maximum likelihood method was used to examine the measurement models. In this study, in addition to measurement models, total/direct structural model analysis was employed. The structural model examined the structural-oriented relationship between the original and extended versions of TPB. For this purpose, the original TPB was first tested by entering three independent variables, namely APMC, PBCPMC, and SNPMC. In The next step, the extended version was tested by adding MNPMC and SCPMC to the original TPB. This is one of the most common ways to develop behavioral patterns.

4. Results

4.1. Correlation Matrix between Independent and Dependent Variables

The correlation relationships of the variables showed that five variables of APMC (r = 0.670; $p > 0.01$), SNPMC (r = 0.465; $p < 0.01$), PBCPMC (r = 0.544; $p < 0.01$), SCPMC (r = 0.547; $p < 0.01$), and MNPMC (r = 0.709; $p < 0.01$) had positive and significant correlations with IPMC. These correlations mean that increasing or strengthening these five variables can lead to enhancing the IPMC. By comparing the correlation values, it can be inferred that the intensity of the correlation of MNPMC and APMC with IPMC is higher than other variables (Table 2).

Table 2. The results of Pearson correlation for the study variables.

Variable Name	IPMC	APMC	SNPMC	PBCPMC	SCPMC	MNPMC
IPMC	1					
APMC	0.670 **	1				
SNPMC	0.465 **	0.461 **	1			
PBCPMC	0.544 **	0.694 **	0.387 **	1		
SCPMC	0.547 **	0.494 **	0.294 **	0.451 **	1	
MNPMC	0.709 **	0.680 **	0.425 **	0.591 **	0.549 **	1

Note: Abbreviations: Intention towards participation in management and conservation (IPMC), Attitude towards participation in management and conservation (APMC), Subjective norms towards participation in management and conservation (SNPMC), Perceived behavioral control on participation management and conservation (PBCPMC), Self-concept about participation in management and conservation (SCPMC), Moral norms of participation in management and conservation (MNPMC). ** Sig. level: 0.01.

4.2. Evaluating Validity and Reliability Results Using Measurement Models

The evaluation of the measurement models of the extended TPB variables showed that the values of the loading factors of all measurement items were greater than or equal to the acceptable value of 0.4 (Table 3). This means that the items having a loading factor less than 0.4 are not loaded on the specific factor and must be eliminated. In fact, the values of the loading factors represent the correlation of individual items with the main factors (variables). The recommendation presented by Nunnally [51] and Maleksaeidi et al. [52] was used to determine this acceptable threshold value. The obtained numerical coefficients for CR and AVE were greater than or equal to the cut off values of 0.7 and 0.5, respectively. It can be inferred that the questionnaire used in this study had composite reliability and convergent validity. Comparing divergent validity indicators (ASV and MSV) with AVE values also revealed that the values of these indicators were less than AVE values. Therefore, it can be concluded that the study instrument had divergent validity. Since Mardia's multivariate skewness and kurtosis coefficients were less than ±1.96, it can be inferred that the research data had a normal distribution (Table 3).

Table 3. Measurement models' estimations and validity and reliability results.

Items/Variables/ Normality Measure	IPMC	APMC	SNPMC	PBCPMC	SCPMC	MNPMC	Skew	Kurtosis
IPMC1	0.75 *						1.302	0.765
IPMC2	0.75						0.625	0.969
IPMC3	0.60						0.374	0.751
IPMC4	0.57						1.502	−0.298
APMC1		0.52 *					0.881	0.531
APMC2		0.60					−1.449	1.123
APMC3		0.66					−0.460	0.741
APMC4		0.77					0.367	0.758
APMC5		0.72					0.811	0.298
APMC6		0.58					1.742	0.453
APMC7		0.70					−0.569	0.961
SNPMC1			0.78 *				0.276	−0.816
SNPMC2			0.78				0.704	0.316
SNPMC3			0.75				−0.557	0.098
PBCPMC1				0.69 *			0.472	−0.318
PBCPMC2				0.66			1.121	0.811
PBCPMC3				0.52			1.025	−0.591
PBCPMC4				0.62			0.745	0.637
SCPMC1					0.75 *		0.907	0.733

Table 3. *Cont.*

Items/Variables/ Normality Measure	IPMC	APMC	SNPMC	PBCPMC	SCPMC	MNPMC	Skew	Kurtosis
SCPMC2					0.80		0.282	−0.808
SCPMC3					0.40		−0.596	−0.652
MNPMC1						0.77 *	1.082	0.770
MNPMC2						0.67	1.602	1.238
MNPMC3						0.61	0.729	0.415
MNPMC4						0.75	0.361	−1.608
CR	0.73	0.71	0.78	0.70	0.72	0.74	-	-
AVE	0.62	0.58	0.71	0.54	0.59	0.63	-	-
MSV	0.45	0.32	0.52	0.24	0.38	0.49	-	-
ASV	0.31	0.41	0.36	0.08	0.27	0.44	-	-

* Fixed item in the confirmatory factor analysis.

4.3. Evaluation of Structural Models and Analyzing the Relationships among Latent Variables

The results of testing the total/direct structural model for the original TPB revealed that the variables such as APMC (β = 0.489; $p > 0.01$), SNPMC (β = 0.175; $p > 0.01$), and PBCPMC (β = 0.135; $p > 0.05$) had positive and significant effects on IPMC (Table 4). This means that the first, second, and third hypotheses were supported by the results of the original TPB. Comparing the standardized effects of these variables reveals that APMC had the strongest effect on IPMC. Independent variables in the original TPB could account for 47.4% of the variance changes in IPMC.

The estimation of the total/direct structural model for the extended TPB models showed that APMC (β = 0.260; $p > 0.01$), SNPMC (β = 0.120; $p > 0.01$), SCPMC (β = 0.169; $p > 0.01$), and MNPMC (β = 0.362; $p > 0.01$) had positive and significant effects on IPMC (Table 4; Figure 3). This means that the first, second, fourth, and fifth hypotheses were supported by the results of the extended TPB. By comparing the effects of these four significant variables, it can be said that both MNPMC and APMC variables had the strongest standardized effects in the extended version of TPB, respectively. This result indicates that these two variables can explain IPMC more than other variables. However, it should be mentioned that SCPMC and SNPMC also play a significant role in explaining variance and directing IPMC. In the total/direct structural model in the extended TPB, the effect of PBCPMC on IPMC was not significant. Therefore, the hypothesis or path of PBCPMC → IPMC was not supported (β = −0.027; n.s.). It should be noted that the independent variables of extended TPB could explain 58.1% of the variance changes in IPMC (Table 4 and Figure 3). The comparison of the total variances explained in original and extended versions of planned behavior demonstrates that the inclusion of the variables such as MNPMC and SCPMC in the original version can improve its explanatory power.

Table 4. The results of estimating standardized effects of independent variables on IPMC in original and extended versions of TPB using the structural model.

Model	Hypothesized Relationship	Unstandardized Coefficients	S.E.	Standardized Coefficients	Sig.	Hypothesis Test
Extended TPB	APMC→IPMC	0.175	0.041	0.260	0.001	Supported
	SNPMC→IPMC	0.177	0.064	0.120	0.006	Supported
	PBCPMC→IPMC	0.033	0.067	0.027	0.625	Un-supported
	SCPMC→IPMC	0.273	0.076	0.169	0.001	Supported
	MNPMC→IPMC	0.387	0.060	0.362	0.001	Supported
Original TPB	APMC→IPMC	0.329	0.041	0.489	0.001	Supported
	SNPMC→IPMC	0.258	0.071	0.175	0.001	Supported
	PBCPMC→IPMC	0.166	0.72	0.135	0.022	Supported

Note: Abbreviations: Intention towards participation in management and conservation (IPMC), Attitude towards participation in management and conservation (APMC), Subjective norms towards participation in management and conservation (SNPMC), Perceived behavioral control on participation management and conservation (PBCPMC), Self-concept about participation in management and conservation (SCPMC), Moral norms of participation in management and conservation (MNPMC).

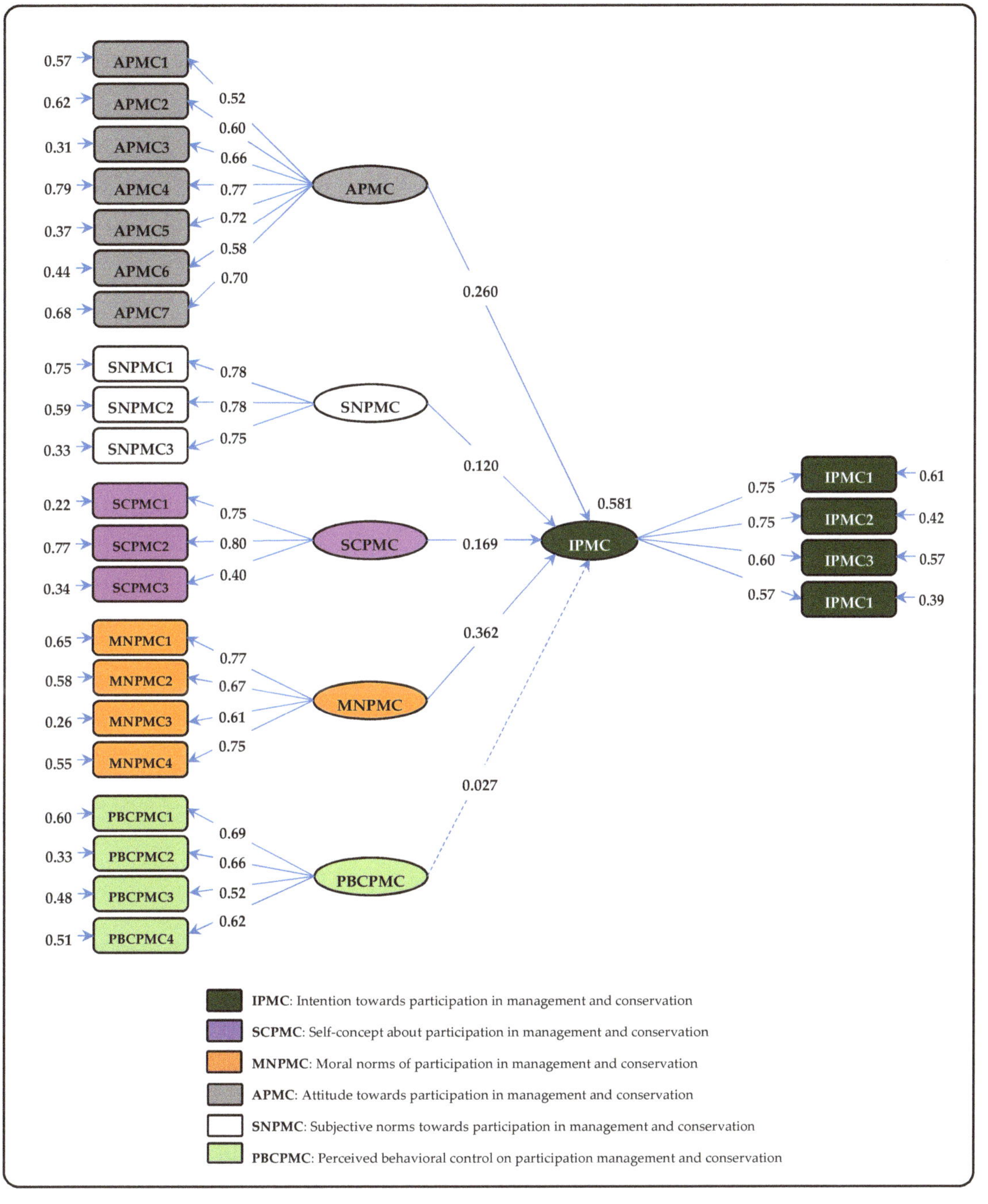

Figure 3. Direct structural model with standardized extended TPB.

4.4. Fit Indices of the Structural Model

The fit of the direct structural models for the original and extended TPB was evaluated using some indices (e.g., CFI, GFI, AGFI, NFI, IFI, RMSEA, and CMIN/DF). In structural equation modeling, CFI, GFI, AGFI, NFI, IFI, RMSEA, and CMIN/DF fit indices are used as standards or criteria to compare the two models. As Table 5 shows, all fit indices are within a plausible range in both models. In other words, both original and extended TPBs had an acceptable fit. However, in deciding on the most appropriate model, each of these fit indices (standards/criteria) must be compared separately for both models. Considering the results in the table, all the reported values of fit indices for the extended TPB are more favorable than the fit indices for the original TPB. Therefore, the extended version of TPB fits better than the original TPB. Overall, it can be concluded that the extended TPB better predicts IPMC than the original TPB. The comparison of R2 values between the two models also confirms this result.

Table 5. Cut-offs and results for fit indices in the original and extended TPB.

Fit Index	Cut-Off	Results for the Present Study	
		Extended TPB	Original TPB
Comparative Fit Index (CFI)	≥0.90	0.912	0.906
Goodness of Fit Index (GFI)	≥0.90	0.953	0.924
Adjusted Goodness of Fit Index (AGFI)	≥0.90	0.947	0.915
Normed Fit Index (NFI)	≥0.90	0.933	0.908
Incremental Fit Index (IFI)	≥0.90	0.984	0.937
Root Mean Square Error of Approximation (RMSEA)	≤0.10	0.076	0.083
Chi-square Normalized by Degrees of Freedom (CMIN/DF)	≤5	1.39	2.41

5. Discussion

The results highlighted that MNPMC had a positive and significant effect on IPMC and this variable has the highest explanatory power in the extended TPB. Based on this finding, it can be concluded that part of the issues related to the conservation and management of wetlands can be improved by creating moral norms among farmers. These findings are consistent with the findings of Yazdanpanah and Forouzani [46] and Han et al. [34]. In many cases, the degradation of wetlands, ecosystems, and natural resources is due to the fact that various stakeholders, such as farmers and local communities, do not consider participation in the management and conservation of wetlands as a moral action. In such conditions, collective interests, altruistic values, and intergenerational and intragenerational equality are not considered in the behavior of individuals. As a result, farmers attempt to maximize their personal profits by maximizing the use of wetlands services. Such anti-social and anti-environmental actions in the long term could lead to further wetlands degradation and natural resources.

According to the results, APMC had a positive and significant impact on IPMC. It can be concluded that having a favorable attitude towards the conservation and management of the wetlands can play a significant role in strengthening the IPMC of farmers. This result has been supported by Valizadeh et al. [53] and Han et al. [34]. The favorable attitude towards the management and conservation of the wetlands is influenced by various prerequisites such as the awareness of non-participation consequences, previous experiences of participation, and outcome expectancy in the field of participation in the conservation and management of the wetlands. Therefore, at this stage, they should be aware of the consequences of their participation or non-participation in the management and conservation of wetlands so that their attitude is based on solid awareness and understanding.

The results revealed that SNPMC had a significant positive effect on IPMC. Based on this finding, we can understand the key role and importance of farmers' social environments in directing farmers' IPMC. Similar findings can be found among the results of

Yazdanpanah and Forouzani [46] and Valizadeh et al. [48]. Such evidence proves that good subjective norms or positive social pressures about wetlands management and conservation increase farmers' intention to participate in wetlands conservation and management. SCPMC also had a positive and significant impact on IPMC. This result is in line with the findings of other researchers see [37–43]. SCPMC refers to a set of socially constructed roles and beliefs about farmers' participation in wetlands conservation and management. Despite the key role of SCPMC in encouraging farmers' behavioral intentions towards participation in wetlands management and conservation, in many conditions, SCPMC of farmers is not at the acceptable level. As a result, they are reluctant to actively participate in sustainable wetlands management activities. In order to strengthen the IPMC using SCPMC, this perspective must be institutionalized among farmers who do not necessarily have to accept and implement all the proposed guidelines for wetlands management and conservation at the same time and in a short period of time. Another strengthening strategy of SCPMC that may be used is self-regulation training in wetlands management and conservation activities. In other words, farmers are educated on how to monitor and direct their conservation activities. This can help strengthen the SCPMC and IPMC.

From a practical point of view, this study provides justifications and insights into the use of the MNPMC, APMC, SNPMC, and SCPMC variables in the policies and programs that intend to encourage farmers and local communities to participate in wetlands management and conservation.

From a theoretical point of view, it should be emphasized that this study contributes to the development of the original TPB. The extended version of TPB has two new variables (MNPMC and SCPMC). However, the results showed that PBCPMC should be removed from the original version of TPB. Many studies deny the effect of PBCPMC on IPMC (see [46,54–57]). The non-significance of the effect of PBCPMC on IPMC in the extended TPB has various reasons for occurring. First, in structural equation modeling, independent variables compete with each other simultaneously to predict the independent variable (IPMC). In general, the greater the number of independent variables in a study is, the more likely that variables may not have a significant effect on the dependent variable. However, by reducing the predictor variables, the non-significant effects of some variables may become significant [58–61]. In this study, the inclusion of the two new variables in the extended TPB model has made the effect of PBCPMC on IPMC non-significant. Second, adding the two new variables to the original TPB can increase the non-causal effects since the introduction of the new variables into the original model will affect the mechanisms and relationships between independent and dependent variables [58,62]. As a result, in the extended TPB, the effect of the independent variables (such as PBCPMC) on the dependent variable (IPMC) may be mediated using new and unknown variables and factors. These modifications in the TPB could pave the way for further research to develop this theory. In addition, it provides new insights into the mechanisms of communication between the variables of TPB theory.

6. Conclusions and Recommendations

Institutions responsible for the planning and sustainable operation of wetlands, such as the Environmental Conservation Organization and the Ministry of Agriculture, must take prompt and purposeful interventions to establish and strengthen the MNPMC. These interventions can be done via participatory and face-to-face discussions with farmers in the farming communities. The executives of these participatory meetings should try to convince the farmers that participation in the management and conservation of the wetlands is a moral action and for the benefit of the agricultural community. This could ultimately lead to strengthening the IPMC. However, interveners should know that enhancing goodwill and trust is the first step in strengthening MNPMC and IPMC.

In order to strengthen the IPMC using attitudinal changes, it is necessary to precisely examine the social context of the farmers' community in participatory activities. In this process, behavioral and attitudinal change agents must first investigate the experiences of the farmers in the field of participation in participatory programs and projects. In many cases, due to unsuccessful experiences of farmers' participation in previous development and partnership programs, they may be reluctant to participate in future programs. In such cases, it is necessary to provide them with tangible guarantees and motivation to ensure that the project will not have a similar fate.

It is also suggested that formal and non-formal farmers' networks, which are generally in the form of various organizations, are identified to be used as levers for the behavioral change of the individuals. In addition, leading, influential, and trusted farmers in the farming community have a high actual and potential ability to direct farmers' IPMC. These farmers can also be used as executive elements of programs to enhance farmers' desire to conserve the wetlands. If leading farmers are used, there is no need for one-to-one communication to strengthen the IPMC, and the social environment is strengthened via social pressures. Moreover, wetlands management programs should be community- and/or region-based. In this way, complex ecosystems can be managed collectively by the (complex) network of farmers.

It is recommended that planners, policy makers, and operators of the wetlands management program use strengthening strategies of SCPMC in the agricultural community. One of the best strategies in this field is to "avoid perfectionism" in participation in the conservation and management of wetlands. In order to strengthen IPMC using SCPMC, it is necessary that some of the easiest and most user-friendly measures of participation in the management and conservation of the wetlands should be adopted and implemented by farmers. After the implementation and completion by farmers, their actions should be encouraged by program supervisors to lead to the repetition and development of behavior.

This study led to four general conclusions. First, extending the TPB by including MNPMC and SCPMC can increase its explanatory power. It seems that these two variables have a positive effect on predicting farmers' intention to participate in wetlands management and conservation. Second, PBCPMC had no impact on IPMC in the extended TPB. Third, MNPMC and APMC variables were the strongest predictors of IPMC in the extended TPB. Fourth, the results of the fitness indices supported the extended TPB.

Directions for Future Research

It should be emphasized that the extended TPB is open for further development in the future. Therefore, researchers can extend this version by including other variables such as knowledge, information, and environmental concern, as well as individual (self) identities and relationships with other farmers and communities. The collected data in this study were analyzed for the whole study area. However, future studies can cluster the results with respect to the geographical division of the study area. This can make it possible to identify the determinants of farmers' intention in areas close to and away from the wetlands more precisely. In this way, the wetlands managers and extension workers can strategize to adjust their plans based on stratified classes or clusters of respondents. In addition, it is worth mentioning that the final validation of the extended TPB requires a cross-validation analysis, which should be addressed in future studies. Overall, due to budget and time constraints, performing such an analysis was not feasible in the current study. Accordingly, this issue is recognized as a limitation for this research which should be addressed by future studies. In the end, it is worth mentioning that the present study focused on analyzing and understanding farmers' intention towards the conservation and management of the wetlands. Futures studies can investigate the intention of other stakeholders including the experts of the Department of Environment, project managers, agricultural extension agents, etc.

Supplementary Materials: The following are available online at https://www.mdpi.com/article/10.3390/land10080860/s1, Table S1: Survey questionnaire.

Author Contributions: Conceptualization, methodology, software, validation, formal analysis, investigation, resources, data curation, writing—original draft preparation, N.V.; writing—review and editing, S.E.B., H.A., A.-H.V., V.T. and A.K.; visualization, supervision, M.B., D.H. and H.A. All authors have read and agreed to the published version of the manuscript.

Funding: This research paper was partly funded by the Strategic Priority Research Program of Chinese Academy of Sciences (Grant No. XDA20060303) and the Chinese Academy of Sciences President's International Fellowship Initiative (PIFI grant no. 2021VCA0004).

Institutional Review Board Statement: Not applicable.

Informed Consent Statement: Not applicable.

Data Availability Statement: Raw data were generated from Shiraz University. We confirm that the data, models, and methodology used in the research are proprietary, and the derived data supporting the findings of this study are available from the first author on request.

Acknowledgments: The authors hereby express their special gratitude to all the respondents who presented the needed data with great patience as well as the surveyors and interviewers who did their best in terms of data collection.

Conflicts of Interest: The authors declare no conflict of interest.

Abbreviations

Intention towards participation in management and conservation (IPMC), Attitude towards participation in management and conservation (APMC), Subjective norms towards participation in management and conservation (SNPMC), Perceived behavioral control on participation management and conservation (PBCPMC), Self-concept about participation in management and conservation (SCPMC), Moral norms of participation in management and conservation (MNPMC).

References

1. Bouahim, S.; Rhazi, L.; Ernoul, L.; Mathevet, R.; Amami, B.; Er-Riyahi, S.; Muller, S.D.; Grillas, P. Combining vulnerability analysis and perceptions of ecosystem services in sensitive landscapes: A case from western Moroccan temporary wetlands. *J. Nat. Conserv.* **2015**, *27*, 1–9. [CrossRef]
2. Ahmad, Z.; Hussain, A.; Shakeel, A. Economics Importance of Wetlands, Their Benefits and Values Case of Pakistan. Munich Personal RePEc Archive (MPRA) 2019. Available online: https://mpra.ub.uni-muenchen.de/95260/ (accessed on 13 November 2020).
3. Shea, A.; Violin, C.R.; Wallace, C.; Forster, B.M. Teaching Water Quality Analysis using a constructed wetlands microcosm in a Non-Science Majors Environmental Science Laboratory. *Pedagog. Res.* **2019**, *4*, em0046. [CrossRef]
4. Chatterjee, K.; Bandyopadhyay, A.; Ghosh, A.; Kar, S. Assessment of environmental factors causing wetland degradation, using Fuzzy Analytic Network Process: A case study on Keoladeo National Park, India. *Ecol. Model.* **2015**, *316*, 1–13. [CrossRef]
5. Stepanenko, V.; Repina, I.; Artamonov, A. Derivation of Heat Conductivity from Temperature and Heat Flux Measurements in Soil. *Land* **2021**, *10*, 552. [CrossRef]
6. Gardner, R.C.; Finlayson, M. Global Wetland Outlook: State of the World's Wetlands and Their Services to People 2018. Secretariat of the Ramsar Convention. Stetson University College of Law Research Paper 2018, No. 2020-5. Available online: https://ssrn.com/abstract=3261606 (accessed on 23 January 2020).
7. Xu, W.; Fan, X.; Ma, J.; Pimm, S.L.; Kong, L.; Zeng, Y.; Li, X.; Xiao, Y.; Zheng, H.; Liu, J.; et al. Hidden loss of wetlands in China. *Curr. Biol.* **2019**, *29*, 3065–3071. [CrossRef]
8. Salazar, A.A.; Arellano, E.C.; Muñoz-Sáez, A.; Miranda, M.D.; Oliveira da Silva, F.; Zielonka, N.B.; Crowther, L.P.; Silva-Ferreira, V.; Oliveira-Reboucas, P.; Dicks, L.V. Restoration and Conservation of Priority Areas of Caatinga's Semi-Arid Forest Remnants Can Support Connectivity within an Agricultural Landscape. *Land* **2021**, *10*, 550. [CrossRef]
9. Shen, W.; Zhang, J.; Zhou, X.; Li, S.; Geng, X. How to Perceive the Trade-Off of Economic and Ecological Intensity of Land Use in a City? A Functional Zones-Based Case Study of Tangshan, China. *Land* **2021**, *10*, 551. [CrossRef]
10. Vélez, J.M.M.; García, S.B.; Tenorio, A.E. Policies in coastal wetlands: Key challenges. *Environ. Sci. Policy* **2018**, *88*, 72–82. [CrossRef]
11. Knippschild, R.; Zöllter, C. Urban Regeneration between Cultural Heritage Preservation and Revitalization: Experiences with a Decision Support Tool in Eastern Germany. *Land* **2021**, *10*, 547. [CrossRef]

12. Ferrario, F.; Beck, M.W.; Storlazzi, C.D.; Micheli, F.; Shepard, C.C.; Airoldi, L. The effectiveness of coral reefs for coastal hazard risk reduction and adaptation. *Nat. Commun.* **2014**, *5*, 1–9. [CrossRef]
13. Narayan, S.; Beck, M.W.; Reguero, B.G.; Losada, I.J.; van Wesenbeeck, B.; Pontee, N.; Sanchirico, J.N.; Ingram, J.C.; Lange, G.; Burks-Copes, K.A. The effectiveness, costs and coastal protection benefits of natural and nature-based defences. *PLoS ONE* **2016**, *11*, e0154735. [CrossRef]
14. Sutton-Grier, A.E.; Sandifer, P.A. Conservation of wetlands and other coastal ecosystems: A commentary on their value to protect biodiversity, reduce disaster impacts, and promote human health and well-being. *Wetlands* **2019**, *39*, 1295–1302. [CrossRef]
15. Cui, B.; He, Q.; Gu, B.; Bai, J.; Liu, X. China's coastal wetlands: Understanding environmental changes and human impacts for management and conservation. *Wetlands* **2016**, *36* (Suppl. 1), S1–S9. [CrossRef]
16. Adekola, O.; Mitchell, G.; Grainger, A. Inequality and ecosystem services: The value and social distribution of Niger Delta wetland services. *Ecosyst. Serv.* **2015**, *12*, 42–54. [CrossRef]
17. Lamsal, P.; Pant, K.P.; Kumar, L.; Atreya, K. Sustainable livelihoods through conservation of wetland resources: A case of economic benefits from Ghodaghodi Lake, western Nepal. *Ecol. Soc.* **2015**, *20*, 1–10. [CrossRef]
18. Rodrigues, J.G.; Conides, A.J.; Rivero Rodriguez, S.; Raicevich, S.; Pita, P.; Kleisner, K.M.; Pita, C.; Lopes, P.F.M.; Alonso Roldani, V.; Ramos, S.S.; et al. Marine and coastal cultural ecosystem services: Knowledge gaps and research priorities. *One Ecosyst.* **2017**, *2*, e12290. [CrossRef]
19. Sandifer, P.A.; Sutton-Grier, A.E.; Ward, B.P. Exploring connections among nature, biodiversity, ecosystem services, and human health and well-being: Opportunities to enhance health and biodiversity conservation. *Ecosyst. Serv.* **2015**, *12*, 1–15. [CrossRef]
20. Nguyen, H.H.; Dargusch, P.; Moss, P.; Aziz, A.A. Land-use change and socio-ecological drivers of wetland conversion in Ha Tien Plain, Mekong Delta, Vietnam. *Land Use Policy* **2017**, *64*, 101–113. [CrossRef]
21. Mokhtari, D. *Participatory Management of Irrigation Water Resources in Iran: Fundamentals and Learning from Experiences*; Ilaf: Shiraz, Iran, 2012.
22. Ancog, R.; Ruzol, C. Urbanization adjacent to a wetland of international importance: The case of Olango Island Wildlife Sanctuary, Metro Cebu, Philippines. *Habitat Int.* **2015**, *49*, 325–332. [CrossRef]
23. Mondal, B.; Dolui, G.; Pramanik, M.; Maity, S.; Biswas, S.S.; Pal, R. Urban expansion and wetland shrinkage estimation using a GIS-based model in the East Kolkata Wetland, India. *Ecol. Indic.* **2017**, *83*, 62–73. [CrossRef]
24. Li, Y.; Shi, Y.; Qureshi, S.; Bruns, A.; Zhu, X. Applying the concept of spatial resilience to socio-ecological systems in the urban wetland interface. *Ecol. Indic.* **2014**, *42*, 135–146. [CrossRef]
25. Iran Department of Environment. *Introduction of the Wetlands of Iran*; Iran Department of Environment: Tehran, Iran, 2020; Unpublished Report.
26. Fishbein, M.; Ajzen, I. *Belief, Attitude, Intention and Behavior: An Introduction to Theory and Research*; Addison-Wesley: Boston, MA, USA, 1975.
27. Park, E.; Lee, S.; Peters, D.J. Iowa wetlands outdoor recreation visitors' decision-making process: An extended model of goal-directed behavior. *J. Outdoor Recreat. Tour.* **2017**, *17*, 64–76. [CrossRef]
28. Jowi, J.O. Deans in Kenyan Universities: Their Leadership Styles and Impacts on Staff Commitment. Ph.D. Thesis, University of Twente, Enschede, The Netherlands, 2018.
29. Gao, L.; Wang, S.; Li, J.; Li, H. Application of the extended theory of planned behavior to understand individual's energy saving behavior in workplaces. *Resour. Conserv. Recycl.* **2017**, *127*, 107–113. [CrossRef]
30. Ajzen, I. *The Theory of Planned Behavior*; Organizational Behavior and Human Decision Processes: San Diego, CA, USA, 1991; Volume 50, pp. 179–211.
31. Kim, E.; Lee, J.A.; Sung, Y.; Choi, S.M. Predicting selfie-posting behavior on social networking sites: An extension of theory of planned behavior. *Comput. Hum. Behav.* **2016**, *62*, 116–123. [CrossRef]
32. Kolvereid, L. Prediction of employment status choice intentions. *Entrep. Theory Pract.* **1996**, *21*, 47–58. [CrossRef]
33. Lortie, J.; Castogiovanni, G. The theory of planned behavior in entrepreneurship research: What we know and future directions. *Int. Entrep. Manag. J.* **2015**, *11*, 935–957. [CrossRef]
34. Han, H.; Meng, B.; Kim, W. Emerging bicycle tourism and the theory of planned behavior. *J. Sustain. Tour.* **2017**, *25*, 292–309. [CrossRef]
35. Ajzen, I. The theory of planned behaviour: Reactions and reflections. *Psychol. Health* **2011**, *26*, 1113–1127. [CrossRef] [PubMed]
36. Viira, A.H.; Põder, A.; Värnik, R. Discrepancies between the Intentions and Behaviour of Farm Operators in the Contexts of Farm Growth, Decline, Continuation and Exit—Evidence from Estonia. *Ger. J. Agric. Econ.* **2014**, *63*, 46–62.
37. Carfora, V.; Caso, D.; Sparks, P.; Conner, M. Moderating effects of pro-environmental self-identity on pro-environmental intentions and behaviour: A multi-behaviour study. *J. Environ. Psychol.* **2017**, *53*, 92–99. [CrossRef]
38. Bagheri, A.; Bondori, A.; Allahyari, M.S.; Damalas, C.A. Modeling farmers' intention to use pesticides: An expanded version of the theory of planned behavior. *J. Environ. Manag.* **2019**, *248*, 109291. [CrossRef] [PubMed]
39. Ajzen, I.; Fishbein, M. Questions raised by a reasoned action approach: Comment on Ogden. *Health Psychol.* **2004**, *23*, 431–434. [CrossRef]
40. Zhang, L.; Ruiz-Menjivar, J.; Luo, B.; Liang, Z.; Swisher, M. Predicting climate change mitigation and adaptation behaviors in agricultural production: A comparison of the theory of planned behavior and the Value-Belief-Norm Theory. *J. Environ. Psychol.* **2020**, 101408. [CrossRef]

41. Whitmarsh, L.; O'Neill, S. Green identity, green living? The role of pro-environmental self-identity in determining consistency across diverse pro-environmental behaviours. *J. Environ. Psychol.* **2010**, *30*, 305–314. [CrossRef]
42. Mobrezi, H.; Khoshtinat, B. Investigating the factors affecting female consumers' willingness toward green purchase based on the model of planned behavior. *Proc. Econ. Financ.* **2016**, *36*, 441–447. [CrossRef]
43. Mancha, R.M.; Yoder, C.Y. Cultural antecedents of green behavioral intent: An environmental theory of planned behavior. *J. Environ. Psychol.* **2015**, *43*, 145–154. [CrossRef]
44. Pilving, T.; Kull, T.; Suškevičs, M.; Viira, A.H. Creating shared collaborative tourism identity in a post-communist environment. *Scand. J. Hosp. Tour.* **2021**, *21*. [CrossRef]
45. Krejcie, R.V.; Morgan, D.W. Determining sample size for research activities. *Educ. Psychol. Meas.* **1970**, *30*, 607–610. [CrossRef]
46. Yazdanpanah, M.; Forouzani, M. Application of the Theory of Planned Behaviour to predict Iranian students' intention to purchase organic food. *J. Clean. Prod.* **2015**, *107*, 342–352. [CrossRef]
47. Tommasetti, A.; Singer, P.; Troisi, O.; Maione, G. Extended theory of planned behavior (ETPB): Investigating customers' perception of restaurants' sustainability by testing a structural equation model. *Sustainability* **2018**, *10*, 2580. [CrossRef]
48. Valizadeh, N.; Bijani, M.; Abbasi, E. Farmers' participatory-based water conservation behaviors: Evidence from Iran. *Environ. Dev. Sustain.* **2021**, *23*, 4412–4432. [CrossRef]
49. Turyahabwe, N.; Kakuru, W.; Tweheyo, M.; Tumusiime, D.M. Contribution of wetland resources to household food security in Uganda. *Agric. Food Secur.* **2013**, *2*, 1–2. [CrossRef]
50. Byrne, B.M. *Structural Equation Modeling with Amos: Basic Concepts, Applications, and Programming (Multivariate Applications Series)*; Taylor and Francis Group: New York, NY, USA, 2010; Volume 396, p. 7384.
51. Nunnally, J.C. *Psychometric Theory 3*; Tata McGraw-Hill Education: New York, NY, USA, 1994.
52. Maleksaeidi, H.; Karami, E.; Zamani, G.H. Farm households' resilience scale under water scarcity. *Mitig. Adapt. Strateg. Glob. Chang.* **2015**, *20*, 1305–1318. [CrossRef]
53. Valizadeh, N.; Bijani, M.; Abbasi, E.; Ganguly, S. The role of time perspective in predicting Iranian farmers' participatory-based water conservation attitude and behavior: The role of time perspective in water conservation behavior. *J. Hum. Behav. Soc. Environ.* **2018**, *28*, 992–1010. [CrossRef]
54. Akbari, M.; Ardekani, Z.F.; Pino, G.; Maleksaeidi, H. An extended model of Theory of Planned Behavior to investigate highly-educated Iranian consumers' intentions towards consuming genetically modified foods. *J. Clean. Prod.* **2019**, *1*, 784–793. [CrossRef]
55. Li, J.; Zuo, J.; Cai, H.; Zillante, G. Construction waste reduction behavior of contractor employees: An extended theory of planned behavior model approach. *J. Clean. Prod.* **2018**, *20*, 1399–1408. [CrossRef]
56. Maleksaeidi, H.; Keshavarz, M. What influences farmers' intentions to conserve on-farm biodiversity? An application of the theory of planned behavior in fars province, Iran. *Glob. Ecol. Conserv.* **2019**, *1*, e00698. [CrossRef]
57. Ataei, P.; Gholamrezai, S.; Movahedi, R.; Aliabadi, V. An analysis of farmers' intention to use green pesticides: The application of the extended theory of planned behavior and health belief model. *J. Rural Stud.* **2021**, *1*, 374–384. [CrossRef]
58. Kalantari, K. *Data Processing and Analysis in Socio-Economic Research*; Design and Landscape Publishing: Tehran, Iran, 2012.
59. Kalantari, K. *Structural Equation Modeling in Socio-Economic Research with LISREL AND SYMPLIS Software*; Design and Landscape Publishing: Tehran, Iran, 2012.
60. Valizadeh, N.; Bijani, M.; Karimi, H.; Naeimi, A.; Hayati, D.; Azadi, H. The effects of farmers' place attachment and identity on water conservation moral norms and intention. *Water. Res.* **2020**, *185*, 116131. [CrossRef]
61. Mohammadi-Mehr, S.; Bijani, M.; Abbasi, E. Factors Affecting the Aesthetic Behavior of Villagers towards the Natural Environment: The Case of Kermanshah Province, Iran. *J. Agric. Sci. Technol.* **2018**, *20*, 1353–1367.
62. Bijani, M.; Ghazani, E.; Valizadeh, N.; Fallah Haghighi, N. Predicting and Understanding of Farmers' Soil Conservation Behavior in Mazandaran Province, Iran. *J. Agric. Sci. Technol.* **2019**, *21*, 1705–1719.

land

MDPI

Article

Changes in Ecosystem Service Value in the 1 km Lakeshore Zone of Poyang Lake from 1980 to 2020

Xinchen Gu [1,2,†], Aihua Long [2,3,*,†], Guihua Liu [4,†], Jiawen Yu [2,3,†], Hao Wang [1,2], Yongmin Yang [2] and Pei Zhang [2]

1 State Key Laboratory of Hydraulic Engineering Simulation and Safety, School of Civil Engineering, Tianjin University, Tianjin 300072, China; gxc@stu.shzu.edu.cn (X.G.); 1016205072@tju.edu.cn (H.W.)

2 State Key Laboratory of Simulation and Regulation of Water Cycle in River Basin, China Institute of Water Resources and Hydropower Research, Beijing 100044, China; yujiawen_415@stu.shzu.edu.cn (J.Y.); yangym@iwhr.com (Y.Y.); zhangpei@iwhr.com (P.Z.)

3 Xinjiang Production and Construction Group Key Laboratory of Modern Water-Saving Irrigation, College of Water Conservancy & Architectural Engineering, Shihezi University, Shihezi 832000, China

4 School of Geography and Environment, Key Laboratory of Poyang Lake Wetland and Watershed Research, Ministry of Education, Jiangxi Normal University, 99 Ziyang Road, Nanchang 330022, China; liugh2013@jxnu.edu.cn

* Correspondence: ahlong@iwhr.com

† These authors contributed equally to this work and should be considered co-first authors.

Abstract: Poyang Lake is a typical lake in the middle and lower reaches of the Yangtze River and is the largest freshwater lake in China. The habitat quality of Poyang Lake has been declining in recent years, leading to a series of ecological problems. An ecological risk evaluation, based on land use, is important in order to promote a coordinated development of land use and the ecological environment. In this paper, land use data from the Poyang Lake basin in the corresponding years are interpreted based on the images from the Landsat satellite mission in seven periods from 1980 to 2020. The lake surface and the 1 km lakeshore zone of Poyang Lake are extracted based on the interpreted land use data. Finally, the ecological service value per unit area of the area is measured by combining it with the Chinese terrestrial ecosystem service value equivalent table, and then with the value of each ecological factor and the value of the changes to land use type. The research results show that: (1) from 1980 to 2000, the lake area of Poyang Lake had an overall decreasing trend (the area slightly increased from 1980 to 1990); from 2000 to 2020, the lake area of Poyang Lake gradually increased (the area slightly decreased from 2015 to 2020). (2) The farmland, forest, grassland and desert areas gradually increased and the wetlands gradually decreased over 40 years; the area of the water body gradually increased from 1980 to 2010, and gradually decreased from 2010 to 2020. (3) The ecosystem service value of the lakeshore zone of Poyang Lake fluctuated around 15,000 × 10^6 Yuan from year to year.

Keywords: Poyang Lake; ecosystem service value; lakeshore zone; Landsat

Citation: Gu, X.; Long, A.; Liu, G.; Yu, J.; Wang, H.; Yang, Y.; Zhang, P. Changes in Ecosystem Service Value in the 1 km Lakeshore Zone of Poyang Lake from 1980 to 2020. *Land* **2021**, *10*, 951. https://doi.org/10.3390/land10090951

Academic Editor: Richard C. Smardon

Received: 5 August 2021
Accepted: 4 September 2021
Published: 8 September 2021

1. Introduction

How to use land use/cover changes to understand changes in complex human–environment systems in an integrated manner—i.e., causes, consequences, and effects—is one of the focal issues in the land use discipline [1]. Many scholars are now beginning to focus on changes in land use landscape patterns from different perspectives and scales [2,3] and are trying to understand land use change and its ecological effects [4–7]. With the correction and refinement of the principles and methods of Ecosystem Service Value (ESV) estimation and the various ecosystem service values proposed by Costanza et al. [8], scholars in various countries have determined the appropriate ESVs for different study areas. Research on the value of ecosystem services has entered a period of rapid development [9]. At the same time, the United Nations Environment Programme (UNEP), the World Bank and other agencies launched the Millennium Ecosystem Assessment (MA) from 2001 to

2005. After the MA, the United Nations organized and implemented The Economics of Ecosystems and Biodiversity (TEEB) project (2007–2010), and the UNEP supported the establishment of the Intergovernmental Science-Policy in 2012. In 2012, the UNEP also supported the establishment of the Intergovernmental Science-Policy Platform on Biodiversity and Ecosystem Services (IPBES). Since then, ecosystems have gone through a rapid development phase and a diversified development phase before reaching the current integrated application phase [10]. In recent years, scholars have also started to use machine learning in order to construct ecosystem service frameworks [11].

Xie Gaodi et al. [12–14], among other scholars in China, completed a table of ecosystem service value equivalent factors in China, based on existing studies, to provide support for the study of the ecosystem value in China. On this basis, Chinese scholars tried to analyze the value of ecosystem service functions under different time frames and spaces. Chen Juncheng et al. [15] assessed the ecological service value of provincial administrative regions in China and analyzed the characteristics of changing spatial differences; Ma Guoxia et al. [16] studied the ecological damage loss in the Chinese ecosystem in 2015, and from that quantified the ecological loss caused by different factors. The study by Ma Guoxia et al. [16] quantified the ecological losses caused by different factors in 2015; Zhang Hao et al. [17] measured the value of arable land development rights and protection compensation based on this theory. Ecosystem services have a very significant impact on human life [18], and studying the value of ecosystem services helps to understand the process of change in ecosystem service functions, which has become a hotspot for ecosystem sustainability research in recent years.

The lakeshore zone is the area at the edge of the lake basin, which is adjacent to the surrounding land [19], and has the function of regulating water quality [20], water quantity [21], and groundwater recharge [22] of the lake. In addition, the lakeshore zone maintains the biodiversity of flora and fauna in the area [23,24] and is critical for food security in nearby subsistence agriculture areas [25,26]. The lakeshore of Poyang Lake influences wetlands maintenance and the biodiversity of Poyang Lake, but previous studies assumed that there was a fixed area of the lakeshore zone [27–29]. In contrast, the lake area of Poyang Lake has experienced large fluctuations caused by the implementation of different water policies in the last 40 years, and the extent of the corresponding lakeshore zone on the lake surface has changed accordingly [30]. In this paper, the distribution of the corresponding 1 km lakeshore zone is divided based on the changes to the lake area in different years from 1980 to 2020. While the remote sensing film of each year cannot fully reflect the overall situation of the lakeshore zone that year, we believe that the long time scale can reflect the inter-annual spatial variation of the lakeshore zone of Poyang Lake, which can provide some reference for the subsequent study of Poyang Lake and the lakeshore area.

The main research contributions of this paper are: (1) an interpretation of the land use data from Poyang Lake basin, based on remote sensing images from the Landsat series satellite from October to November from 1980 to 2020 over seven periods, and we conducted a change analysis of Poyang Lake and the lakeshore zone, based on the land use data from Poyang Lake over seven periods. (2) The area of Poyang Lake, the area of the lakeshore zone, and the land use type of the lakeshore zone from 1980 to 2020 are analyzed. (3) At the end of the article, the ecosystem service value table of the Poyang Lake basin is also calculated and the changes to the ESV of the riparian zone of Poyang Lake from 1980 to 2020 are evaluated.

2. Materials and Methods

2.1. Study Site

Poyang Lake is located on the south bank of the Yangtze River, north of Jiangxi Province, in three prefecture-level cities of Jiujiang, Nanchang and Shangrao. The geographical range of the main lake body is 115°49′ E~116°46′ E, and 28°24′ N~29°46′ N, and is connected to the lower reaches of five major rivers, including Ganjiang River, Fu River,

Xinjiang River, Rao River and Xiushui River (Figure 1). Poyang Lake is connected to the Yangtze River via its mouth under the lake, and is an overwater (exchange of nutrients in lake waters with different rivers) throughput type (water in the lake is fed by rivers and outflow) seasonal shallow freshwater lake, which has a water level change that is influenced twice as much by the five major rivers and the Yangtze River [31]. The groundwater flow direction in the lakeshore zone of Poyang Lake is mainly from the surrounding hilly area to the downstream lake area where the terrain is relatively flat, and the groundwater flows towards the river and lake area in general [32].

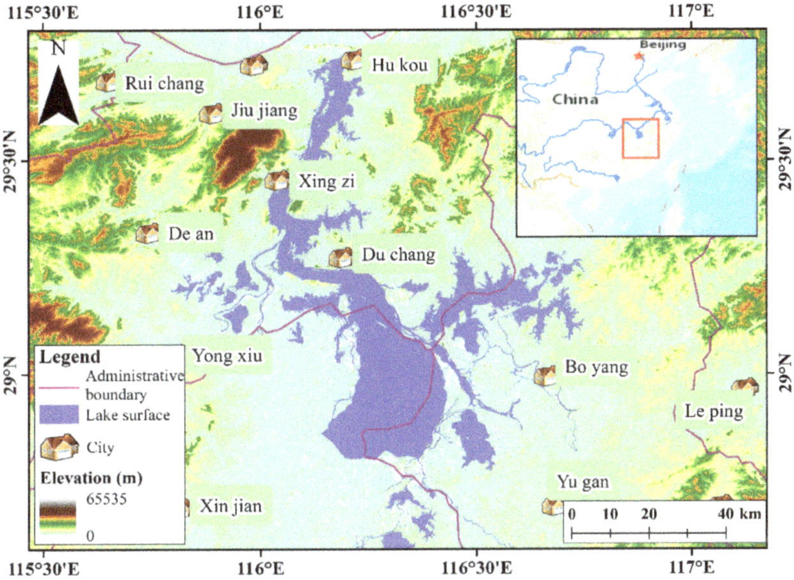

Figure 1. Location of Poyang Lake.

Poyang Lake and the adjacent area is a basin that is composed of different landform types, such as mountains, hills, plains and lakes. The elevation of the lake basin is generally high in the south and low in the north; the maximum elevation difference can be up to 13 m, and the average elevation difference between the south and north is about 6.5 m [33]. Taking Songmen Mountain between Duchang County and Wucheng Town of Yongxiu County as the boundary, Poyang Lake is divided into two lakes in the north and south; the southern part is wide and shallow, which is the main lake area; the northern part is narrow and deep, which is the waterway into the Yangtze River. The geomorphology of the lake is combined by a waterway, continental beach, island, inner lake and branching port. The waterway includes five rivers that lead into the lake and Poyang Lake leads into the river waterway. The distribution is strongly influenced by the water level. Due to the influence of human activities, most of the branches and harbors were turned into closed water bodies for aquaculture, or were transformed into paddy fields for rice cultivation.

The biodiversity and ecological conservation of the Yangtze River basin has been the priority area of the World Wide Fund for Nature or World Wildlife Fund, and the ecological wetlands of Poyang Lake were listed in the "List of Wetlands of International Importance" in 1992, which is the only representative of China to join the "World Network of Lakes for Life". Poyang Lake is responsible for various ecological functions, such as flood control and water storage, climate regulation and pollution degradation. Poyang Lake is an important storage lake of the Yangtze River, and the annual average water volume into the river accounts for about 15.6% of the runoff of the Yangtze River. The sustainable stability of water quantity and quality of Poyang Lake is directly related to the water security of the surrounding area of Poyang Lake and even the middle and lower reaches of the Yangtze River.

Throughout history, China has seen several southward migrations of populations in the Central Plains, and a southward migration of the population lived nearby the water, The migration led to an increase in the population around Poyang Lake, and the seasonally submerged beach was reclaimed to a large extent, which is called "polder farming". With the increase in population and demand for arable land, the activity of "paddock farming" around Poyang Lake was very active in the early 1970s; after the 1980s, people gradually realized the harm caused by excessive paddock farming in the Poyang Lake, and paddock farming was subsequently prohibited. However, due to the influence of the Yangtze River Three Gorges Storage Project, the low water level of Poyang Lake was persistently low, and since 2008, the climax of paddock farming, paddock city farming and paddock land farming in the areas along the lake began. The official approval of the "Poyang Lake Ecological Economic Zone Plan" by the State Council of China in December 2009 marks the point at which the construction of Poyang Lake Ecological Economic Zone was formally upgraded to a national strategy. The Poyang Lake Ecological Economic Zone is an economic development zone with Poyang Lake as the core, Poyang Lake City Cluster as the base, the low-carbon economic development pioneer area as the target, and ecological civilization and coordinated economic and social development as the strategic concept.

2.2. Data Source

We obtained Landsat-MSS, Landsat-TM/ETM, and Landsat-8 remote sensing data from the Geospatial Data Cloud Platform of the Computer Network Information Center of the Chinese Academy of Sciences (http://www.gscloud.cn/, accessed on 31 August 2021) for October–November 1980, 1990, 2000, 2005, 2010, 2015, and 2020 for the Poyang Lake basin. Among them, Landsat-MSS remote sensing image data were mainly used for the 1980 land use interpretation and Landsat-TM/ETM remote sensing image data were mainly used for the remote sensing interpretation of the 1990, 2000, 2005 and 2010 data periods, while Landsat8 remote sensing image data were used for the 2015 and 2020 land use data. All Landsat satellite mission remote sensing image data were filtered appropriately, based on cloudiness. Then, ENVI 4.5/ArcGIS 10.2 was used for the preliminary interpretation of remote sensing images, and land use classification was carried out using the secondary land use classification standard issued by the Chinese Academy of Sciences (Appendix A, Table A1). We used the computer supervised classification method, based on the maximum likelihood method, to classify land use in the Poyang Lake region. The advantages of the computer supervised classification method are higher classification accuracy, based on training samples, and faster speed, as compared to the traditional classification method, and finally, the seven-phases of the Poyang Lake Basin land use data could be obtained.

The DEM data used in this paper were obtained from the geospatial data cloud platform of the Computer Network Information Center of Chinese Academy of Sciences (http://www.gscloud.cn, accessed on 31 August 2021), which was processed from the data from ASTER GDEM version 1, which is a digital elevation data product with a global spatial resolution of 30 m.

2.3. Research Methodology

Costanza et al. [8,34] classified ecosystem services into 17 types, and in this paper, the 17 types of ecosystem services proposed by Costanza et al. [35] were reclassified into 11 types, based on the results obtained by Xie Gaodi et al. [35]. We reclassified the 17 categories of ecosystem services found by Costanza into 11 categories, with the following criteria (Appendix A, Table A2): (1) climate regulation includes climate regulation and disturbance regulation in the Costanza system; (2) soil conservation includes erosion control and sediment retention, and soil formation in the Costanza system; (3) biodiversity includes pollination, biological control, refugia, and genetic resources in the Costanza system; and (4) the aesthetic landscape includes recreation and culture in the Costanza system.

According to Costanza's findings [8], the amount of economic value obtained from one ecological service value equivalent factor was 446.58 Yuan/hm^2 (in 1997, the exchange rate between the US dollar and the Chinese yuan was 1:8.27); however, Xie Gaodi et al. [12,14,35] argued that the calculation of ecological service value equivalent per unit area of the Chinese ecosystems should set the ecological service value equivalent of farmland failure production as 1, and then the value of other ecological services provided by the ecosystem should be determined. The size of the other ecological services provided by the ecosystem was determined (Appendix A, Table A3). The calculation method of the ecosystem service value is as follows.

(1) Firstly, the ecological service value per unit area of the secondary ecosystem in Poyang Lake basin is calculated as (Table 1):

$$P_{ij} = \frac{1}{7}k \cdot b \cdot c \cdot a_{ij} \tag{1}$$

where P_{ij} is the value of the ecosystem service of class i per unit area of the class j secondary land use type; $i = 1, 2, \ldots$ 11 represent different types of ecosystem service values of gas regulation, climate regulation, etc.; $j = 1, 2, \ldots$ 14 represent secondary land use types of dryland, paddy field, grassland, meadow, etc.; a_{ij} represents class i ecosystem of the class j secondary land use type service equivalent factor [14]; b represents the unit area grain yield; c is the average grain purchase price in the region in that year; k is the correction coefficient of the equivalent factor; and $\frac{1}{7}k \cdot b \cdot c$ is the corrected unit ecosystem value equivalent in the Poyang Lake basin.

(2) Combine secondary land use types into primary land use types according to the specific research scale of this paper.

Since the land use classification standard used for remote sensing interpretation is the secondary land use classification standard issued by the Chinese Academy of Sciences (Appendix A, Table A1), this paper converts both classification standards into six primary land use types according to the table of ecological service values per unit area of the primary ecosystem in Poyang Lake Basin (Table 2) before analysis.

According to the geomorphological characteristics of the lakeshore zone of Poyang Lake [28], we combined the land use types of the lakeshore zone of Poyang Lake; the average value of dry land and grassland was taken as the ecological service value of the primary use type farmland; the ecological service value of the shrubs with the highest area share were taken as the ecological service value of the forest; the ecological service value of the meadow was taken as the ecological service value of the grassland; wetlands were kept unchanged; the ecological service value of bare land was taken as the ecological service value of the desert; the ecological service value of the water system was taken as the ecological service value of the watershed. Finally, the unit to calculate the ecosystem value service was converted to Yuan/km^2, and the table of the ecological service value per unit area of the ecosystem in the Poyang Lake basin level after conversion is shown as follows (Table 2):

Table 1. Table of ecological service value per unit area of the secondary ecosystem in Poyang Lake basin (Yuan/hm^2).

Ecosystem Classification		Provisioning Services			Regulating Services				Supporting Services			Cultural Services
Primary Classification	Secondary Classification	Food Production	Raw Material Production	Water Supply	Gas Regulation	Climate Regulation	Environmental Purification	Hydrological Regulation	Soil Conservation	Maintenance of Nutrient Cycles	Biodiversity	Aesthetic Landscape
Farmland	Dryland	3063.47	1441.63	72.08	2414.73	1297.47	360.41	973.10	3712.20	432.49	468.53	216.24
	Paddy field	4901.55	324.37	−9478.73	4000.53	2054.33	612.69	9803.10	36.04	684.78	756.86	324.37
Forest	Coniferous	792.90	1874.12	973.10	6126.94	18,272.69	5370.08	12,037.63	7424.40	576.65	6775.67	2955.35
	Mixed coniferous	1117.26	2558.90	1333.51	8469.59	25,336.68	7172.12	12,650.32	10,307.67	792.90	9370.61	4108.65
	Broad-leaved	1045.18	2378.69	1225.39	7820.85	23,426.52	6955.87	17,083.34	9550.81	720.82	8685.83	3820.32
	Shrub	684.78	1549.75	792.90	5081.75	15,245.26	4613.22	12,073.67	6199.02	468.53	5658.41	2486.82
Grassland	Grass	360.41	504.57	288.33	1838.08	4829.47	1585.80	3532.00	2234.53	180.20	2018.28	901.02
	Scrub	1369.55	2018.28	1117.26	7100.04	18,777.26	6199.02	13,767.59	8649.79	648.73	7856.89	3459.92
	Meadow	792.90	1189.35	648.73	4108.65	10,884.32	3604.08	7965.02	5009.67	396.45	4577.18	2018.28
Wetland	Wetland	1838.08	1802.04	9334.57	6847.75	12,974.69	12,974.69	87,326.86	8325.42	648.73	28,364.11	17,047.30
Desert	Desert	36.04	108.12	72.08	396.45	360.41	1117.26	756.86	468.53	36.04	432.49	180.20
	Bare ground	0.00	0.00	0.00	72.08	0.00	360.41	108.12	72.08	0.00	72.08	36.04
Waters	Water system	2883.26	828.94	29,877.82	2775.14	8253.34	20,002.64	368,481.14	3351.79	252.29	9190.40	6811.71
	Glacial snow	0.00	0.00	7784.81	648.73	1946.20	576.65	25,697.09	0.00	0.00	36.04	324.37

Table 2. Table of ecological service value per unit area of the ecosystem in the Poyang Lake basin level (Yuan/km^2).

Ecosystem Level Classification	Provisioning Services			Regulating Services				Supporting Services			Cultural Services	Total
	Food Production	Raw Material Production	Water Supply	Gas Regulation	Climate Regulation	Environmental Purification	Hydrological Regulation	Soil Conservation	Maintenance of Nutrient Cycle	Biodiversity	Aesthetic Landscape	
Farmland	398,251	88,300	−470,332	320,763	167,590	48,655	538,810	187,412	55,863	61,269	27,031	1,423,612
Forests	68,478	154,975	79,290	508,175	1,524,526	461,322	1,207,367	619,902	46,853	565,841	248,682	5,485,411
Grassland	79,290	118,935	64,873	410,865	1,088,432	360,408	796,502	500,967	39,645	457,718	201,828	4,119,463
Wetlands	183,808	180,204	933,457	684,775	1,297,469	1297,469	8,732,686	832,542	64,873	2,836,411	1,704,730	18,748,424
Desert	0	0	0	7208	0	36,041	10,812	7208	0	7208	3604	72,081
Waters	288,326	82,894	2,987,782	277,514	825,334	2,000,264	36,848,114	335,179	25,229	919,040	681,171	45,270,847

(3) Calculate the value of the ecosystem services in the lakeshore zone.

$$ESV = \sum_{m=1}^{6} P_m \cdot A_m \quad (2)$$

where $m = 1, 2, \ldots 6$ represent the primary land use types, such as farmland, forest and grassland, respectively; P_m represents the ecological service value of the m class primary land use types, and the conversion relationship between P_m and P_{ij} is shown in Table 2 ($P_j = \sum_{i=1}^{11} P_{ij}$, and P_j represents the ecological service value of the j class secondary land use types). We converted the secondary land use ecosystem service value (P_j) into the primary land use ecosystem service value (P_m), based on the situation of the lakeshore zone of Poyang Lake); A_m is the area size of the m class land use types in the lakeshore zone, unit hm^2; and ESV is the total ecological service value of the lakeshore zone (units are in Yuan).

Based on the results obtained in Xie Gaodi's study [14,36], the economic value of 1 ecosystem value equivalent in China was determined to be 3406.50 Yuan/hm^2, and this paper calculated the grain yield per unit area using the National Bureau of Statistics (http://www.stats.gov.cn/tjsj/zxfb/201512/t20151208_1286449.html, accessed on 31 August 2021), published by Jiangxi Province in 2015; the grain yield per unit area of cultivated land was 5798.6 kg/hm^2, while the average grain yield per unit area of the cultivated land in China in the same period was 5482.9 kg/hm^2, and the correction coefficient was calculated as 1.058, i.e., $k = 1.058$. Then, the value equivalent of one ecosystem in the Poyang Lake basin was determined to be 3604.08 Yuan/hm^2. In this paper, the corresponding secondary ecosystem service value of Poyang Lake was calculated based on the corrected ecosystem service value equivalent and different land use areas of the corresponding lakeshore zones (Tables 1 and 2).

3. Results

3.1. Spatial and Temporal Changes of Poyang Lake and Lakeshore Zone

According to the results from the remote sensing image interpretation, the lake area of Poyang Lake had a generally increasing trend from 1980 to 2020, the lake area decreased from 1990 to 2000, and the lake area decreased from 2015 to 2020. From Figure 2a, the lake area of Poyang Lake had an increasing trend, year-by-year, from 1980 to 2020. Due to the increase in the lake area, all the waters of the lake started to be connected, and the lake integrity of Poyang Lake was enhanced. It can be seen that the decrease in the lake level of Poyang Lake firstly leads to the degradation of the lake surface in the southwest area of Poyang Lake, which is the most significant area, and this reflects the changes to the lake area of Poyang Lake. The lake area increased by 109.96 km^2 from 1980 to 2000, which was an increase in the lake area by 7.46% in 20 years. The lake area increased by 920.74 km^2 from 2000 to 2020, which was an increase in the lake area by 58.13% in 20 years. The total lake area increased by 1030.43 km^2 in 40 years, and the lake area obviously increased (Figure 2).

According to the results from the remote sensing interpretation, the area of the 1 km range lakeshore zone of the lake in 1980 was 1215.38 km^2, the area of the 1 km range lakeshore zone of the lake in 1990 was 1370.22 km^2, the area of the 1 km range lakeshore zone of the lake in 2000 was 1076.20 km^2, and the area of the 1 km range lakeshore zone of the lake in 2005 was 1399.76 km^2. The area of the 1 km extent lakeshore zone in 2010 was 1584.66 km^2, the area of the 1 km extent lakeshore zone in 2015 was 1494.19 km^2, and the area of the 1 km extent lakeshore zone in 2020 was 1457.13 km^2. The area of the lakeshore zone decreased by 139.18 km^2 between 1980 and 2000, a decrease by −11.45%.

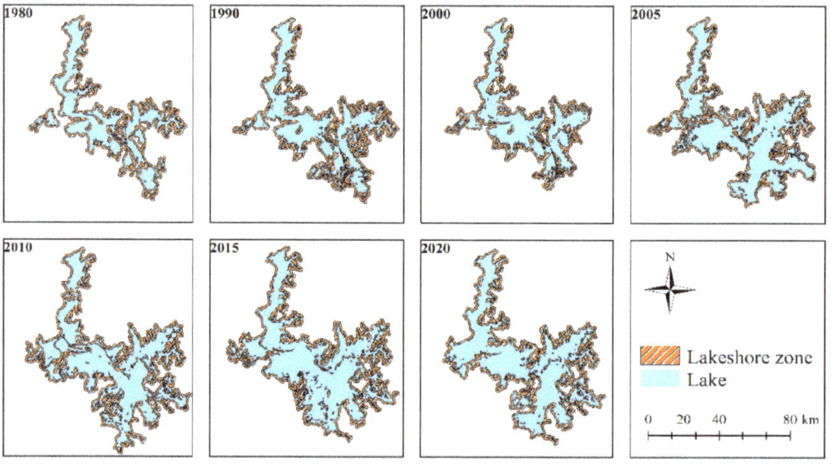

(**a**)

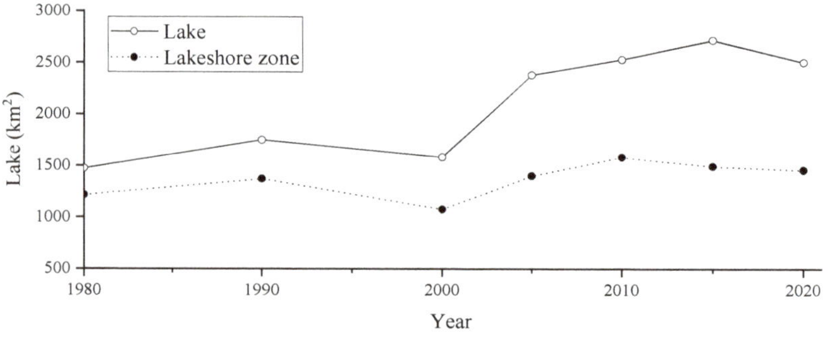

(**b**)

Figure 2. (**a**) Spatial changes and (**b**) area changes of Poyang Lake and the lakeshore zone from 1980 to 2020.

The area change of Poyang Lake and the area change of the 1 km lakeshore zone of the lake were highly consistent; both reached the lowest value in 2000 and then started to increase. Over 40 years, the average change rate of the lake area was 25.76 km^2/year, and the average change rate of the 1 km lakeshore zone of the lake was 6.04 km^2/year.

3.2. Changes of Various Land Use Types in the Lakeshore Zone of Poyang Lake

The area of each land use type at the first level of the lakeshore zone of Poyang Lake from 1980 to 2020 was analyzed by counting the area of each land use type at the first level of the lakeshore zone of Poyang Lake from 1980 to 2020; the area of each land use type at the first level of the lakeshore zone of Poyang Lake from 1980 to 2020 shows an increasing trend from 1980 to 1990, and a decreasing trend from 1990 to 2000 (Table 3). From 2000 to 2015, the area of farmland and forest shows a continuously increasing trend, the area of grassland and wetlands show a decreasing trend, the area of desert shows a decreasing and then increasing trend, and the area of water shows an increasing and then decreasing trend. From 2015 to 2020, the area of farmland, forest and water shows a decreasing trend, and the area of grassland, wetlands and desert shows an increasing trend; over 40 years, except the area of wetland, which decreased, the area of all other types increased (Figure 3). The area of all types of land use increased during the 40-year period, except for the area of wetlands. The area changes and trends of each land use type were analyzed using a linear fit, as shown in Figure 3. The proportion of various types of land uses in the lakeshore zone of Poyang Lake remained the same in 2000, but its area was greatly reduced compared to other years. This is due to the fact that the lake area of Poyang Lake shrank to a minimum in 2000, resulting in the lowest area of the Poyang Lake riparian zone in history.

Table 3. The values of the lake area of Poyang Lake and the area of the land use type of the lakeshore zone of Poyang Lake from 1980 to 2020 (km^2).

Year	1980	1990	2000	2005	2010	2015	2020
Lake area (km^2)	1473.47	1750.00	1583.43	2381.27	2529.60	2718.23	2503.90
Lakeshore zone area (km^2)	1215.38	1370.22	1076.20	1399.76	1584.66	1494.19	1457.13
Farmland	430.83	500.30	351.16	648.61	724.59	794.51	728.49
Forests	103.08	107.94	91.13	123.08	114.18	132.90	128.60
Grassland	48.58	60.86	42.02	79.63	70.37	69.46	71.59
Wetlands	529.69	555.66	460.77	374.37	290.90	224.26	311.03
Desert	27.22	37.50	25.43	53.80	48.29	59.14	59.95
Waters	75.97	107.97	105.68	120.26	336.34	213.92	157.47

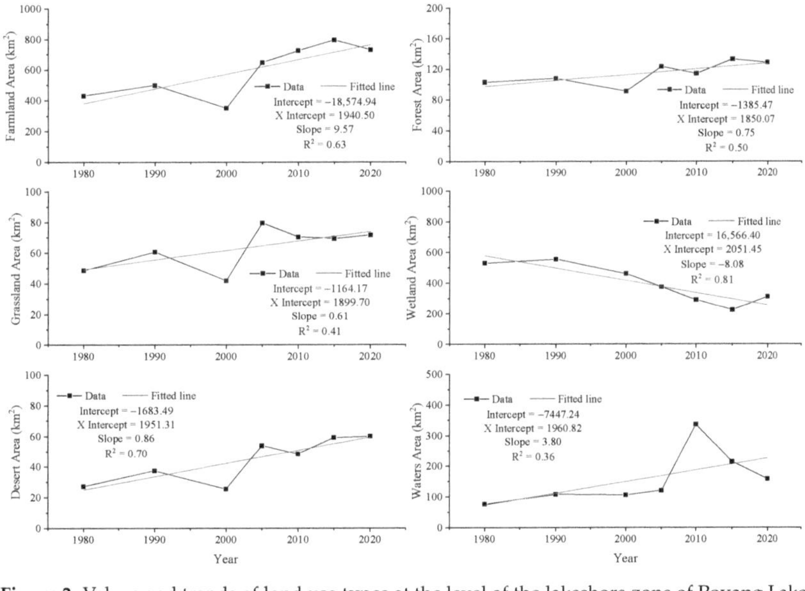

Figure 3. Values and trends of land use types at the level of the lakeshore zone of Poyang Lake from 1980 to 2020.

The farmland area generally showed an increasing trend from 1980 to 2020: the lowest value of 351.16 km^2 was reached in 2000 and the highest value of 794.51 km^2 was reached in 2015. Through linear fitting, it was found that the farmland area in the lakeshore zone increased at a rate of 9.57/year over 40 years. It is worth noting that the area of farmland in the lakeshore zone was 728.49 km^2 in 2020, which is a decrease of 66.02 km^2, indicating a decreasing trend in the area of farmland. The forest area generally showed an increasing trend between 1980 and 2020; the lowest value of 91.13 km^2 was reached in 2000 and the highest value of 132.90 km^2 was reached in 2015. Through linear fitting, it was found that the forest area in the lakeshore zone increased at a rate of 0.75/year over the 40-year period, an insignificant increase. It is noteworthy that the forest area in the lakeshore zone shows a small fluctuation between 2005 and 2020, indicating that its area tends to be stable. The grassland area generally shows an increasing trend between 1980 and 2020; the lowest value of 42.02 km^2 was in 2000 and the highest value of 79.63 km^2 was in 2015. There is a high similarity between the area of grassland and agricultural land during the 40-year period. From 1980 to 2020, the wetlands area showed a generally decreasing trend; the lowest value of 224.26 km^2 was in 2015 and the highest value of 555.66 km^2 was in 1990. The area of wetlands in the lakeshore zone declined at an average rate of 8.08/year over the 40-year period, and the area of wetlands rebounded in 2020, as compared to that in 2015. The water area showed an overall increasing trend between 1980 and 2020; the lowest value of 75.97 km^2 was in 1980 and the highest value of 336.34 km^2 was in 2010. The linear fit shows that the water area in the lakeshore zone increased at an average rate of 0.86/year over 40 years. The average rate of increase in the water area in the lakeshore zone was 3.80/year over the 40 years.

3.3. Changes of Ecosystem Service Value in the Lakeshore Zone of Poyang Lake

Based on the area of each primary land use type in the lakeshore zone from 1980 to 2020 (Table 4), combined with Table 2, changes in the value of various ecosystem services were calculated, as shown in the following table (Table 4).

Table 4. Values of ESV in the lakeshore zone of Poyang Lake from 1980 to 2020 (10^6 Yuan).

Classification	1980	1990	2000	2005	2010	2015	2020
Farmland	613.34	712.23	499.91	923.37	1031.54	1131.08	1037.09
Forests	565.44	592.09	499.90	675.13	626.33	729.01	705.41
Grassland	200.13	250.71	173.12	328.03	289.87	286.13	294.93
Wetlands	9930.93	10,417.78	8638.73	7018.90	5453.83	4204.47	5831.32
Desert	1.96	2.70	1.83	3.88	3.48	4.26	4.32
Waters	3439.30	4887.74	4784.42	5444.39	15,226.19	9684.26	7128.61
Total	14,751.11	16,863.26	14,597.91	14,393.70	22,631.24	16,039.21	15,001.68

From 1980 to 2020, the ecosystem value service of the lakeshore zone of Poyang Lake showed an increasing trend from 1980 to 1990, and a decreasing trend from 1990 to 2000. From 2000 to 2010, the ecosystem value service of farmland, forest and water areas showed a continuously increasing trend, the ecosystem value service of grassland and desert showed an increasing trend first and then a decreasing trend, and the ecosystem value service of wetlands showed a decreasing trend. During the period 2015–2020, the ecosystem value services of farmland, forest and watershed showed a decreasing trend, and grassland, wetlands and desert showed an increasing trend; during the 40-year period, except for the decrease in ecosystem value services of wetland, all other types of ecosystem value services increased. The trend of the total ecosystem value services is shown in Figure 4.

The overall trend of the ecosystem service value (ESV) of the lakeshore zone of Poyang Lake shows an increasing trend, but the increasing trend is not obvious. Among them, the ESV in 2010 is the highest in the lakeshore zone of Poyang Lake, reaching 22,631.24 × 10^6 Yuan, which may be related to the fact that the water area in the lakeshore zone of Poyang Lake in 2010 was the largest from 1980 to 2020 (the water area contributes a significant part of the ESV). The ESV in the lakeshore zone of Poyang Lake decreased slightly from 2015 to 2020.

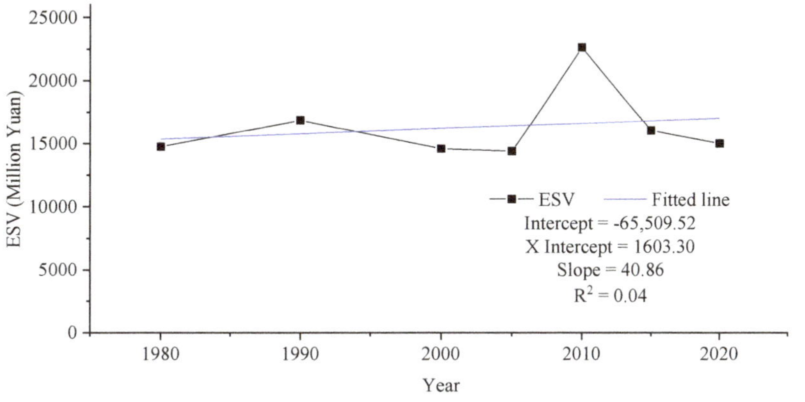

Figure 4. Values and trends of ESV in the lakeshore zone of Poyang Lake from 1980 to 2020.

4. Discussion

Human activities and changes in natural conditions directly cause changes in land use land cover, which in turn affects the changes in the ecological environment [2,37,38]. With the development of "enclosing the lake and making fields" of Poyang Lake, the wetlands area is reduced, vegetation is destroyed, soil erosion is serious, and biodiversity is seriously threatened [39,40]. According to a survey, 119 species of wetlands plants and 126 species of fishery resources in the Poyang Lake in the 1960s reduced to 102 and 118 species, respectively, in the 1980s, and the reed and dipterocarp clusters everywhere were gradually replaced by mossy grass clusters with short plants [41,42].

The results of this study show that the lake area of Poyang Lake decreased obviously between 1990 and 2000, and the lake area decreased from 1370.22 km^2 in 1990 to 1076.20 km^2 in 2000. Meanwhile, about 40% of the land use types in the 1 km lakeshore zone are farmland, which is probably due to the "enclosing lake for farming". It is likely that the area of the Poyang Lake has been decreasing year after year due to the "enclosing lake for farming". The study of domestic related scholars [43] shows that the utilization of the wetlands of Poyang Lake is mainly to enclose the lake in order to make fields for agricultural production, which is consistent with the results of this study. Meanwhile, the area of wetlands increased in 2020, as compared to 2015, which indicates that the proposal and construction of "Poyang Lake Eco-Economic Zone" has begun to show effect [43], and this reflects that the ecological service value generated by wetlands accounts for about 40% of the ecological service value generated by all the lakeshore zones in 2020. The land use types that changed more significantly between 1980 and 2020 were agricultural land, wetlands, and waters. Among them, agricultural land had the highest change in area, increasing at a rate of 9.57/year, followed by wetlands, decreasing at the rate of 8.08/year. it is worth noting that the wetlands area in the lakeshore zone of Poyang Lake has continued to decrease from 1990 to 2015, which is consistent with some current findings [44–46].

Between 2000 and 2020, the lake area rebounded and the area of the lakeshore zone increased accordingly. The year 2010 benefited from the increase in the area of water bodies and the ecosystem service value (ESV) reached 22,631.24 $\times$ 10^6 Yuan in 2010, which was the highest in 40 years. The area of water bodies in the lakeshore zone decreased in 2020, as compared to 2015, which is the main reason for the lower ESV in 2020, as compared to 2015. This is the main reason why the ESV of the lakeshore zone in 2020 was lower than the ESV in 2015. Except for 2010, the ecological service value of the riparian zone of Poyang Lake grew slowly and remains around 15,000 $\times$ 10^6 Yuan all year round, which indicates that the ecosystem service value of the riparian zone of Poyang Lake fluctuates within the normal range (1500–3000 km^2) under the change of lake area and spatial location and area of the riparian zone of Poyang Lake. This manuscript differs from previous papers in the valuation of ecosystem services in the riparian zone of Poyang Lake in that we used the interannual dynamic riparian zone to more accurately quantify the magnitude of ecosystem services provided by the riparian zone to the Poyang Lake region. Few papers have been published on the ecosystem service value of the riparian zone of Poyang Lake. Previous studies [27,32,47–49] used the multi-year mean water level of Poyang Lake in order to determine the area of the surface of the Poyang Lake and thus the riparian zone area, and the ESV results obtained from those studies may be inaccurate.

Since the implementation of the Yangtze River Protection Law in March 2021, Poyang Lake has entered a ten-year ban on fishing, which promotes the implementation of the "catching big protection and not big development" in the Yangtze River basin; the previously artificially closed water body has gradually broken the dike, and the hydrological connectivity of the lake area tends to be in the natural state. The local wetlands protection policy has achieved remarkable results and should be continued, as well as the ecological restoration of previously developed agricultural land and construction sites. The local decision makers could consider the following two aspects: (1) ensure that the total amount of wetlands in the Poyang Lake area will not be reduced and the ecological function will not be diminished. That is, consider the need for wetlands protection and biodiversity protection, especially the need to maintain the function of various aquatic biological reserves, and establish an ecological protection alert line in the Poyang Lake area. (2) Speed up the biodiversity reply in the basin, establish a mechanism for establishing aquatic life race restoration, and study the influence of human interference on ecosystem degradation. Use certain anthropogenic methods to accelerate the ecological restoration process of the Poyang Lake region.

In the future, if the farmland type in the lakeshore zone of Poyang Lake is further transformed into the wetlands or water body, and the land use types such as building land, forest and grassland are reasonably allocated, the ecological service value of lakeshore zone will be further improved, the ecological effect of the lakeshore zone will be more stable, and the quality of the lakeshore zone of Poyang Lake will be further improved.

5. Conclusions

In this paper, based on the land use data corresponding to a 30 m resolution, based on 7-Landsat satellite image interpretation from 1980 to 2020, the changes in the ecosystem service value of the Poyang Lake surface and the 1 km lakeshore zone are assessed, in combination with the ecological service value per unit area of different terrestrial ecosystems in Poyang Lake basin, and the conclusions are as follows.

(1) The lake area of Poyang Lake has shown a decreasing trend from 1980 to 2000; the lake area of Poyang Lake has gradually increased from 2000 to 2020. This indicates that the current lake area of Poyang Lake has recovered.

(2) The area of farmland, forest, grassland and desert has gradually increased and the area of wetlands has gradually decreased over 40 years. The area of the water body gradually increased from 1980 to 2010, while the area of water body gradually decreased from 2010 to 2020.

(3) The ecosystem service value of the lakeshore zone of Poyang Lake fluctuates around $15{,}000 \times 106$ Yuan from year to year.

Author Contributions: Conceptualization, X.G. and A.L.; methodology, G.L.; software, J.Y.; validation, X.G., A.L. and H.W.; formal analysis, X.G.; investigation, X.G.; resources, H.W.; data curation, H.W.; writing—original draft preparation, X.G.; writing—review and editing, A.L., Y.Y. and P.Z.; visualization, H.W.; supervision, A.L.; project administration, J.Y.; funding acquisition, G.L. All authors have read and agreed to the published version of the manuscript.

Funding: This research was funded by the National Key Research and Development Program of China (Project Numbers: 2017YFC0404301 and 2016YFA0601602) and the National Natural Science Foundation of China (Project Number: 41961004).

Institutional Review Board Statement: Not applicable.

Informed Consent Statement: Not applicable.

Data Availability Statement: The data set is provided by Geospatial Data Cloud site, Computer Network Information Center, Chinese Academy of Sciences (http://www.gscloud.cn/, accessed on 31 August 2021).

Conflicts of Interest: The authors declare no conflict of interest.

Appendix A

Table A1. Chinese Academy of Sciences secondary land use classification system.

No. (Level 1)	Name	No. (Level 2)	Name	Description
1	Farmland			Refers to the planting of crops cab land, including ripe cultivated land, newly opened land, recreational land, rotational rest land, grass field rotation crop land; to plant crops which are mainly agricultural fruit, agricultural mulberry, and agricultural forestry land; cultivated for more than three years of the beach and sea shoals.
		11	Water Field	Refers to the arable land with water guarantee and irrigation facilities, which can be irrigated normally in normal years and used for growing rice, lotus root and other aquatic crops, including the arable land where rice and dryland crops are rotated.
		12	Dryland	Cultivated land without irrigation sources and facilities, growing crops by natural precipitation; cultivated land with water sources and irrigation facilities, which can be irrigated normally in one year; cultivated land mainly for growing vegetables; recreational land and rotational land with normal crop rotation.
2	Forests			Refers to forestry land such as growing trees, shrubs, bamboos, and coastal mangrove land.
		21	With woodland	Refers to natural forests and plantations with a denseness of >30%. Including timber forests, economic forests, protective forests and other mature woodlands.
		22	Shrubland	Refers to short woodlands and scrub woodlands with densities > 40% and heights below 2 m.
		23	Open woodland	Refers to forest land with 10–30% tree densities.
		24	Other woodland	Refers to the non-forested plantations, trails, nurseries and various types of gardens (orchards, mulberry gardens, tea gardens, hot crop forests, etc.).
3	Grassland			Refers to all kinds of grassland with herbaceous plants growing mainly and covering more than 5%, including scrub grassland mainly for grazing and open forest grassland with less than 10% depression.
		31	High-cover grassland	Refers to natural grassland, improved grassland and mowed grassland covering > 50%. This type of grassland has a good moisture condition and dense grass cover growth.
		32	Grassland with medium cover	Natural grassland and improved grassland with >20–50% cover, where one strand has insufficient moisture and the grass cover is sparse.

Table A1. *Cont.*

No. (Level 1)	Name	No. (Level 2)	Name	Description
		33	Low-cover grassland	Refers to natural grassland with a cover of 5–20%. This kind of grassland lacks moisture and the grass cover is sparse and poorly used for grazing.
4	Waters			Natural terrestrial waters and water facilities.
		41	River and canal	Refers to the land below the perennial water level of naturally formed or artificially dug rivers and their main trunks. Artificial canal including embankment.
		42	Lakes	Refers to the land below the perennial water level of naturally formed waterlogged areas.
		43	Reservoir ponds	Refers to the land below the perennial water level of artificially constructed water storage areas.
		44	Permanent glacial snow	Includes land covered by glaciers and snow all year round.
		45	Mudflats	Refers to the tidal inundation zone between the high and low tide levels of the coastal high tide.
5	Urban and rural, industrial and mining, residential land			Refers to urban and rural settlements and the land outside of them for industry, mining, transportation, etc.
		51	Urban land	The land in large, medium and small cities and built-up areas above the county town.
		52	Rural settlements	Refers to the rural settlements independent of the towns.
		53	Other construction use	Land for large industrial areas, oil fields, salt fields, quarries, etc., as well as traffic roads, airports and special land.
6	Unused land			Land that is currently unused, including land that is difficult to use.
		61	Sandy land	Refers to land with a sand-covered surface and vegetation cover of less than 5%, including deserts, excluding deserts in water systems.
		62	Gobi	Refers to land where the ground surface is dominated by gravel and the vegetation cover is less than 5% cab land.
		63	Saline land	Refers to the land where salt and alkali gather on the surface, vegetation is sparse, and only strong salt-tolerant plants can grow.
		64	Swampy land	Refers to land with flat and low-lying terrain, poor drainage, chronically wet, seasonally waterlogged or perennially waterlogged, and wet plants growing on the surface.
		65	Bare land	Refers to land with surface soil cover and vegetation cover of less than 5%.
9		99	Marine	Marine

Table A2. Classification of ecosystem services [8,14].

Primary Classification	Secondary Classification	Comparison with Constaza Classification
Provisioning services	Food production	Food production (13)
	Raw material production	Raw material (14)
	Water supply	Water supply (5)
Regulating services	Gas regulation	Gas regulation (1)
	Climate regulation	Climate regulation (2), Disturbance regulation (3)
	Environmental purification	Waste treatment (9)
	Hydrological regulation	Water regulation (4)
Supporting services	Soil conservation	Erosion control and sediment retention (6), Soil formation (7)
	Maintenance of nutrient cycle	Nutrient cycling (8)
	Biodiversity	Pollination (10), Biological control (11), Refugia (12), Genetic resources (15)
Cultural services	Aesthetic landscape	Recreation (16), Cultural (17)

Table A3. Ecosystem service equivalent value per unit area [14].

Ecosystem Classification		Provisioning Services			Regulating Services				Supporting Services			Cultural Services
Primary Classification	Secondary Classification	Food Production	Raw Material Production	Water Supply	Gas Regulation	Climate Regulation	Environmental Purification	Hydrological Regulation	Soil Conservation	Maintenance of Nutrient Cycles	Biodiversity	Aesthetic Landscape
Farmland	Dryland	0.85	0.40	0.02	0.67	0.36	0.10	0.27	1.03	0.12	0.13	0.06
	Paddy field	1.36	0.09	−2.63	1.11	0.57	0.17	2.72	0.01	0.19	0.21	0.09
Forest	Coniferous	0.22	0.52	0.27	1.70	5.07	1.49	3.34	2.06	0.16	1.88	0.82
	Mixed coniferous	0.31	0.71	0.37	2.35	7.03	1.99	3.51	2.86	0.22	2.60	1.14
	Broad-leaved	0.29	0.66	0.34	2.17	6.50	1.93	4.74	2.65	0.20	2.41	1.06
	Shrub	0.19	0.43	0.22	1.41	4.23	1.28	3.35	1.72	0.13	1.57	0.69
Grassland	Grass	0.10	0.14	0.08	0.51	1.34	0.44	0.98	0.62	0.05	0.56	0.25
	Scrub	0.38	0.56	0.31	1.97	5.21	1.72	3.82	2.40	0.18	2.18	0.96
	Meadow	0.22	0.33	0.18	1.14	3.02	1.00	2.21	1.39	0.11	1.27	0.56
Wetland	Wetlands	0.51	0.50	2.59	1.90	3.60	3.60	24.23	2.31	0.18	7.87	4.73
Desert	Desert	0.01	0.03	0.02	0.11	0.10	0.31	0.21	0.13	0.01	0.12	0.05
	Bare ground	0.00	0.00	0.00	0.02	0.00	0.10	0.03	0.02	0.00	0.02	0.01
Waters	Water system	0.80	0.23	8.29	0.77	2.29	5.55	102.24	0.93	0.07	2.55	1.89
	Glacial snow	0.00	0.00	2.16	0.18	0.54	0.16	7.13	0.00	0.00	0.01	0.09

References

1. Huajun, T.; Wenbin, W.; Peng, Y.; Youqi, C.; Verburg, P.H. Recent progresses of land use and land cover change (lucc) models. *Acta Geogr. Sin.* **2009**, *64*, 456–468.
2. García-Álvarez, D.; Lloyd, C.D.; van Delden, H.; Olmedo, M.T.C. Thematic resolution influence in spatial analysis. An application to land use cover change (lucc) modelling calibration. *Comput. Environ. Urban Syst.* **2019**, *78*, 101375. [CrossRef]
3. Bingfang, W.; Quanzhi, Y.; Zhangzhen, Y.; Zongming, W.; Xinfang, Y.; Ainong, L.; Ronghua, M.; Jinliang, H.; Jinsong, C.; Cun, C.; et al. Land cover changes of china from 2000 to 2010. *Quat. Sci.* **2014**, *34*, 723–731.
4. Li, W.; Binggeng, X. The variation differences of cultivated land ecological security between flatland and mountainous areas based on lucc. *PLoS ONE* **2019**, *14*, e0220747.
5. Lei, Z.; Bingfang, W.; Xiaosong, L.; Qiang, X. Classification system of china land cover for carbon budget. *Acta Ecol. Sin.* **2014**, *34*, 7158–7166.
6. Yupeng, L.I.; Yaning, C.; Zhi, L.I. Effects of land use and cover change on surface wind speed in china. *J. Arid. Land* **2019**, *11*, 345–356.
7. Rubo, Z.; Meizhen, L.; Jianzhou, G.; Zhuo, W. Spatiotemporal heterogeneity and influencing mechanism of ecosystem services in the pearl river delta from the perspective of lucc. *J. Geogr. Sci.* **2019**, *29*, 831–845.
8. Costanza, R.; dArge, R.; deGroot, R.; Farber, S.; Grasso, M.; Hannon, B.; Limburg, K.; Naeem, S.; ONeill, R.V.; Paruelo, J.; et al. The value of the world's ecosystem services and natural capital. *Nature* **1997**, *387*, 253–260. [CrossRef]
9. Winkler, R. Valuation of ecosystem goods and services part 1: An integrated dynamic approach. *Ecol. Econ.* **2006**, *59*, 82–93. [CrossRef]
10. Nan, Y.; Shuai, W.; Yanxu, L. Ecosystem service value assessment: Research progress and prospects. *Chin. J. Ecol.* **2021**, *40*, 233–244.
11. Willcock, S.; Martinez-Lopez, J.; Hooftman, D.A.P.; Bagstad, K.J.; Balbi, S.; Marzo, A.; Prato, C.; Sciandrello, S.; Signorello, G.; Voigt, B.; et al. Machine learning for ecosystem services. *Ecosyst. Serv.* **2018**, *33*, 165–174. [CrossRef]
12. Gaodi, X.; Chunxia, L.; Yunfa, L.; Du, Z.; Shuangcheng, L. Ecological assets valuation of the tibetan plateau. *J. Nat. Resour.* **2003**, 189–196.
13. Gaodi, X.; Caixia, Z.; Changshun, Z.; Yu, X.; Chunxia, L. The value of ecosystem services in china. *Resour. Sci.* **2015**, *37*, 1740–1746.
14. Gaodi, X.; Caixia, Z.; Leiming, Z.; Wenhui, C.; Shimei, L. Improvement of the evaluation method for ecosystem service value based on per unit area. *J. Nat. Resour.* **2015**, *30*, 1243–1254.
15. Juncheng, C.; Tianhong, L. Changes of spatial variations in ecosystem service value in china. *Acta Sci. Nat. Univ. Pekin.* **2019**, *55*, 951–960.
16. Guoxia, M.; Xiafei, Z.; Fei, P.; Ying, Z. Cost of ecological degradation accounting in china in 2015. *Sci. Geogr. Sin.* **2019**, *39*, 1008–1015.
17. Hao, Z.; Yaya, J.; Bo, W.; Shuyi, F.; Futian, Q. Compensation for cultivated land protection of shaanxi province based on calculation of cultivated land development rights. *Trans. Chin. Soc. Agric. Eng.* **2018**, *34*, 256–266.
18. Li, L.; Wenbo, Z.; Yanhong, L.; Lianqi, Z.; Shuaibo, X.; Xiaoyan, F. Ecosystem service value gains and losses of qihe river basin based on topographic gradient characteristics. *Res. Soil Water Conserv.* **2019**, *26*, 287–295.
19. Smardon, R. Wetlands and sustainability. *Water* **2014**, *6*, 3724–3726. [CrossRef]
20. de Souza, A.L.T.; Fonseca, D.G.; Liborio, R.A.; Tanaka, M.O. Influence of riparian vegetation and forest structure on the water quality of rural low-order streams in se brazil. *For. Ecol. Manag.* **2013**, *298*, 12–18. [CrossRef]
21. Helfenstein, J.; Kienast, F. Ecosystem service state and trends at the regional to national level: A rapid assessment. *Ecol. Indic.* **2014**, *36*, 11–18. [CrossRef]
22. Bruijnzeel, L. Hydrological functions of tropical forests: Not seeing the soil for the trees? *Agric. Ecosyst. Environ.* **2004**, *104*, 185–228. [CrossRef]
23. Naiman, R.J.; Decamps, H.; Pollock, M. The role of riparian corridors in maintaining regional biodiversity. *Ecol. Appl. A Publ. Ecol. Soc. Am.* **1993**, *3*, 209–212. [CrossRef]
24. Surasinghe, T.D.; Baldwin, R.F. Importance of riparian forest buffers in conservation of stream biodiversity: Responses to land uses by stream-associated salamanders across two southeastern temperate ecoregions. *J. Herpetol.* **2015**, *49*, 83–94. [CrossRef]
25. Hiraoka, M. Floodplain farming in the peruvian amazon. *Geogr. Rev. Jpn. Ser. B* **1985**, *58*, 1–23. [CrossRef]
26. Swanson, C.; Bohlman, S. Cumulative impacts of land cover change and dams on the land–water interface of the tocantins river. *Front. Environ. Sci.* **2021**, *9*, 120. [CrossRef]
27. Dongming, X.; Guohua, J. Riparian landscape change in poyang lake. *Acta Ecol. Sin.* **2016**, *36*, 5548–5555.
28. Huang, W.; Mao, J.; Zhu, D.; Lin, C. Impacts of land use and land cover on water quality at multiple buffer-zone scales in a lakeside city. *Water* **2020**, *12*, 47. [CrossRef]
29. Shen, R.; Lan, Z.; Chen, Y.; Leng, F.; Jin, B.; Fang, C.; Chen, J. The effects of flooding regimes and soil nutrients on lakeshore plant diversity in a pristine lake and a human managed lake in subtropical china. *J. Freshw. Ecol.* **2019**, *34*, 757–769. [CrossRef]
30. Haisheng, C.; Xueling, Z.; Dehai, Z. Research on land utilization and deterioration of poyang lake area. *Yangtze River* **2006**, 86–89, 131.

31. Haiyan, Z. Study on Land Use Changes and Its Effects on Eco-Environment in the Poyang Lake Region. Ph.D. Thesis, Nanjing Agricultural University, Nanjing, China, 2011.
32. Yunliang, L.; Guizhang, Z.; Jing, Y.; Qi, Z. Interactions between groundwater and lake water of riparian zone in the typical area of poyang lake. *Trop. Geogr.* **2017**, *37*, 522–529.
33. Li, Y.; Zhang, Q.; Yao, J.; Werner, A.D.; Li, X. Hydrodynamic and hydrological modeling of the poyang lake catchment system in china. *J. Hydrol. Eng.* **2014**, *19*, 607–616. [CrossRef]
34. Costanza, R.; de Groot, R.; Sutton, P.; van der Ploeg, S.; Anderson, S.J.; Kubiszewski, I.; Farber, S.; Turner, R.K. Changes in the global value of ecosystem services. *Glob. Environ. Chang.-Hum. Policy Dimens.* **2014**, *26*, 152–158. [CrossRef]
35. Gaodi, X.; Lin, Z.; Chunxia, L.; Yu, X.; Cao, C. Expert knowledge based valuation method of ecosystem services in china. *J. Nat. Resour.* **2008**, *23*, 911–919.
36. Kasimu, Y.; Shengtian, Y.; Simayi, Z. Impact of land use change on ecosystem service value in ebinur lake basin, xinjiang. *Trans. Chin. Soc. Agric. Eng.* **2019**, *35*, 260–269.
37. Gu, X.; Yang, G.; He, X.; Zhao, L.; Li, X.; Li, P.; Liu, B.; Gao, Y.; Xue, L.; Long, A. Hydrological process simulation in manas river basin using cmads. *Open Geosci.* **2020**, *12*, 946–957. [CrossRef]
38. Smardon, R. International wetlands policy and management issues. *Natl. Wetl. Newsl.* **2015**, *37*, 10–16.
39. Gan, X.; Chen, F. Study on influence factors of famers' willingness to accept ecological compensation living in wetland area based on the investigation for 322 households in poyang lake wetland. In Proceedings of the 2016 2nd International Conference on Economy, Management, Law and Education (Emle 2016), Moscow, Russia, 15–17 December 2016; Volume 20, pp. 100–104.
40. Niu, D.; Guo, X.; Chen, C.; Peng, B. Perspectives on restoration strategies for poyang lake wetlands. In Proceedings of the Eplww3s 2011: 2011 International Conference on Ecological Protection of Lakes-Wetlands-Watershed and Application of 3s Technology, Nanchang, China, 25–26 June 2011; Volume 1, pp. 371–375.
41. Changgen, S. Study on the ecological effects of agricultural activities in poyang lake region. *Rural. Eco-Environ.* **1996**, *4*, 11–14.
42. Guoqin, H. On ecological security and ecological construction of poyang lake district. *Sci. Technol. Rev.* **2006**, *24*, 73–78.
43. Qiwu, H.; Bo, Y.; Ying, L.; Qin, W.; Zhonggang, Z.; Wushan, L. Analysis on changes of man-land relationship and associated driving force in poyang lake region. *Resour. Environ. Yangtze Basin* **2010**, *19*, 628–633.
44. Tang, M. Study on the Influence of Urbanization Process on the Ecological Environment of Poyang Lake Waters. Ph.D. Thesis, Shanghai Normal University, Shanghai, China, 2019.
45. You, H.; Fan, H.; Xu, L.; Wu, Y.; Wang, X.; Liu, L.; Yao, Z.; Yan, B. Effects of water regime on spring wetland landscape evolution in poyang lake between 2000 and 2010. *Water* **2017**, *9*, 467. [CrossRef]
46. Wang, Y.; Molinos, J.G.; Shi, L.; Zhang, M.; Wu, Z.; Zhang, H.; Xu, J. Drivers and changes of the poyang lake wetland ecosystem. *Wetlands* **2019**, *39*, S35–S44. [CrossRef]
47. Li, X.W.; Yu, X.B.; Jiang, L.G.; Li, W.Y.; Liu, Y.; Hou, X.Y. How important are the wetlands in the middle-lower yangtze river region: An ecosystem service valuation approach. *Ecosyst. Serv.* **2014**, *10*, 54–60. [CrossRef]
48. Liu, H.; Zheng, L.; Wu, J.; Liao, Y.H. Past and future ecosystem service trade-offs in poyang lake basin under different land use policy scenarios. *Arab. J. Geosci.* **2020**, *13*, 46. [CrossRef]
49. Zhao, X.F.; Tong, P.H. Ecosystem services valuation based on land use change in a typical waterfront town, poyang lake basin, china. In *Progress in Environmental Protection and Processing of Resource, Pts 1–4*; Tang, X., Zhong, W., Zhuang, D., Li, C., Liu, Y., Eds.; Trans Tech Publications: Baech, Switzerland, 2013; Volume 295–298, pp. 722–725.

land

MDPI

Perspective

Obligations of Researchers and Managers to Respect Wetlands: Practical Solutions to Minimizing Field Monitoring Impacts

Jessica A. Bryzek [1], Krista L. Noe [1], Sindupa De Silva [1], Andrew MacKenzie [1], Cindy L. Von Haugg [2], Donna Hartman [1], Jordan E. McCall [2], Walter Veselka IV [1] and James T. Anderson [2,*]

1 School of Natural Resources, West Virginia University, 1145 Evansdale Drive, Morgantown, WV 26506, USA; jab00051@mix.wvu.edu (J.A.B.); kn0054@mix.wvu.edu (K.L.N.); sd0131@mix.wvu.edu (S.D.S.); as0038@mix.wvu.edu (A.M.); Donna.Hartman@mail.wvu.edu (D.H.); Walter.Veselka@mail.wvu.edu (W.V.IV)

2 James C. Kennedy Waterfowl and Wetlands Conservation Center, Belle W. Baruch Institute of Coastal Ecology and Forest Science, Clemson University, P.O. Box 596, Georgetown, SC 29442, USA; cvonhau@clemson.edu (C.L.V.H.); jmccal4@g.clemson.edu (J.E.M.)

* Correspondence: jta6@clemson.edu; Tel.: +1-304-276-8956

Abstract: Research and field monitoring can disturb wetland integrity. Adoption of ethical field practices is needed to limit monitoring induced stressors such as trampling, non-native seed and invertebrate dispersal, and disease and fungal spread. We identify a linear pathway of deterioration highlighting stressors that can progress to cumulative impacts, consequences, and losses at the site scale. The first step to minimize disturbance is to assess and classify the current ecosystem quality. We present a tiered framework for wetland classification and link preventative measures to the wetland tier. Preventative measures are recommended at various intensities respective to the wetland tier, with higher tiered wetlands requiring more intense preventative measures. In addition, preventative measures vary by time of implementation (before, during, and after the wetland visit) to mitigate impacts at various temporal scales. The framework is designed to increase transparency of field monitoring impacts and to promote the adoption of preventative measures. Implementing preventative measures can build accountability and foster a greater appreciation for our roles as researchers and managers in protecting wetlands.

Keywords: cleaning; efficacy; ethics; researcher impacts; wetland decontamination

Citation: Bryzek, J.A.; Noe, K.L.; De Silva, S.; MacKenzie, A.; Von Haugg, C.L.; Hartman, D.; McCall, J.E.; Veselka, W., IV; Anderson, J.T. Obligations of Researchers and Managers to Respect Wetlands: Practical Solutions to Minimizing Field Monitoring Impacts. *Land* **2022**, *11*, 481. https://doi.org/10.3390/land11040481

Academic Editor: Richard C. Smardon

Received: 4 March 2022
Accepted: 24 March 2022
Published: 26 March 2022

1. Call to Action for Wetland Researchers

Wetlands hold special significance to researchers and managers for a multitude of personal, professional, and public-service reasons [1]. The importance of wetlands, and their local and landscape functions, have historically been underappreciated by society [2] (pp. 15,16). Although views are shifting as researchers disseminate information on the values and ecosystem functions of wetlands such as reducing flood damage, providing clean water [3–5], preserving biodiversity, and mitigating global climate change [6–8]. These shifting attitudes in public perception are partially a result of ongoing research and education. Moreover, the importance of understanding these wetland functions in the face of climate change presents managers with an obligation to prevent further degradation, to a practical extent, during research and field monitoring events. To aid managers in encouraging researchers and monitoring personnel to maintain the ecological integrity of a wetland, we propose a conceptual framework that includes a tiered approach to classify wetland sensitivity, with guidelines for preventative measures recommended at various intensities and times of implementation to protect wetland integrity.

Activities of wetland managers and researchers should be held to a higher standard than the public's because the scientific community has an obligation to cause minimal negative impacts to the areas they conserve and study. Despite the recommended ethical field practices within the field of ecology [9–15], there is no specific guidance for wetlands.

As more research recognizes the multitude of wetland ecosystem services [16], the adoption of ethical field practices becomes a moral responsibility for researchers and managers. Alternatively, others have proposed a Universal Declaration of the Rights of Wetlands that recognizes the inherent rights of wetlands to exist unaltered from human presence [6]. Moreover, the Ramsar Convention strategic plan proposes the vision that "Wetlands are conserved, wisely used, restored and their benefits are recognized and valued by all" [17]. Our paper builds off these concepts and provides a roadmap of proactive steps that managers and researchers can follow to conserve wetland integrity. We briefly: (1) review literature on researcher- and monitoring-induced stressors and subsequent impacts and consequences to the wetland system; (2) conduct an abbreviated synthesis of successful strategies to counter these impacts; and (3) propose a tiered hierarchical approach, based on landscape function and ecological importance, to allow managers to determine practical preventative measures to better ensure protection for monitoring impacts in these wetland systems.

2. Linear Pathway of Deterioration

Researcher-induced stressors and disturbances should be understood by managers and mitigated when possible. These researcher- and monitoring-induced stressors lead to a predictable and linear pathway of deterioration, progressing from stressors to impacts to consequences to losses at the site scale. Impacts describe direct results from stressors, while consequences are the effects of these impacts. The result of these consequences leads to losses that describe partial or complete deterioration of a physical capacity or function. We describe monitoring and research-induced disturbances as four potential stressor categories: (1) physical trampling; (2) non-native seed dispersal; (3) non-native invertebrate dispersal; and (4) disease and fungal spread. These singular stressors of introduction alter structural features of wetlands and result in amplified ecosystem impacts, consequences, and losses (Figure 1).

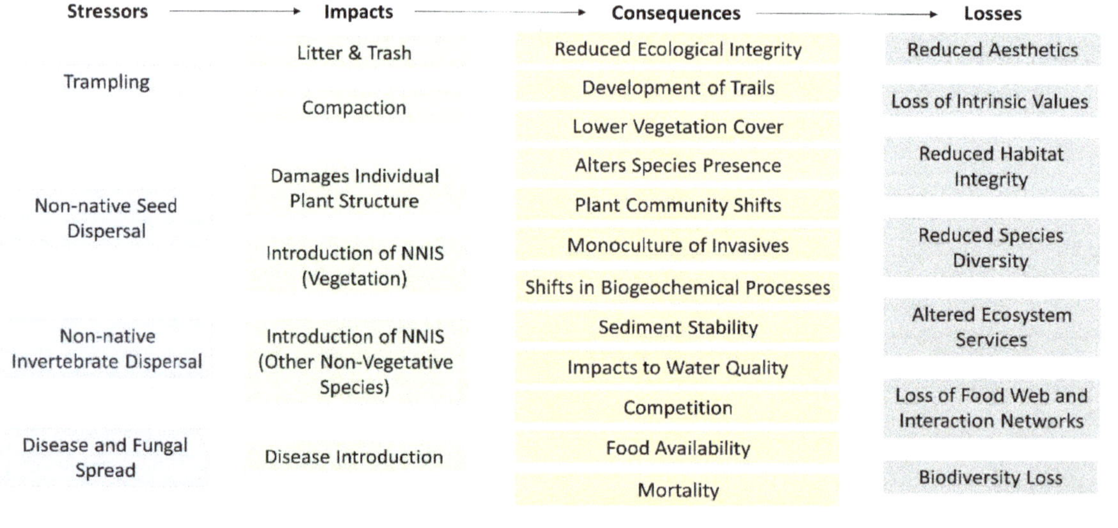

Figure 1. Researcher-induced stressors degrade ecosystem attributes down a linear pathway, progressing from impacts to consequences to ecosystem losses. NNIS = Non-native invasive species.

Physical impacts from field research are broad and can impact characteristics of wetland soil, hydrology, and vegetation [18,19]. Repeated trampling impacts vegetation height with more intensive trampling limiting vegetation cover; however ultimately declines in species richness are apparent [20]. Animal trails, which are susceptible to repetitive use like repetitive human disturbance, display higher compacted soil, more standing water, and distinctive vegetation communities different from surrounding areas [18]. While submerged community vegetation is particularly sensitive to trampling, emergent communities are particularly vulnerable to the formation of single file trails since they are much easier to walk through [21]. Generally, intensity of trampling correlates with damage to vegetation but depends on vegetation type [20,21].

To access remote study areas, often indicative of pristine or best-case conditions, the use of off-highway vehicles (i.e., airboats, motorboats, all-terrain vehicles, etc.) negatively alters and degrades the vegetation community [18,22]. Submerged and shoreline vegetation communities can become altered with repeated boat traffic, resulting wakes [23], and use of motorized vehicles resulting in the formation of deep ruts with fewer plant communities [22]. Even continued foot traffic can result in trampling of vegetation, changing the soil compaction, and subsequent hydrology to influence the vegetative structure [18]. The impacted areas typically contain fewer species, consisting of less cover [24]. Changes to wetland function fluctuates based on the intensity of the trampling [25] and individual species recovery occurs at different rates [20,26], taking as long as 15 years for recovery [20].

These small-scale disturbances can shift wetland characteristics and provide a mode for colonization of invasive species as the consistent trampling and soil compactions create pockets of disturbed microhabitats [27]. While the soil compaction is not necessarily a precursor to invasive species, it creates a transportation corridor for invasive hitchhikers carried inadvertently by researchers and others into potentially uncolonized wetlands. These invasive species may be spread via field gear and waders [28], boats [29], or even vehicles [30]. They may require specific studies and comparisons on their distribution from region to region in order to adequately assess their control [31]. The resulting invasive plants, such as reed canarygrass (*Phalaris arundinacea* L. var. *picta*) and phragmites (*Phragmites australis* Cav. Trin), can form dense monocultures [32,33] leading to a loss of native grasses [32]. These habitat changes can create disruptions to soil biota [34], wildlife communities [35–37], and insect communities on the landscape [38].

These ecosystem-disruptive invasive species are not limited to plants. New Zealand mudsnails (*Potamopyrgus antipodarum* [Gray]) outcompete native snails [39] and other macroinvertebrates within the same trophic level [40]. Other species, such as killer shrimp (*Dikerogammarus villosus* [Sowinsky]) reduce amphipod diversity of both native and other exotic species [41], and impact fish and anuran populations, preying upon larval populations [42]. The literature is replete with numerous other examples.

Researchers themselves have the potential to spread several pathogens and fungi that can have devastating effects on the surrounding ecosystems [43,44]. For example, there are two species of chytrid fungus: *Batrachochytrium dendrobatidis* (Bd), which have a global distribution, and *Batrachochytrium salamandrivorans* (Bsal), which is morphologically like Bd but currently known to only exist in Asia and Europe [45]. Both can lead to localized population crashes for amphibian communities [46–48], are believed to be spread by direct contact among frogs or through infected water [47,48], and impact over 350 amphibian host species [46]. In addition, ranaviruses are another type of disease that can spread through contact or ingestion of exposed animals [49] or exposure to infected soil and water [50]. These pathogens can lead to losses in endemic site-level biodiversity [51]. In 2015, 175 species of fish, amphibians, and reptiles were known to have been infected by viruses in the Ranavirus genus [49]. Ranavirus has led to mass die-offs in amphibians, reptiles, and fish [49,50,52], and is believed to have been spread worldwide due to the international pet trade [49,52]. The spread of Ranavirus can be deterred by disinfecting equipment and attire [53]. The spread of pathogenic bacteria and fungi is an ongoing problem within wetlands, as outbreaks of infectious diseases are occurring more frequently [51].

3. Identify Successful Intervention Strategies

Because the severity of researcher-induced impacts is dependent on timing, frequency, magnitude, and intensity, managers can suggest regulations for study design. Traversing on more durable surfaces such as rock or stone can decrease damage to vegetation [21]. In situations where this is not practical, assessing the vulnerability of vegetation to trampling based on morphological characteristics is possible [21,26] and follows a general trend of resistance with graminoids being the most resistant, and shrubs being the least resistant [20,24]. The rate of recovery for trampled vegetation increased when trampling was limited to single trails as opposed to large, trampled areas [24].

Decontamination procedures exist to limit the introduction and spread of invasive and non-native species (Table 1, Supplementary Materials Table S1). While some treatment methods may be most effective at targeting a specific invasive, bleach (Sodium hypochlorite) is often used as a universal decontaminant. Bleach has shown to be effective at eliminating aquatic invasives such as the spiny water flea (*Bythotrephes longimanus* [Leydig]) and the bloody red mysid shrimp (*Hemimysis anomala* [Sars]) [54], as well as didymo [55]. In addition, bleach is an effective treatment for both species of Chytrid fungus [56,57] and Ranavirus [53]. One notable exception is bleach does not kill New Zealand mudsnails [54].

Table 1. Methods used to control the spread of invasive plants, invertebrates, and diseases in wetlands. Concentrations, durations, target organisms, and references are found in Supplementary Materials Table S1.

Disease		**Invasive Vegetation**
(Chytrid Fungus (C), Ranavirus (R), Snake Fungal Disease (S))	**Aquatic Invertebrates**	**(Aquatic (A), Seeds (S))**
Air dry C,R	Air dry	Air dry A
Alcohol C	Alcohol	Alcohol A
Biocidal C	Bleach and water	Bleach and water A, S
Bleach and water C,R, S	Chlorine bleach	Chlorine bleach A
Chloramine-T C	Freezing	Freezing A, S
Chlorine bleach C, R	Hot water bath	Hot water bath A, S
Dettol medical C	Rinse/power wash	Rinse/power wash A
Disolol C	Steam	Steam A, S
F10 C	Virasure	
Hibiscrub C	Virkon Aquatic	
Hot water bath C, R	Virkon S®	
Kickstart C		
Nolvasan® C, R		
Potassium permanganate solution C		
QUAT-128 C		
Safe4 C		
Sodium Chloride C		
UV light R		

While bleach is effective for targeting invasives and diseases, some biota may require targeting in different ways specific to the species of invasive or disease. Other successful intervention strategies include treatments with hot water [58], air drying [29], steam treatments [59,60], and other chemicals such as Virkon Aquatic and Virasure [28,59,60]. To increase the efficacy of treatment, decontamination of clothing, boots, transportation, and all field gear is recommended [28,60].

4. Classify and Prioritize Ecosystem Sensitivity

To develop pragmatic and sensible protection measures for wetland condition monitoring or research, managers should take into consideration ecosystem sensitivity on the rate of recovery from disturbances. Our framework recognizes three categories representing a hierarchy of sensitivity characteristics that is indicative of wetland quality and significance. Our practical recommendations recognize that the most sensitive and important wetlands should have stringent safeguards to protect and maintain the exemplary functional and ecological integrity and valuable ecosystem services provided on the landscape. Whereas other wetlands fall on the spectrum of productivity and are subject to one of two lower protective tiers for minimizing the opportunity for researcher impacts. Conditions and characteristics of wetlands must be taken into consideration when identifying ecosystem sensitivity (Table 2) [61]. We note that not all conditions need to be unanimous in determining the appropriate level of protection, rather this is intended to be a guide to consider important factors in the decision. It ultimately relies on the professional judgment of the resource manager to make an informed decision that is best for protecting their wetland ecosystem.

Table 2. Classification criteria and ranking criteria for wetlands.

Ranking Criteria and Definition	A Tier	B Tier	C Tier
Rank of T&E Species: The presence and rank of threatened and endangered species, considering both global and state ranks.	Globally significant	Regional	Not present (to our knowledge)
Biodiversity: Natural assemblages of species that exist in a stable state and support ecosystem functions.	High	Moderate	Low
Ecosystem Services: Assess the functions of the wetland at their small- and large-scale roles.	Significant and unique	Moderate and multiple	Minimal or singular
Availability of Management Actions: Ownership factors influencing current and long-term management strategies such as grazing, as well as the availability of conservation resources and investments.	International, national, or regional	Regional or private	Private
Current Quality: Describes the wetland on a spectrum of natural/pristine to degraded/destroyed.	Large and/or intact	Intact or threatened	Low and/or degraded
Immediacy/Extent of Threats: Assess the scale and intensity of anthropogenic impact. Scale describes the distribution and extent of threats, and intensity describes their severity.	Minimal	Minimal and threatened	Present and extensive
Public Interest: Refers to how much the public is involved, interested, and aware of the wetland.	High	High or moderate	Low
Recovery Potential: Recognizes the disturbed and degraded state and approximates the investment of resources needed for the wetland to recover.	Low (not much to recover)	Medium (could benefit from some recovery)	High
Monitoring Difficulty: Characteristics that describe the accessibility and feasibility of access, as well as potential temporal and spatial variability difficulties.	Difficult	Difficult or moderate	Moderate or low

A Tier: Global and Regional Significance

A-Tier wetlands are globally or regionally significant and represent examples of functioning natural wetlands in an undisturbed state. These wetlands are large and/or intact wetlands with existing conservation investments under federal management, including Ramsar sites, or wetlands within a country's national park system or other protected and managed lands. A-Tier wetlands support globally endangered and threatened species, are hotspots for biodiversity, and are producers of ecosystem services based on their predominantly undisturbed state. They often provide important habitat for an imperiled species, at least temporarily (e.g., migration, stopover, breeding habitat, etc.), for some duration of the year.

B Tier: High Quality Wetlands

B-Tier wetlands comprise high quality wetlands that exist in a stable, natural state with limited signs of human impairment or are managed to support specific target organisms such as waterfowl. This tier could likely support regional and national endangered and threatened species. These may be wetlands owned by a government agency or exemplary wetlands either on private property or owned and managed by other non-government organizations and nonprofits. While these wetlands may not be among the most exemplary on the landscape, they do house locally important species, provide many ecosystem services, and are important to the overall biodiversity on the landscape. However, a notable difference from A-Tier is that they are threatened on the landscape in terms of nearby encroachment or loss.

C Tier: Low Quality Wetlands

C-Tier wetlands include low quality wetlands that are typically privately owned and have been substantially impacted by humans, which has limited or altered their functional capacity. To our knowledge, they do not currently support national or regional endangered or threatened species. Biodiversity at these wetlands is usually low and ecosystem services may be minimal or driven towards a particular function to serve human needs or infrastructure (e.g., stormwater interception and sewage overflow wetlands). These wetlands would certainly benefit from restorative actions such as wetland enhancement or restoration to provide a more diverse suite of landscape services. At the lowest end of the C-Tier spectrum, wetlands are entirely constructed or engineered and exist only to support human infrastructure (sediment ponds).

5. Recommended Preventative Measures

We propose managers instill the levels of preventative measures in research protocols reflected based on the Tiered classification. These levels of protection ascend as the recommendations provide a compelling rationale for increasing the protection to preserve the natural state (Figure 2). This phased approach to intervention requires preventative actions before, during, and after the wetland visit to mitigate impacts at various temporal scales. Pre-planning the site visit is vital to sustaining the integrity of research by proactively mitigating anticipated impacts based on the known quality of the wetland ecosystem. In addition, decontamination of clothing and field gear after the site visit is essential to limit the spread of non-native vegetation and invertebrates (Figure 3). The motivation for a tiered and ranked approach is to recognize the limitation on time and resources. This paper provides a rationale for managers to encourage the formation of specific protocols incorporating these universal, minimally intensive measures to protect the integrity of wetlands. It provides context to field staff to minimize the chance of perceived resentment as changes are implemented.

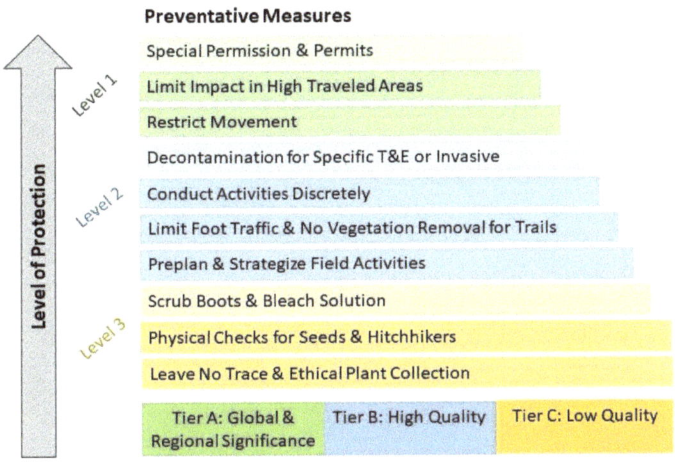

Figure 2. Recommended preventative measures for each level based on the wetland tier, where Tier A is the highest quality wetlands, Tier C is the lowest quality wetlands, and Level 1 affords the highest level of protection and includes all preventive measures included under Levels 1, 2, and 3.

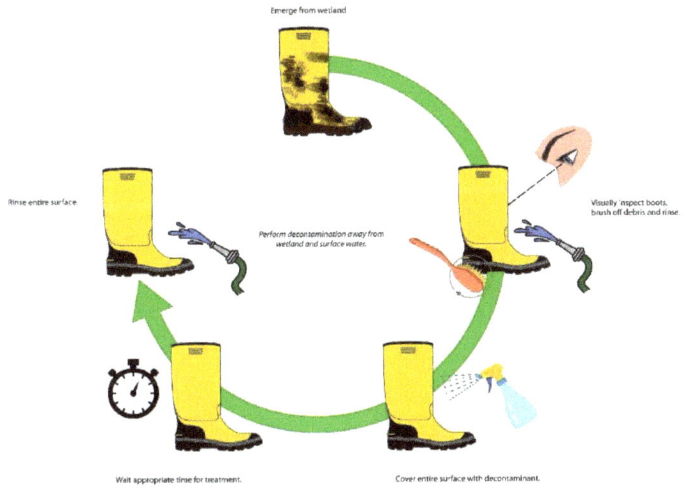

Figure 3. Decontamination steps for boots and equipment before and after entering a wetland. Treatment time varies depending on target organisms but for bleach, which is useful in most situations, a minimum of 5 min is suggested at 10% concentration (Table 1, Supplementary Materials Table S1). Decontamination procedures should occur about 200 m away from the wetland to avoid inadvertent contamination.

We believe managers understand the importance of their respective resources, and the following are intended for guidance in the development of specific protocols pertaining to research and field-monitoring staff. Generically speaking, we have divided these measures based on time of implementation including before, during, and after the site visit (Table 3).

Table 3. Managers and researchers should coordinate procedures for all activities before, during, and after site visits. These activities should reflect all preventative measures for the appropriate Tier and those below it.

Tier	Preventative Measures		
	Before Visit	**During Visit**	**After Visit**
A: Global and Regional Significance	• Plan and define steps to reduce intensity, frequency, and magnitude of study design. • Set limitations on date, duration, and purpose of visit. • Tier B preventative measures.	• Use planks or tarps in areas of high activity to minimize trampling. • Tier B preventative measures.	• List decontamination procedures at each site in a special permission permit. • Tier B preventative measures.
B: High Quality Wetlands	• Coordinate access points and travel routes to limit trampling. • Establish plan for specific threats and decontamination procedures. • Obtain scientific collection permits. • Tier C preventative measures.	• Restrict foot traffic to single trails when possible and limit use of multiple trails to decrease intensity and spatial distribution of impacts. • Avoid cutting vegetation to create trails. • Increase efforts to minimize disturbance in areas that are publicly visible. • Place soil plugs on a tarp upon excavation and return to its original layers. • Limit collecting multiple plant specimens for identification. • Tier C preventative measures.	• Organize and document plant vouchers and specimens collected. • Tier C preventative measures.
C: Low Quality Wetlands	• Identify goals and objectives. • Identify invasive and T&E species presence.	• Practice 'Leave No Trace' and conduct activities discreetly. • Minimize use of mechanized/motorized equipment unless used for specific restoration or management action. Transportation vehicle use should be prohibited in the wetland. • Back fill soil pits and do not leave open holes. • Use biodegradable materials to mark points of interest. • Follow ethical plant collection guidelines and limit intensity of harvest.	• Physically check for attached seeds or macroinvertebrates on boots, clothing, and equipment. • Scrub equipment and boots (including tread) with bristle brush. • Spray bleach solution at 5% and set for 10 minutes to eliminate wildlife diseases and invasives. • Follow decontamination guidelines for all invasives present. Bleach is not effective for certain invasive species (i.e. New Zealand mudsnail, faucet snail, Asian clam, spiny water flea eggs). • Clean and dispose of decontamination equipment and solvents away from wetland and surface waters. • Retrieve long-term monitoring equipment and markers. • Ensure efficiency and accuracy of data storage.

6. Conclusions: What's at Stake?

If managers and researchers fail to take precautions, a cascading scale of implications can occur ranging from site-specific to broader-scale inclusions. Researcher-induced stressors lead to a linear pathway of degradation, progressing from stressors to impacts to consequences and losses at the site scale. Impacts can potentially skew data collection and lead to biased results and misrepresentation of ecosystem attributes. In addition, research is often conducted in remote areas where the researcher is not part of the local community. Managers and researchers have a duty to collect accurate and representative ecological attributes, but also to act as ambassadors to the local populations, peers, and the next generation of researchers and field staff in demonstrating the importance and value of the site, and by extension, research through their actions. A failure to convey this reverence and importance to the resource does the field of science a disservice and may contribute to a cultural loss of trust in the scientific process and those that conduct research (Figure 4).

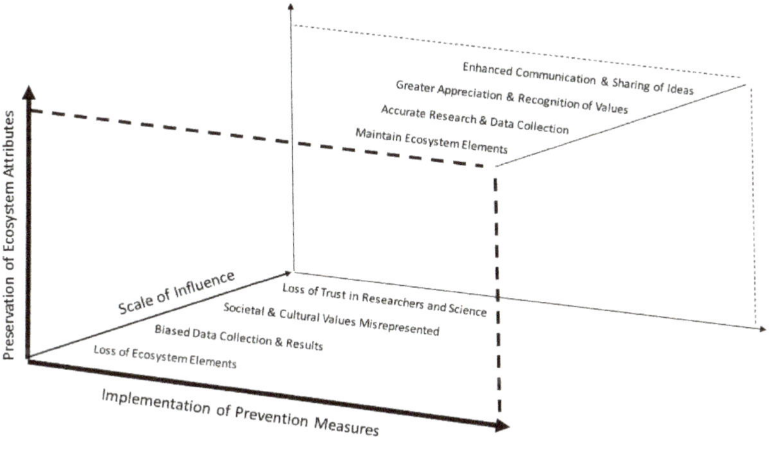

Figure 4. Consequences from research-induced impacts result in a cascading scale, ranging from local disturbances to broader societal consequences. When no preventative measures are taken, researchers' fail to fulfill their duty at the site level, and this can lead to a loss of trust in science at the societal scale. Embracing preventative measures allows ecosystem attributes to persist and leads to greater appreciation of wetlands and enhanced communication between stakeholders.

Science is grounded in observations and gains strength through collaboration and sharing of ideas. At the foundational level, researchers must accept the inherent rights of wetland ecosystems to exist unaltered from human presence [6], especially researcher-induced impacts. Wetlands exist singularly within the natural world, and the researcher is a visitor who does not remain. Our role should be to design unbiased studies that capture the best representation of ecosystem processes. It is incredibly important to control what we can and limit direct stresses to the wetland ecosystem. Researchers should feel empowered to reduce impacts and limit disturbance to preserve ecosystem integrity, increase credence in the scientific community, and foster a greater appreciation for the intrinsic value of wetlands. When preventative measures are implemented, ecosystem attributes are retained, creating a better perspective and representation of wetlands, while also protecting their integrity and the integrity of the researcher.

Supplementary Materials: The following supporting information can be downloaded at: https://www.mdpi.com/article/10.3390/land11040481/s1, References [62–64] are cited in the supplementary materials, Table S1: Table of cleaning and decontamination protocols.

Author Contributions: Conceptualization, J.A.B., K.L.N., S.D.S., A.M., D.H., W.V.IV, and J.T.A.; investigation, J.A.B., K.L.N., S.D.S., A.M., C.L.V.H., J.E.M., and W.V.IV; writing—original draft preparation, J.A.B., K.L.N., S.D.S., A.M., C.L.V.H., D.H., J.E.M., W.V.IV, and J.T.A.; writing—review and editing, J.A.B., K.L.N., S.D.S., A.M., C.L.V.H., D.H., J.E.M., W.V.IV, and J.T.A.; visualization, J.A.B., K.L.N., S.D.S., A.M., C.L.V.H., and J.E.M.; supervision, D.H., W.V.IV, and J.T.A.; project administration, D.H. and J.T.A.; funding acquisition, J.T.A. All authors have read and agreed to the published version of the manuscript.

Funding: The APC was funded by the James C. Kennedy Waterfowl and Wetlands Conservation Center at Clemson University.

Institutional Review Board Statement: Not applicable.

Informed Consent Statement: Not applicable.

Data Availability Statement: Not applicable.

Conflicts of Interest: The authors declare no conflict of interest.

References

1. Wilcox, D.A. *History of Wetland Science: A Perspective from Wetland Leaders*; Wilcox, D.A., Ed.; Amazon Print-on-Demand: Middletown, DE, USA, 2020.
2. Mitsch, W.J.; Gosselink, J.G. *Wetlands*, 4th ed.; John Wiley & Sons Inc: Hoboken, NJ, USA, 2007; Volume 99, pp. 15–16.
3. Kadlec, R.H.; Kadlec, J.A. *Wetlands and Water Quality. Wetland Functions and Values: The State of Our Understanding*; Greeson, P.E., Clark, J.R., Clark, J.E., Eds.; American Water Resources Association: Minneapolis, MN, USA, 1979; pp. 436–456.
4. Whigham, D.F.; Chitterling, C.; Palmer, B. Impacts of Freshwater Wetlands on Water Quality: A Landscape Perspective. *J. Environ. Manag.* **1988**, *12*, 663–671. [CrossRef]
5. Acreman, M.; Holden, J. How Wetlands Affect Floods. *Wetlands* **2013**, *33*, 773–786. [CrossRef]
6. Davies, G.T.; Finlayson, C.M.; Pritchard, D.E.; Davidson, N.C.; Gardner, R.C.; Moomaw, W.R.; Okuno, E.; Whitacre, J.C. Towards a Universal Declaration of the Rights of Wetlands. *Mar. Freshwater Res.* **2021**, *72*, 593. [CrossRef]
7. Taillardat, P.J.; Thompson, B.S.; Garneau, M.; Trottier, K.; Friess, D.A. Climate change mitigation potential of wetlands and the cost-effectiveness of their restoration. *Interface Focus* **2020**, *10*, 20190129. [CrossRef] [PubMed]
8. Sinthumule, N.I. An analysis of communities' attitudes towards wetlands and implications for sustainability. *Glob. Ecol. Conserv.* **2021**, *27*, e01604. [CrossRef]
9. Crozier, G.K.D.; Schulte-Hostedde, A.I. Towards Improving the Ethics of Ecological Research. *Sci. Eng. Ethics* **2015**, *21*, 577–594. [CrossRef] [PubMed]
10. Farmer, M.C. Setting Up an Ethics of Ecosystem Research Structure Based on the Precautionary Principle. *ILAR J.* **2013**, *54*, 58–62. [CrossRef]
11. Farnsworth, E. *Guidelines for Ethical Field Research on Rare Plant Species*; New England Wild Flower Society: Framingham, MA, USA, 2005.
12. Costello, M.J.; Beard, K.H.; Corlett, R.T.; Cumming, G.S.; Devictor, V.; Loyola, R.; Maas, B.; Miller-Rushing, A.J.; Pakeman, R.; Primack, R.B. Field Work Ethics in Biological Research. *Biol. Conserv.* **2016**, *203*, 268–271. [CrossRef]
13. Minteer, B.A.; Collins, J.P. From Environmental to Ecological Ethics: Toward a Practical Ethics for Ecologists and Conservationists. *Sci. Eng. Ethics* **2008**, *14*, 483–501. [CrossRef]
14. Parris, K.M.; McCall, S.C.; McCarthy, M.A.; Minteer, B.A.; Steele, K.; Bekessy, S.; Medvecky, F. Assessing Ethical Trade-offs in Ecological Field Studies. *J. Appl. Ecol.* **2010**, *47*, 227–234. [CrossRef]
15. Wallace, M.C.; Curzer, H.J. Moral Problems and Perspectives for Ecological Field Research. *ILAR J.* **2013**, *54*, 3–4. [CrossRef] [PubMed]
16. Xu, X.; Chen, M.; Yang, G.; Jiang, B.; Zhang, J. Wetland Ecosystem Services Research: A Critical Review. *Glob. Ecol. Conserv.* **2020**, *22*, e01027. [CrossRef]
17. Ramsar Convention. *The 4th Strategic Plan 2016–2024*; Ramsar Convention Secretariat: Gland, Switzerland, 2016; Available online: https://www.ramsar.org/document/the-fourth-ramsar-strategic-plan-2016-2024 (accessed on 17 March 2022).
18. Koning, C.O. Vegetation Patterns Resulting from Spatial and Temporal Variability in Hydrology, Soils, and Trampling in an Isolated Basin Marsh, New Hampshire, USA. *Wetlands* **2005**, *25*, 239–251. [CrossRef]
19. Ross, P.M. Macrofaunal Loss and Microhabitat Destruction: The Impact of Trampling in a Temperate Mangrove Forest, NSW Australia. *Wetl. Ecol. Manag.* **2006**, *14*, 167–184. [CrossRef]
20. Hill, R.; Pickering, C. Differences in Resistance of Three Subtropical Vegetation Types to Experimental Trampling. *J. Environ. Manag.* **2009**, *90*, 1305–1312. [CrossRef]
21. Rees, J.; Tivy, J. Recreational Impact on Scottish Lochshore Wetlands. *J. Biogeogr.* **1978**, *5*, 93. [CrossRef]
22. Taylor, B.R.; Raney, S. Correlation Between ATV Tracks and Density of a Rare Plant (*Drosera filiformis*) in a Nova Scotia Bog. *Rhodora* **2013**, *115*, 158–169. [CrossRef]

23. Sagerman, J.; Hansen, J.P.; Wikström, S.A. Effects of boat traffic and mooring infrastructure on aquatic vegetation: A systematic review and meta-analysis. *Ambio* **2020**, *49*, 517–530. [CrossRef]
24. Arnesen, T. Vegetation Dynamics Following Trampling in Rich Fen at Sølendet, Central Norway; a 15 Year Study of Recovery. *Nord. J. Bot.* **1999**, *19*, 313–327. [CrossRef]
25. Hsu, C.; Chen, C.; Hsieh, H. Effects of sediment compaction on macroinfauna in a protected coastal wetland in Taiwan. *Mar. Ecol. Prog. Ser.* **2009**, *375*, 73–83. [CrossRef]
26. Pescott, O.L.; Stewart, G.B. Assessing the Impact of Human Trampling on Vegetation: A Systematic Review and Meta-Analysis of Experimental Evidence. *PeerJ* **2014**, *2*, e360. [CrossRef] [PubMed]
27. Kettenring, K.M.; Whigham, D.F.; Hazelton, E.L.G.; Gallagher, S.K.; Weiner, H.M. Biotic Resistance, Disturbance, and Mode of Colonization Impact the Invasion of a Widespread, Introduced Wetland Grass. *Ecol. Appl.* **2015**, *25*, 466–480. [CrossRef] [PubMed]
28. Stockton, K.A.; Moffitt, C.M. Disinfection of Three Wading Boot Surfaces Infested with New Zealand Mudsnails. *N. Am. J. Fish. Manag.* **2013**, *33*, 529–538. [CrossRef]
29. Mohit, S.; Johnson, T.; Arnott, S. Recreational Watercraft Decontamination: Can Current Recommendations Reduce Aquatic Invasive Species Spread? *Manag. Biol. Invasions* **2021**, *12*, 148–164. [CrossRef]
30. Banha, F.; Marques, M.; Anastácio, P.M. Dispersal of Two Freshwater Invasive Macroinvertebrates, *Procambarus clarkii* and *Physella acuta*, by off-Road Vehicles: Dispersal of Invasive Macroinvertebrates by off-Road Vehicles. *Aquat. Conserv. Mar. Freshw. Ecosyst.* **2014**, *24*, 582–591. [CrossRef]
31. Stinca, A.; Musarella, C.M.; Rosati, L.; Laface, V.L.A.; Licht, W.; Fanfarillo, E.; Wagensommer, R.P.; Galasso, G.; Fascetti, S.; Esposito, A.; et al. Italian Vascular Flora: New Findings, Updates and Exploration of Floristic Similarities between Regions. *Diversity* **2021**, *13*, 600. [CrossRef]
32. Sinks, I.A.; Borde, A.B.; Diefenderfer, H.L.; Karnezis, J.P. Assessment of methods to control invasive reed canarygrass (*Phalaris arundinacea*) in tidal freshwater wetlands. *Nat. Areas J.* **2021**, *41*, 172–185. [CrossRef]
33. Hazelton, E.L.; Downard, R.; Kettenring, K.M.; McCormick, M.K.; Whigham, D.F. Spatial and temporal variation in brackish wetland seed banks: Implications for wetland restoration following Phragmites control. *Estuaries Coast.* **2018**, *41*, 68–84. [CrossRef]
34. Reinhart, K.O.; Callaway, R.M. Soil biota and invasive plants. *New Phytol.* **2006**, *170*, 445–457. [CrossRef]
35. Whitt, M.B.; Prince, H.H.; Cox, R.R., Jr. Avian use of purple loosestrife dominated habitat relative to other vegetation types in a Lake Huron wetland complex. *Wilson Bull.* **1999**, *111*, 105–114.
36. Spyreas, G.; Wilm, B.W.; Plocher, A.E.; Ketzner, D.M.; Matthews, J.W.; Ellis, J.L.; Heske, E.J. Biological consequences of invasion by reed canary grass (*Phalaris arundinacea*). *Biol. Invasions* **2010**, *12*, 1253–1267. [CrossRef]
37. Nagy, C.; Aschen, S.; Christie, R.; Weckel, M. Japanese stilt grass (*Microstegium vimineum*), a nonnative invasive grass, provides alternative habitat for native frogs in a suburban forest. *Urban Habitats* **2011**, *6*, 1–10.
38. Bezemer, T.M.; Harvey, J.A.; Cronin, J.T. Response of native insect communities to invasive plants. *Annu. Rev. Entomol.* **2014**, *59*, 119–141. [CrossRef] [PubMed]
39. Lysne, S.; Koetsier, P. Comparison of desert valvata snail growth at three densities of the invasive New Zealand mudsnail. *West. N. Am. Nat.* **2008**, *68*, 103–106. [CrossRef]
40. Hall, R.O.; Dybdahl, M.F.; VanderLoop, M.C. Extremely high secondary production of introduced snails in rivers. *Ecol. Appl.* **2006**, *16*, 1121–1131. [CrossRef]
41. Dick, J.T.A.; Platvoet, D. Invading predatory crustacean *Dikerogammarus villosus* eliminates both native and exotic species. *Proc. Royal Soc. B.* **2000**, *267*, 977–983. [CrossRef]
42. Warren, D.A.; Bradbeer, S.J.; Dunn, A.M. Superior predatory ability and abundance predicts potential ecological impact towards early-stage anurans by invasive 'Killer Shrimp' (*Dikerogammarus villosus*). *Sci. Rep.* **2021**, *11*, 4570. [CrossRef]
43. Thomas, M.; Samuel, K.A.; Kurian, P. Rodentborne Fungal Pathogens in Wetland Agroecosystem. *Braz. J. Microbiol.* **2012**, *43*, 247–252. [CrossRef]
44. Shelley, V.; Kaiser, S.; Shelley, E.; Williams, T.; Kramer, M.; Haman, K.; Keel, K.; Barton, H. Evaluation of Strategies for the Decontamination of Equipment for Geomyces destructans, the Causative Agent of the White-Nose Syndrome (WNS). *J. Cave Karst Stud.* **2013**, *75*, 1–10. [CrossRef]
45. Ossiboff, R.J.; Towe, A.E.; Brown, M.A.; Longo, A.V.; Lips, K.R.; Miller, D.L.; Carter, E.D.; Gray, M.J.; Frasca, S. Differentiating *Batrachochytrium dendrobatidis* and *B. salamandrivorans* in Amphibian Chytridiomycosis Using RNAScope® in Situ Hybridization. *Front. Vet. Sci.* **2019**, *6*, 304. [CrossRef]
46. Fisher, M.C.; Garner, T.W.J.; Walker, S.F. Global Emergence of *Batrachochytrium dendrobatidis* and Amphibian Chytridiomycosis in Space, Time, and Host. *Annu. Rev. Microbiol.* **2009**, *63*, 291–310. [CrossRef] [PubMed]
47. Kolby, J.E.; Ramirez, S.D.; Berger, L.; Griffin, D.W.; Jocque, M.; Skerratt, L.F. Presence of Amphibian Chytrid Fungus (*Batrachochytrium dendrobatidis)* in Rainwater Suggests Aerial Dispersal Is Possible. *Aerobiologia* **2015**, *31*, 411–419. [CrossRef]
48. Feldmeier, S.; Schefczyk, L.; Wagner, N.; Heinemann, G.; Veith, M.; Lötters, S. Exploring the Distribution of the Spreading Lethal Salamander Chytrid Fungus in Its Invasive Range in Europe—A Macroecological Approach. *PLoS ONE* **2016**, *11*, e0165682. [CrossRef] [PubMed]
49. Gray, M.J.; Chinchar, V.G. (Eds.) . *Ranaviruses;* Springer International Publishing: Cham, Switzerland, 2015. [CrossRef]
50. Harp, E.M.; Petranka, J.W. Ranavirus in wood frogs (*Rana sylvatica*): Potential sources of transmission within and between ponds. *J. Wildl. Dis.* **2006**, *42*, 307–318. [CrossRef]

51. Schmeller, D.S.; Courchamp, F.; Killeen, G. Biodiversity Loss, Emerging Pathogens and Human Health Risks. *Biodivers. Conserv.* **2020**, *29*, 3095–3102. [CrossRef]
52. Brunner, J.; Olson, A.; Rice, J.; Meiners, S.; Le Sage, M.; Cundiff, J.; Goldberg, C.; Pessier, A. Ranavirus Infection Dynamics and Shedding in American Bullfrogs: Consequences for Spread and Detection in Trade. *Dis. Aquat. Org.* **2019**, *135*, 135–150. [CrossRef]
53. Bryan, L.; Baldwin, C.; Gray, M.; Miller, D. Efficacy of Select Disinfectants at Inactivating Ranavirus. *Dis. Aquat. Org.* **2009**, *84*, 89–94. [CrossRef]
54. De Stasio, B.T.; Acy, C.N.; Frankel, K.E.; Fritz, G.M.; Lawhun, S.D. Test of disinfection methods for invasive snails and zooplankton: Effects of treatment methods and contaminated materials. *Lake Reserv. Manag.* **2019**, *35*, 156–166. [CrossRef]
55. Root, S.; O'Reilly, C.M. Didymo Control: Increasing the Effectiveness of Decontamination Strategies and Reducing Spread. *Fisheries* **2012**, *37*, 440–448. [CrossRef]
56. Van Rooji, P.; Pasmans, F.; Coen, Y.; Martel, A. Efficacy of chemical disinfectants for the containment of the salamander chytrid fungus *Batrachochytrium salamandrivorans*. *PLoS ONE* **2017**, *12*, e0186269. [CrossRef]
57. Johnson, M.L.; Berger, L.; Phillips, L.; Speare, R. Fungicidal effects of chemical disinfectants, UV light, desiccation and heat on the amphibian chytrid *Batrachochytrium dendrobatidis*. *Dis. Aquat. Org.* **2003**, *57*, 255–260. [CrossRef] [PubMed]
58. Anderson, L.G.; Dunn, A.M.; Rosewarne, P.J.; Stebbing, P.D. Invaders in Hot Water: A Simple Decontamination Method to Prevent the Accidental Spread of Aquatic Invasive Non-Native Species. *Biol. Invasions* **2015**, *17*, 2287–2297. [CrossRef] [PubMed]
59. Bradbeer, S.J.; Coughlan, N.E.; Cuthbert, R.N.; Crane, K.; Dick, J.T.A.; Caffrey, J.M.; Lucy, F.E.; Renals, T.; Davis, E.; Warren, D.A.; et al. The Effectiveness of Disinfectant and Steam Exposure Treatments to Prevent the Spread of the Highly Invasive Killer Shrimp, *Dikerogammarus villosus*. *Sci. Rep.* **2020**, *10*, 1919. [CrossRef] [PubMed]
60. Coughlan, N.E.; O'Hara, S.; Crane, K.; Dick, J.T.A.; MacIsaac, H.J.; Cuthbert, R.N. Touch Too Much: Aquatic Disinfectant and Steam Exposure Treatments Can Inhibit Further Spread of Invasive Bloody-Red Mysid Shrimp *Hemimysis anomala*. *Wetl. Ecol. Manag.* **2020**, *28*, 397–402. [CrossRef]
61. Elzinga, C.L.; Salzer, D.W.; Willoughby, J.W. *Measuring & Monitoring Plant Populations*; US Department of the Interior, Bureau of Land Management: Washington, DC, USA, 1998; pp. 25–38.
62. Phillott, A.; Speare, R.; Hines, H.; Skerratt, L.; Meyer, E.; McDonald, K.; Cashins, S.; Mendez, D.; Berger, L. Minimising Exposure of Amphibians to Pathogens during Field Studies. *Dis. Aquat. Org.* **2010**, *92*, 175–185. [CrossRef] [PubMed]
63. Gottschalk, S.D.; Karol, K.G. Survivability of Starry Stonewort Bulbils Using Commonly Available Decontamination Strategies. *J. Aquat. Plant Manag.* **2020**, *58*, 19–25.
64. Gold, K.; Reed, P.; Bemis, D.; Miller, D.; Gray, M.; Souza, M. Efficacy of Common Disinfectants and Terbinafine in Inactivating the Growth of Batrachochytrium Dendrobatidis in Culture. *Dis. Aquat. Org.* **2013**, *107*, 77–81. [CrossRef]

MDPI
St. Alban-Anlage 66
4052 Basel
Switzerland
Tel. +41 61 683 77 34
Fax +41 61 302 89 18
www.mdpi.com

Land Editorial Office
E-mail: land@mdpi.com
www.mdpi.com/journal/land

www.ingramcontent.com/pod-product-compliance
Ingram Content Group UK Ltd.
Pitfield, Milton Keynes, MK11 3LW, UK
UKHW060123010826
14089UKWH00006B/119